Control Systems for Heating, Ventilating and Air Conditioning

Control Systems for Heating, Ventilating and Air Conditioning

Fourth Edition

Roger W. Haines
Consulting Engineer

VNR VAN NOSTRAND REINHOLD COMPANY
New York

Library of Congress Catalog Card Number 87-2144
ISBN 0-442-23141-5

Printed in the United States of America

Van Nostrand Reinhold Company Inc.
115 Fifth Avenue
New York, New York 10003

Van Nostrand Reinhold Company Limited
Molly Millars Lane
Wokingham, Berkshire RG11 2PY, England

Van Nostrand Reinhold
480 La Trobe Street
Melbourne, Victoria 3000, Australia

Macmillan of Canada
Division of Canada Publishing Corporation
164 Commander Boulevard
Agincourt, Ontario M1S 3C7, Canada

16 15 14 13 12 11 10 9 8 7 6 5 4 3 2

Library of Congress Cataloging-in-Publication Data

Haines, Roger W.
Control systems for heating, ventilating, and air conditioning.

Bibliography: p.
Includes index.
1. Heating—Control. 2. Air conditioning—Control.
I. Title.
TH7466.5H34 1987 697 87-2144
ISBN 0-442-23141-5

Preface to the Fourth Edition

There are two reasons why we have a new edition every four or five years.

The first is that technology changes. Chapter 10, on computer-based controls, has had to be almost completely rewritten. Fundamentals don't change, but the tools available to us do change. Evaluation and proper use of those tools makes it even more imperative that we understand fundamentals. Many of our control problems stem from the use of new devices as a solution to problems that are, in fact, control design errors. New gadgets, for example, Direct Digital Controls (DDC), will not solve basic problems and may even compound them. None-the-less, you will find an extensive discussion of DDC because I think it is the probable "future" in HVAC control. But it must be applied with a good understanding of fundamentals.

The second reason is that I keep learning and need to pass on my new and improved understanding to my readers. Thus you will find a number of small but important revisions, a dissertation on control "modes," and a much more detailed discussion of how electronic control devices work. There are a few places where I have corrected what I now perceive to be errors. I apologize for these.

I have been much encouraged by the acceptance of this book in the past, and I hope that this new edition will be helpful. Thank you for your support.

Roger W. Haines

Preface to the First Edition

This book is intended for the guidance of the engineer who is designing a heating, ventilating or air conditioning system and wants a simple, practical explanation of how best to control that system. It could also be used as a supplemental text in a college or technical school course on refrigeration and air conditioning.

It does not include mathematical analyses of control systems, response factors, Fourier transforms and the like. These are adequately and thoroughly covered in a number of up-to-date college-level texts.

What is presented here is an elementary but comprehensive explanation of control system theory, control hardware, and both simple and complex control systems. There are also discussions on supervisory controls and the use of computers in control systems.

Throughout, the reader should be aware of the interrelationship between the HVAC controls, the HVAC system, the electrical power system and the building. There have been examples of control systems which failed to control because inherent deficiencies in the building or HVAC system made it uncontrollable; or of control failures because the designer failed to make his intent clear to others. It is hoped that the reader will be helped to avoid some of these pitfalls.

Roger W. Haines

January, 1971

Acknowledgments

Nothing gets done without help and I've had some of the best. First and foremost, Frank Bridgers and Don Paxton provided my basic training in consulting engineering and control design. My debt to them is large. Many friends in ASHRAE, particularly those on TC 10.1 (Controls) gave encouragement and help. Tom Guiterrez, a colleague at Collins Radio, served as sounding board, critic, research assistant and advisor. Don Bahnfleth, editorial director of Heating, Piping and Air Conditioning magazine read the manuscript and gave much good advice. My wife encouraged and helped me as she has so often, besides typing the manuscript.

Professor Clark Pennington and Dr. Joseph Gartner read and criticized the manuscript and their comments were most helpful.

Several manufacturers graciously gave permission to reproduce parts of their literature.

Roger W. Haines

Contents

Control Systems for Heating, Ventilating and Air Conditioning

1 Control Theory and Terminology

1.1 INTRODUCTION

The purpose of this book is to discuss the design of control systems for heating, ventilating and air conditioning systems. Its intent is to help you, the reader, develop an understanding of controls and control systems, and the air conditioning systems to which they are applied. Out of this you should develop a philosophy of design which will enable you to cope not only with the basic systems discussed here, but with the unusual and special requirements which continue to arise as air conditioning becomes more sophisticated.

The term "heating, ventilating and air conditioning" (HVAC) covers a wide range of equipment, from, for example, a kerosene stove, to the large and sophisticated complex of equipment required for the World Trade Center in New York City.

"Control" likewise may vary from the handwheel adjustment of the kerosene stove wick to the elaborate, computerized system in the World Trade Center.

This book will discuss as many as possible of the various HVAC systems in use today, together with the methods of control which may be applied to them. The word "together" should be emphasized. The HVAC system, its control system and the building in which they are installed are inseparable parts of a whole. They interact with one another in many ways, so that neglect of any element may cause a partial or complete loss of controllability.

Unfortunately, there have at times been examples of this neglect, with resultant regret on the part of all concerned. It is possible to design good HVAC systems and controls at a reasonable cost. No HVAC system is better than its controls and the building in which it is installed. As designers we have a duty to provide the owner with the best possible system within budget limits—not necessarily the cheapest. The cheapest may be the most expensive in the long run, in operating cost as well as owner dissatisfaction. The "best" system is one that will provide the required degree of comfort for the application with the least expenditure of energy. That degree of comfort is, of course, a function of the application; that is, we should expect closer control of temperature in a hotel bedroom than in the same hotel's kitchen.

1.2 WHAT IS "CONTROL"?

While many HVAC control systems appear to be, and are, complicated, the most elaborate system may be reduced to a few fundamental elements. Let's get back to that kerosene stove. We are cold, so we strike a match and light the wick (after checking the fuel supply). We turn the wick up high. After a while it gets warm. We sense this and turn the handwheel to lower the wick and get less heat. Or, we may turn it clear down, shutting off the heat entirely.

Here are all the elements of a closed-loop control system. The "controlled variable" is the air temperature in the room. The "process plant" is the stove. The wick is the "controlled device." The sensor and controller are represented by the person in the room. You will note that a man is not a really sensitive controller. Nonetheless he performs the basic function of the sensor-controller, which is to measure the controlled variable, compare it with a "set point" (here, the personal sensation of comfort) and adjust the controlled device.

Notice that only three elements are necessary for a control system: sensor, controller and controlled device. These need only a process plant to give them meaning. In this text the definition of the process plant is limited to heating, ventilating and air conditioning systems.

1.3 ELEMENTARY CONTROL SYSTEM

Figure 1–1 illustrates an elementary control system. This shows air flowing through a heating coil in a duct. The sensor measures the temperature of the air downstream of the coil and passes the information to the

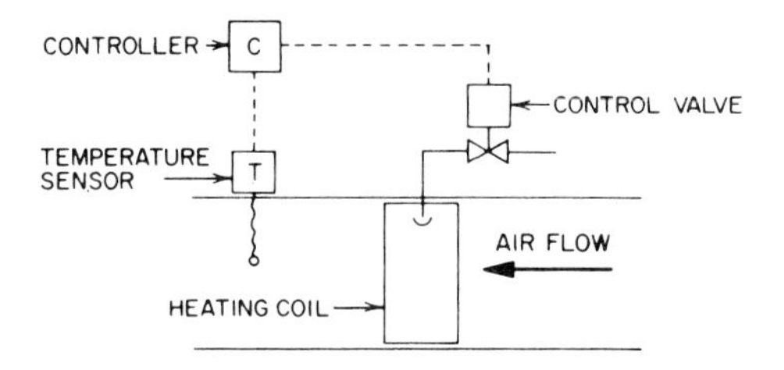

Figure 1-1 Elementary control system.

controller. The controller compares the air temperature with a set point and sends a signal to open or close the hot-water valve (the controlled device) as required to maintain a correspondence between the air temperature and the set point. This is a "closed-loop" system, in which the change in temperature caused by a change in valve position (and/or load) will be sensed and additional adjustments will be made as necessary. The air temperature is known as the "controlled variable."

Most control systems fall in the "closed-loop" classification but "open-loop" systems are sometimes used. In an open-loop system the sensor is not directly affected by the action of the controlled device. A homely example of such a system is the electric blanket, where the thermostat senses room and not blanket temperature.

Keep in mind that, regardless of apparent complexity, all control systems may be reduced to these essential elements. Most complications occur in an attempt to obtain "better" control; that is, to maintain the controlled variable as close to the set point as possible. One of the cardinal rules of control system design is to "keep it simple," and avoid piling relays on resets on multiple sensors.

1.4 PURPOSES OF CONTROL

It is usually thought that the purpose of the automatic control system is to provide control of the temperature and/or humidity in a space. But these are not the only functions which the system can serve:

It can also control the relative pressure between two spaces, a very useful attribute in preventing the spread of contamination.

Safety Controls prevent operating of equipment when in an unsafe condition. They can also trigger visual or audible alarms to alert operating personnel to those conditions.

The HVAC system operates most economically when equipment capacity is closely matched to load, and this can be better accomplished by an automatic control system than manually.

A completely automatic system with changeover controls, interlocks, and internal monitoring and compensating controls minimizes human intervention, and, therefore, the chance of human error.

It can be seen, then, that the purposes of control systems are many and varied. Now, how are the required functions carried out?

1.5 CONTROL ACTION

To satisfy the need for various kinds of control response, there are available a number of types of control actions. These may be broadly classified as follows.

1.5.1 Two-Position or On-Off Action (Figure 1–2)

This is the simplest and most obvious. One example of two-position action is a relay which is open or closed, with no intermediate position. An example of on-off action is a thermostat starting and stopping a ventilating fan motor. Any two-position controller needs a *differential* to prevent "hunting," or too-rapid cycling. This differential is the difference between the setting at which the controller operates to one position and the setting at which it changes to the other. In a thermostat this is expressed in degrees of temperature. The *differential* setting of any controller is usually somewhat less than the *operating differential* of the HVAC system due to the lag of the instrument and the system.

1.5.2 Timed Two-Position Action

This is used to reduce the operating differential by artificially shortening "on" or "off" time in anticipation of system response. A heating

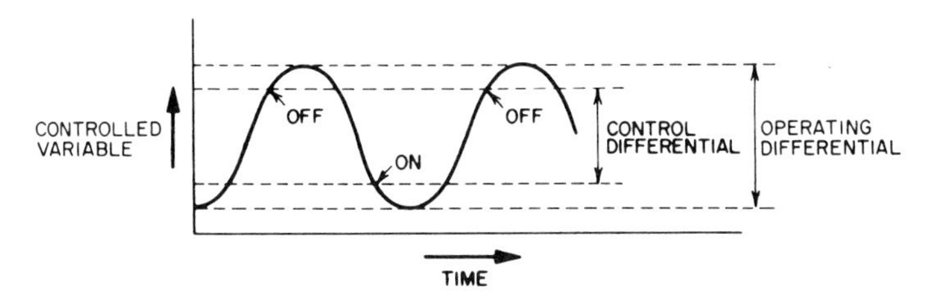

Figure 1-2 Two-position control (low limit).

thermostat may be provided with a small internal heater which is energized during "on" periods, thereby giving a false signal to the thermostat. This is called "heat anticipation."

1.5.3 Floating Action (Figure 1–3)

This term refers to a controlled device which can stop at any point in its stroke and can be reversed without completing its stroke. The controller must have a "dead spot" or neutral zone in which it sends no signal but allows the device to "float" in a partly open position. For good operation this system requires a rapid response in the controlled variable; otherwise it will not stop at an intermediate point.

1.5.4 Modulating Control (Figure 1-4)

Modulating means that the output of the controller can vary infinitely over the range of the controller. In this situation the controlled device will seek a position corresponding to its own range and the controller output. Some new terms are encountered:

Throttling range is the amount of change in the controlled variable required to run the actuator of the controlled device from one end of its stroke to the other.
Set point is the controller setting and is the desired value of the controlled variable.
Control point is the actual value of the controlled variable. If the control point lies within the throttling range of the controller it is said to be in control. When it exceeds the throttling range it is said to be out of control.
Offset or *control point shift* is the difference between the set point and the control point. This is sometimes called drift, droop, or deviation. The amount of offset theoretically possible is determined by the throttling range, but this value may be exceeded in out-of-control situations.

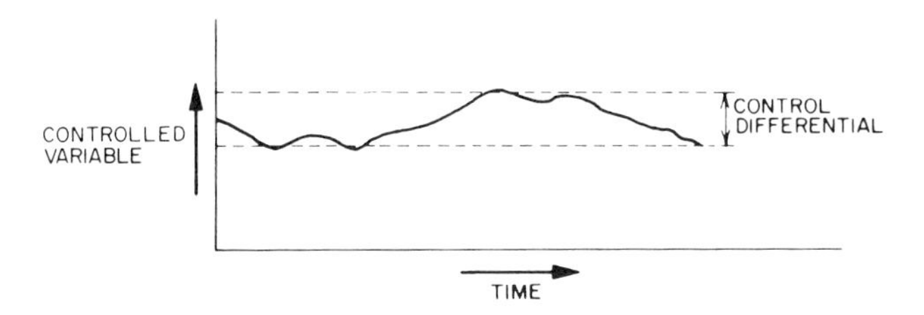

Figure 1-3 Floating action.

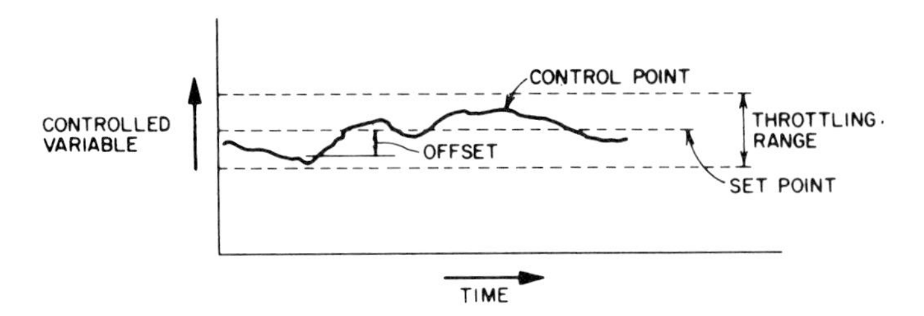

Figure 1-4 Modulating control.

1.5.5 Control Modes—Proportional

There are three control *modes* encountered in modulating control. The first and simplest of these is *proportional* control. This is the control mode used in most pneumatic and older electric systems for HVAC. The mathematical expression for proportional control is:

$$O = A + K_p e \qquad (1\text{–}1)$$

where: O = controller output.

A = a constant equal to the value of the controller output with no error.

e = the *error*, equal to the difference between the set point and the measured value of the controlled variable.

K_p = proportional gain constant.

This means that the output of a proportional controller is equal to a constant plus the error multiplied by another constant which is called the *gain*.

The proportional gain is related inversely to the throttling range. For example, in a pneumatic temperature controller the output will range from 3 to 13 psi (10 psi range). If the throttling range is 10°F then the gain will be the ratio of 10 over 10 (1.0), meaning that the controller output signal will change 1 psi for every degree of error. If the throttling range is reduced to 4°F the gain will increase to 10 over 4, or 2.5 (psi per degree). Increasing the gain will make the controller more responsive, but too high a gain may make the system *unstable*, causing it to oscillate continuously or *hunt* around the set point. Decreasing the gain will improve stability but decrease response and sensitivity.

Figure 1–5 is a graphic illustration of proportional control. This shows the control point—the actual value of the controlled variable—plotted over time. If the HVAC system is started after a prolonged shutdown the variable will be out of control. It will, therefore, be driven so rapidly toward the set point that it will cross the set point value before the system can respond. The return swing will again cross the set point value, and so forth. If the system is stable, it will settle out after a few cycles. If the system is unstable the system will continue to oscillate indefinitely (Figure 1–6).

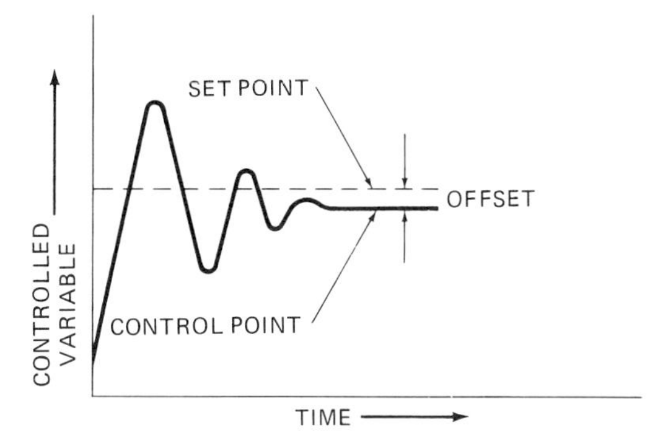

Figure 1-5 Proportion control—stable.

It should be noted that there will always be an offset with proportional control, since the error needed to generate the controller output will produce only enough capacity to match the load on the system. The offset will be greater with low values of gain and at light-load conditions. It will also be affected by *system gains*. The existence of the continuous offset affects system accuracy, comfort and energy consumption. This will be discussed further in later chapters.

With most controllers, gain adjustment requires only a screwdriver. With computer-driven systems, gain is a number in software. (See Chapter 10.)

1.5.6 Control Modes—Proportional plus Integral

Historically, in HVAC control practice, this control mode has been referred to as *proportional with reset*. The correct term is *proportional plus integral*, usually abbreviated PI.

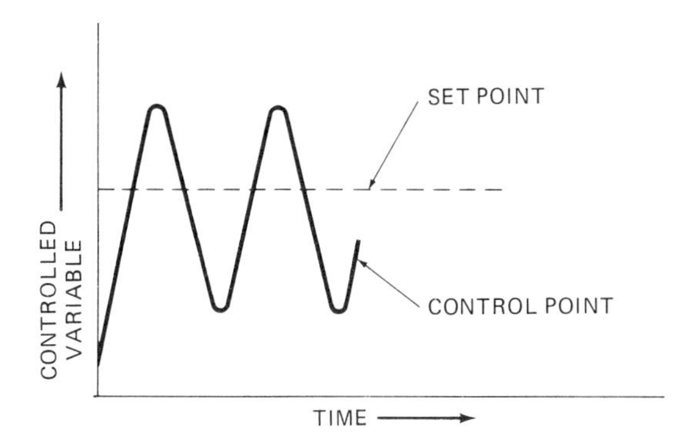

Figure 1-6 Proportional control—unstable.

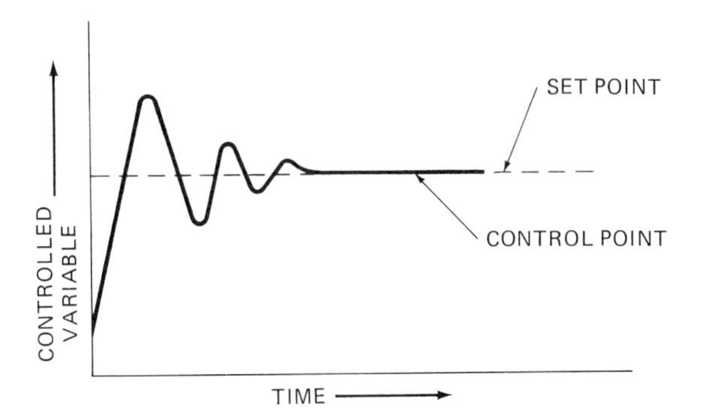

Figure 1-7 Proportional plus integral control.

Mathematically another term is added to the control equation:

$$O = A + K_p e + K_i \int edt \qquad (1\text{–}2)$$

where K_i = integral gain constant.

The added term means that the output of the controller is now affected by the error signal integrated over time and multiplied by the integral gain constant. The integral gain constant is a function of time and may be written as x/t, where x is the number of times the value of variable is sampled per unit time—also known as the *reset rate*. Since the error is being integrated over time, the value of the term will increase as the reset rate is increased. Note that the sign of the error may be positive or negative; therefore the "increase" may be plus or minus.

The effect of this term is that the controller output will continue to change as long as any error persists, and the control offset will be eliminated (Figure 1–7).

While PI mode has long been used in the process control industry it is relatively new to HVAC. Some pneumatic controllers and most electronic controllers use PI mode. Computers can be programmed for any mode.

1.5.7 Control Modes—Derivative

For derivative control mode, still another term is added to the control equation:

$$O = A + K_p e + K_i \int edt + K_d \frac{de}{dt} \qquad (1\text{–}3)$$

where K_d = derivative gain constant.

The derivative term provides additional controller output related to the rate of change of the controlled variable. A rapid rate of change in the error will increase the absolute value of the derivative term. A small rate of change will decrease the value. Offset, per se, has no effect, so long as the offset is constant.

Because most HVAC systems have a relatively slow response to changes in controller output, the use of derivative mode may tend to *overcontrol.* Only when system response is very rapid should PID mode be considered. Most electronic controllers are available in PID mode.

1.6 ENERGY SOURCES FOR CONTROL SYSTEMS

The above-described control actions may be accomplished by various means. Control systems may be electric, electronic, pneumatic, fluidic, hydraulic, self-contained or combinations of these. Chapters 2, 3, 4 and 5 will discuss the control elements in some detail. A brief description here will serve to introduce the next chapters.

1.6.1 Electric Systems

Electric systems provide control by starting and stopping the flow of electricity or varying the voltage and current by means of rheostat or bridge circuits.

1.6.2 Electronic Systems

These systems use very low voltages (24 V or less) and currents for sensing and transmission, with amplification by electronic circuits or servo-mechanisms as required for operation of controlled devices.

1.6.3 Pneumatic Systems

Pneumatic systems usually use low-pressure compressed air. Changes in output pressure from the controller will cause a corresponding position change at the controlled device.

1.6.4 Hydraulic Systems

These are similar in principle to pneumatic systems but use a liquid or gas rather than air. These systems are usually closed, while pneumatic systems are open (some air is wasted).

1.6.5 Fluidic Systems

Fluidic systems use air or gas and are similar in operating principles to electronic as well as pneumatic systems.

1.6.6 Self-Contained System

This type of system incorporates sensor, controller and controlled device in a single package. No external power or other connection is required. Energy needed to adjust the controlled device is provided by the reaction of the sensor with the controlled variable.

1.7 MEASUREMENT

You will have noted that all the control actions described above depend first on *measurement* of a controlled variable. This matter of accurate and rapid measurement is the most serious problem in the control industry. While it is essential for proper control, it is very difficult to obtain an accurate and instantaneous reading, especially if the property being measured is fluctuating or changing very rapidly.

It is, therefore, necessary to analyze very carefully what is actually being measured, how it may vary with time in operation and how great a degree of accuracy is necessary in the measurement.

Thermostats will be affected by the presence or absence of air motion (drafts), the temperature of the surfaces on which they are mounted (if greatly different from the air temperature), the mass of the sensing element and the presence of radiant effects from windows or hot surfaces. For a residence or an office a variation of one or two degrees on either side of the set point may be acceptable. For a standards laboratory (Chapter 9) a variation of $\pm 0.5°F$ may be unacceptable. And, of course, while the thermostat in a standards lab may show a variation of only 0.2° to 0.1°, across the room, 20 ft away, the variation may very well be 0.5° or even 1°.

A pressure sensor which is located at a point of turbulence (such as a turn or change of pipe size) in the fluid can never provide accurate or consistent readings. For this purpose a long straight run of duct or pipe is generally required. Straightening vanes can be used where long straightaways are not possible.

Another difficulty which may arise in measurement or control is delay due to the distance over which the signal must be transmitted. Pneumatic signals will travel only at sonic speeds and are subject to fluid friction losses. Electric signals may become seriously attenuated by resistance in long lines.

Several books have been written on sensors and measuring problems. (See, for example, references 2 and 8 in the Bibliography.) It is not the purpose of this book to go into such detail; but we hope you will study carefully the location of all sensors and their relationship to the rest of the system.

1.8 SYMBOLS AND ABBREVIATIONS

In the back of the book is a list of the symbols and abbreviations used herein. Most of these are "typical" for the HVAC control industry but it should be noted that there is no industry standard. The nearest thing to such a standard is the recommended symbol list in the ASHRAE HANDBOOK. The possibility of using ISA (Instrument Society of America) symbols has been advanced by some people. For the present, however, it appears that, if an industry standard is developed, it will use something similar to the ASHRAE symbols.

All temperatures shown are in degrees Fahrenheit.

1.9 PSYCHROMETRICS

Many of the discussions of control systems use psychrometric chart examples and it is assumed that the reader has a basic knowledge of psychrometrics. For those readers who lack this background a short chapter on the use of psychrometric charts is included (Chapter 11).

1.10 RELATIONSHIPS

Throughout this book it will be evident that HVAC control systems do not exist by and for themselves. There is a symbiotic relationship among the building (or space), the HVAC system and the control system. Typically, when the environment is not properly controlled the blame is placed on the control system. But, in many cases, the real culprit is the HVAC system or the building.

The building must be properly designed to allow the degree of environmental control required. That is, to take an extreme case, a warehouse cannot be used as a clean room. Most examples are more subtle than that, but many such examples exist.

The HVAC system must be designed to provide the degree of environmental control required. A simple HVAC unit with one or two cfm per square foot and a 20-degree delta T will not be satisfactory for a hospital operating suite. Again, most design errors are not that gross, but many such errors exist. For example, refer to the discussion of *system gains* in Chapter 5. Only when the building and HVAC system are properly designed for the required service can the control designer provide a control system which will produce the desired level of control. Thus, the custom—which is still followed in many places—of calling in the control designer *after* the building and HVAC system have been designed is not conducive to a satisfactory solution. All the elements need to be considered together. The control designer must have a thorough understanding of these potential problems and be involved from the beginning of the design process.

With this in mind, it becomes apparent that the actual control devices used are subordinate to the function. The device must only have the required degree of accuracy and function in the desired manner. Computer-based controllers are better tools, sometimes, but even these sophisticated devices cannot solve the problems posed by inadequate building or HVAC system design.

1.11 SUMMARY

This chapter has discussed the elements of a control system, the basic types of control action, and the energy sources commonly used for controls. Many control systems use combinations of these energy types. Interface between unlike types of energy is provided by relays or transducers. The next chapters discuss the various types of control elements in detail. The chapters following the element descriptions will consider how these may be combined to make control *systems* of varying degrees of complexity to perform specific functions.

2 Pneumatic Control Devices

2.1 INTRODUCTION

Chapter 1 dealt with the fundamentals of control circuits. This chapter and the three following will consider various types of control devices: pneumatic, electric, electronic and fluidic. Succeeding chapters will discuss the use of these devices in control systems.

No attempt will be made here to provide an exhaustive catalog of control instruments. Rather, some of the basic principles of operation and general classifications available will be considered. With this as background, it will then be possible for the reader to evaluate a catalog description of a control device with a reasonable degree of understanding.

For simplicity, the chapters which follow classify control devices by energy type: pneumatic, electric, electronic and fluidic. This is followed by a discussion of flow control devices (valves and dampers) since these are essentially independent of the actuator.

Schematic diagrams are used, which avoid details that may obscure the situation and are peculiar to one manufacturer. These details are readily available from the manufacturer's service manuals.

2.2 PNEUMATIC CONTROL DEVICES

Pneumatic controls are powered by compressed air, usually 15 to 20 psig pressure, although higher pressures are occasionally used for operating very large valves or dampers. Pneumatic devices are inherently modulating, since air pressure may easily and simply be provided with infinite variation over the control range. Because of their simplicity and low cost, pneumatic controls are frequently found on commercial and industrial installations using more than 8 or 10 devices.

Pneumatic devices available include sensors, controllers, actuators, relays and transducers. These are described in some detail in the paragraphs that follow. The principles described form the basis of most manufacturers' designs although, of course, each has variations.

2.2.1 Definitions

With the exception of the first two the following definitions are peculiar to pneumatic devices.

1. Direct-acting: A controller is direct-acting when an increase in the level of the sensor signal (temperature, pressure, etc.) results in an increase in the level of the controller output (in a pneumatic system this would be an increase in output air pressure).
2. Reverse-acting is the opposite of direct-acting; that is, an increase in the level of the sensor signal results in a decrease in the level of the controller output.
3. SCFM: Standard cubic feet per minute. This refers to air at standard atmospheric pressure of 14.7 psia and a temperature of 70°F. For ease of comparison most air compressors are rated in SCFM.
4. Psia: Pounds per square inch, absolute pressure.
5. Psig: Pounds per square inch, gauge. At standard atmosphere, psig + 14.7 = psia.
6. SCIM: Standard cubic inches per minute. Similar to SCFM, but SCIM is usually used to describe pneumatic device air consumption. One SCFM equals $1728(12^3)$ SCIM.
7. Manufacturers, in their literature, will often use the terms "sensitivity" or "proportional band." These terms are synonymous with "gain."

Other terms such as bleed, non-bleed, submaster, are more easily described along with the device description.

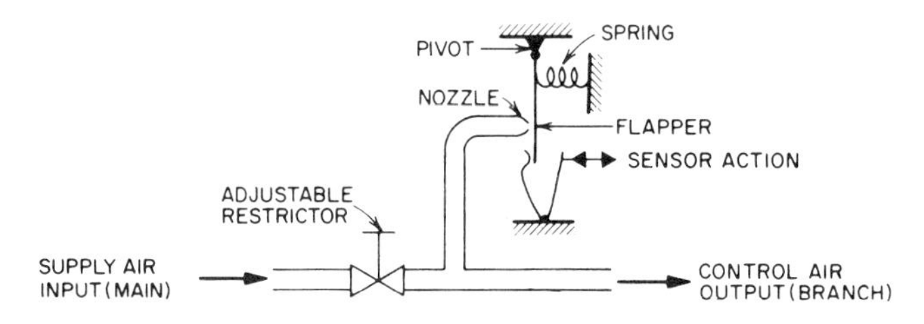

Figure 2-1 Bleed-type controller.

2.2.2 Bleed-Type Controllers

The bleed-type controller (Figure 2–1) is the simplest type of pneumatic control device. The control air ouput pressure is a function of the amount of air flowing through the nozzle; as the nozzle flow is restricted the output pressure rises. Nozzle flow is controlled by varying the position of a flapper valve, in accordance with the sensor requirement. The device may be direct- or reverse-acting, depending on the sensor linkage, and is inherently proportional. For two-position action a relay is required. The output signal may go directly to an actuator, or may be used as input to a relay-controller.

2.2.3 Relay-Type Controllers

Relay-type controllers may be either non-bleed or pilot-operated bleed-type. The non-bleed controller uses air only when exhausting the line to the controlled device.

Figure 2–2 is a schematic diagram of a non-bleed controller. A positive movement from the sensor (increase in temperature or pressure) will cause an inward movement of the diaphragm and a downward movement on the right end of the lever, which raises the other end of the lever and allows the supply air valve to open. This increases the output pressure and the pressure in the valve chamber. As the chamber pressure increases it acts on the diaphragm to offset the pressure from the sensor (negative feedback), so that when equilibrium is attained the supply valve closes and the balanced pressure becomes the output to the controlled device. This is, therefore, a proportional direct-acting controller with negative feedback. A further positive sensor movement will cause a rebalancing at some higher output pressure. A negative sensor movement will decrease the pressure on the right end of the lever, allowing the exhaust valve to open until the chamber and sensor

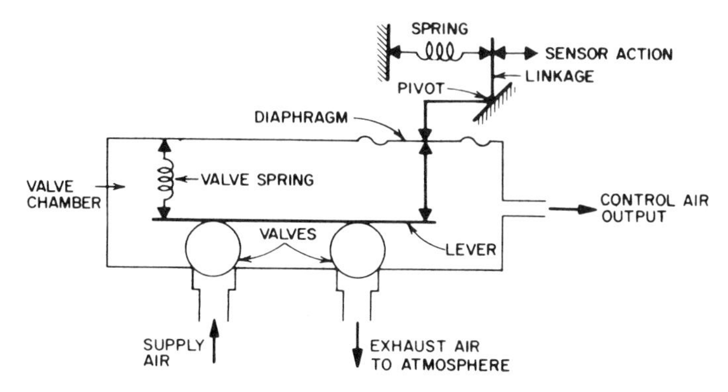

Figure 2-2 Non-bleed controller.

pressures are again in balance. A simple reversal of the linkage will change the controller to reverse-acting.

The bleed-type piloted controller utilizes a reduced-airflow bleed-type pilot circuit combined with an amplifying non-bleed relay to produce a sensitive, fast-acting control device. The controller can be adjusted to produce a large change in output for a small change in pilot pressure, and can be provided with negative feedback for proportional action or positive feedback for two-position action.

The proportional arrangement is as shown in Figure 2–3. The orifice plate is provided to restrict the flow of air to the pilot chamber. The control port may be partially or completely restricted by the flapper valve, which is operated by the sensor.

In Figure 2–4 the various operating conditions are shown: In Figure 2–4(A) the control port is open, the pilot chamber pressure is essentially zero, the

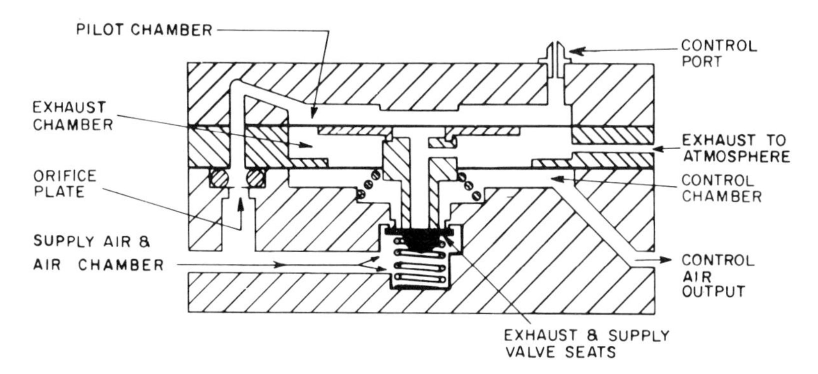

Figure 2-3 Proportional relay controller, pilot-bleed type. *(Courtesy Johnson Service Company.)*

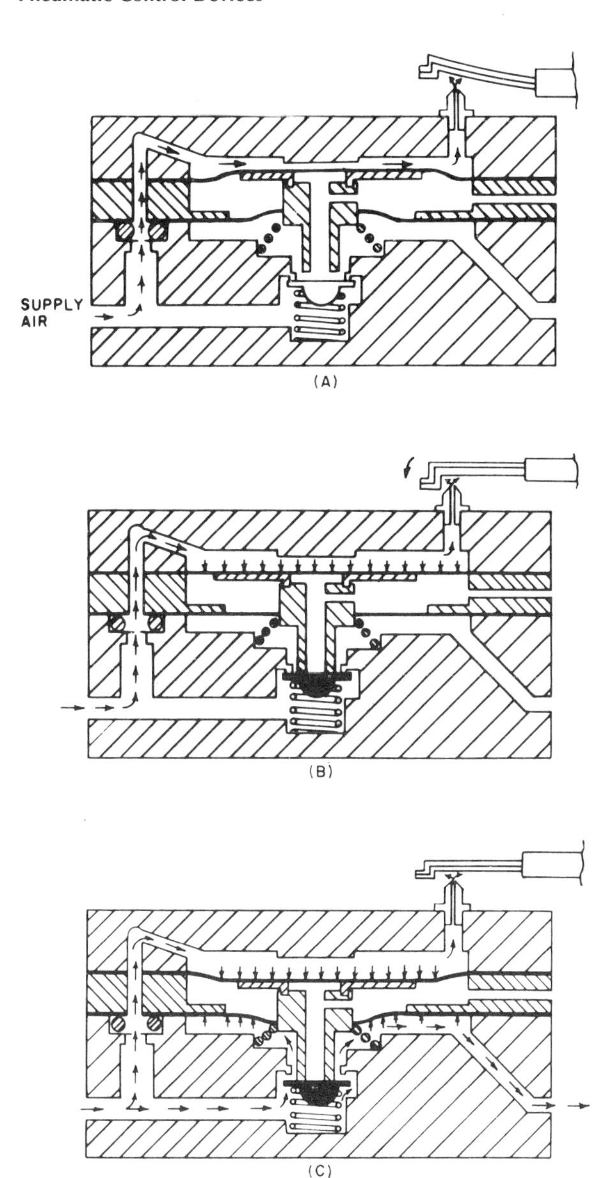

Figure 2-4 Operation of proportional relay. *(Courtesy Johnson Service Company.)* (A) When the control port is open, the exhaust valve between the control and exhaust chambers is open. Thus, air in the control chamber is at zero gage or atmospheric pressure. The supply valve is held closed by a spring and supply pressure. (B) When the sensing element moves closer to the control port, pressure begins to increase in the pilot chamber. At 3 psig, pilot pressure overcomes the force of the opposing spring and closes the exhaust seat. (C) As pilot pressure continues to increase, it forces the pilot diaphragm down and opens the supply port. This allows supply air to flow to the control line.

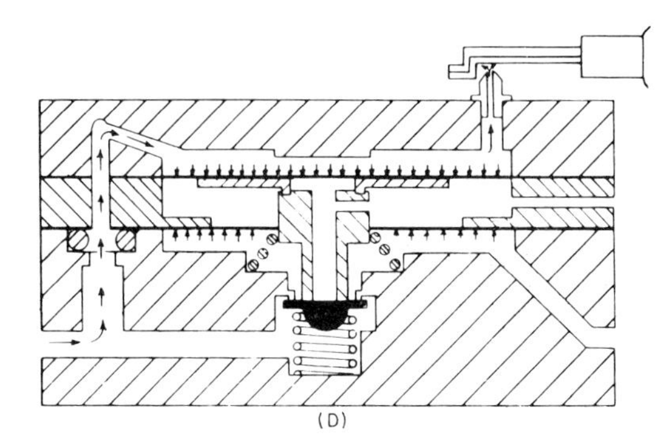

Figure 2-4 (Continued) (D) Pressure now increases in the control chamber, and acts against the control diaphragm to oppose pilot pressure. When the total of forces in each direction is equal, the supply valve is closed and the controller is balanced.

exhaust port is open and output is zero. In Figure 2–4(B) the flapper valve has been partially closed and the pressure begins building up. At some initial pressure, usually 3 psig, the spring allows the exhaust seat to close. In Figure 2–4(C), the pressure continues to increase, pushing the pilot diaphragm down and opening the supply valve. Air flows into the control chamber and to the output line. As the pressure in the control chamber increases the pilot pressure is opposed (Figure 2–4(D)) (negative feedback) until the pressures balance and the supply valve closes.

The two-position controller (Figure 2–5) is similar, but with some important differences in valve and spring arrangement. Figure 2–6 shows the operation: In Figure 2–6(A) the control port is open, and spring pressure holds

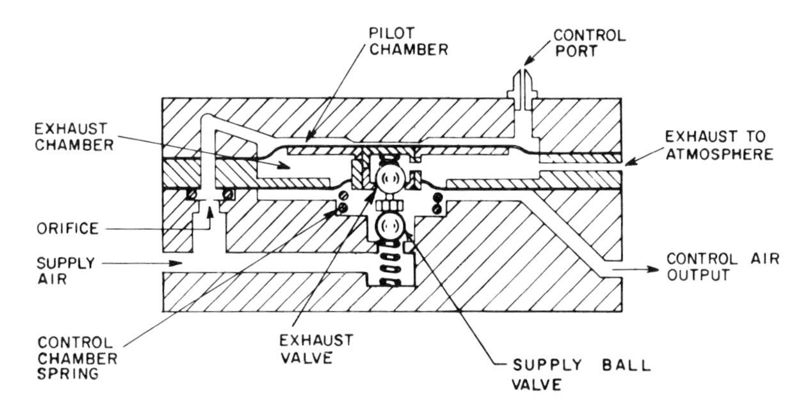

Figure 2-5 Two-position relay controller, pilot-bleed type. (*Courtesy Johnson Service Company.*)

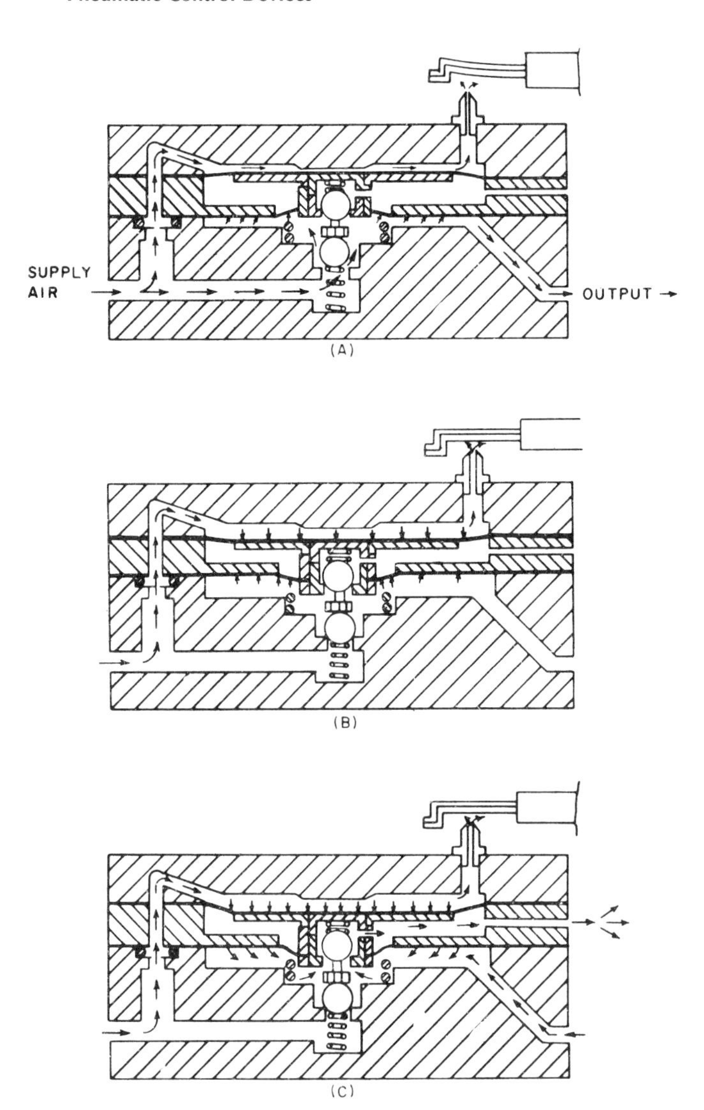

Figure 2-6 Operation of a two-position relay. (*Courtesy Johnson Service Company.*) (A) When the control port is open, supply air flows through the supply chamber and out the control chamber and the controlled devices. Spring and air pressure combine to hold the supply ball valve open and the exhaust valve closed. (B) As the control port is closed by the element, pressure in the pilot chamber increases. When pilot pressure is greater than the opposing forces, the pilot diaphragm moves, seating the supply ball valve. (C) Further movement of the pilot diaphragm opens the exhaust ball valve, so that control air is exhausted. This action reduces the forces opposing the pilot pressure (positive feedback) and causes the exhaust valve to open fully and remain open until pilot pressure is reduced (by opening the control port). The sequence is then reversed.

the supply port open and exhaust port closed. This is a reverse-acting controller. As the control port is partially closed, pilot pressure increases, forcing the pilot diaphragm down and closing the supply valve (Figure 2–6(B)). A further increase in pilot pressure opens the exhaust valve, causing a decrease in pressure in the control chamber. This decreases the force opposing the pilot pressure (positive feedback) and allows the exhaust valve to open completely and remain open until pilot pressure is reduced (Figure 2–6(C)).

Controllers of these types are provided with a broad sensitivity (gain) adjustment. For example, the gain of a temperature controller may be adjusted from as little as 1 psi per degree to as much as 10 psi per degree. (The latter would probably result in an unstable control loop.)

The mounting of the sensing element determines whether the controller is direct- or reverse-acting. Some devices may be changed in the field; others are fixed and cannot be changed.

2.2.4 Sensor-Controller Systems

Sensor-controller systems are used extensively in present HVAC control practice. Single or dual sensor inputs may be used. A single input sensor-controller is shown in Figures 2–7(A), and 2–7(B) and 2–7(C). The system operates as follows:

In the balanced condition, Figure 2–7(A), the following conditions exist:

1. The input diaphragm H force acting on the main lever I (which pivots about J) is balanced by the force of the set point adjustment T and the feedback diaphragm N force, which acts through the proportional-band lever L and adjustment K.
2. Exhaust nozzle P and main nozzle Q are both closed. In this condition, the pressure in the controller branch-line is at a value proportional to the requirements at the sensor.

On an increase in temperature, Figure 2–7(B), the brass tube B expands, tending to close flapper C against the nozzle V. This results in an increased pressure in the sensor-line G and a force on the input diaphragm H, which rotates the main lever I clockwise. The proportional band adjustment K rotates the proportional band lever L counterclockwise about fixed pivot M. Feedback diaphragm N is forced up, and flapper O rotates about nozzle P opening nozzle Q. Main air enters the feedback chamber W, increasing the branch-line pressure. As the pressure in the feedback chamber increases, the feedback diaphragm N is forced down until both nozzle P and Q are closed again, and the system is rebalanced at an increased branch-line pressure.

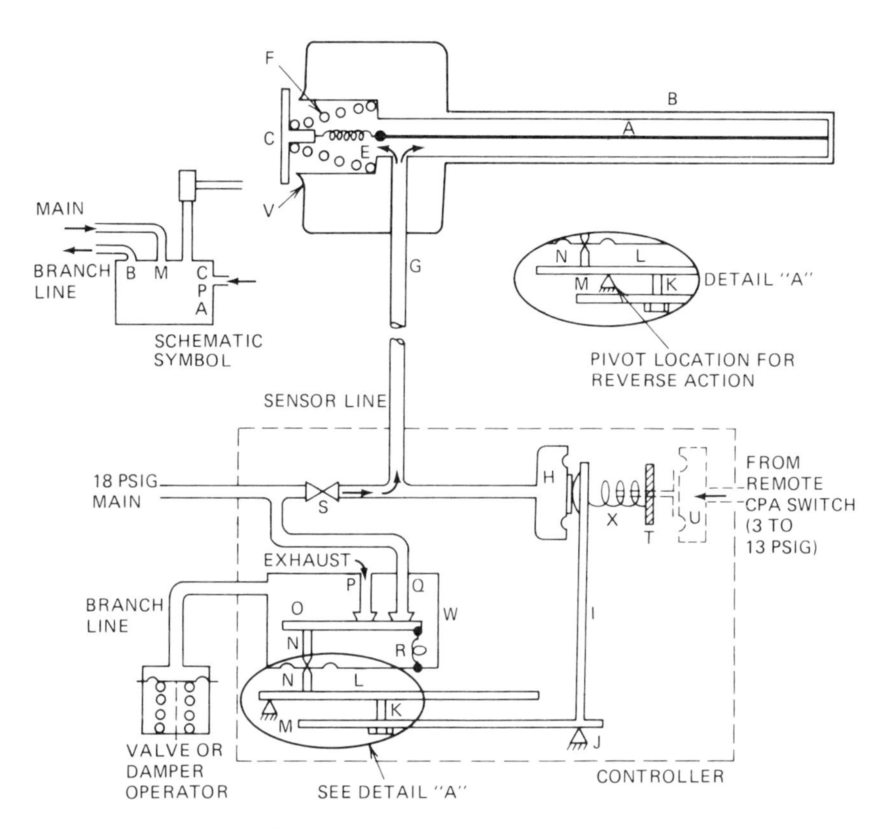

Figure 2-7(A) Diagram of a single-input, sensor-controller system in a balanced condition.

On a decrease in temperature, Figure 2–7(C), the brass tube B contracts, tending to open the flapper C. This results in a decrease in pressure in the sensor line G and a decreased force on input diaphragm H. The net force on feedback diaphragm N now unbalances the lever system. Flapper O rotates counterclockwise following feedback diaphragm N and opening exhaust port P. Feedback diaphragm N, in turn, forces the proportional band lever L to rotate clockwise, and the main lever I and proportional band adjustment rotates counterclockwise. Air is exhausted from the controller branch line and the feedback chamber W. As the pressure in the feedback chamber decreases, the feedback diaphragm N moves up until both nozzles P and Q are closed again, and the system is rebalanced at a decreased branch-line pressure.

For the dual-input controller shown in Figure 2–8, an additional arm is added to the main lever I to provide compensation or reset in the system.

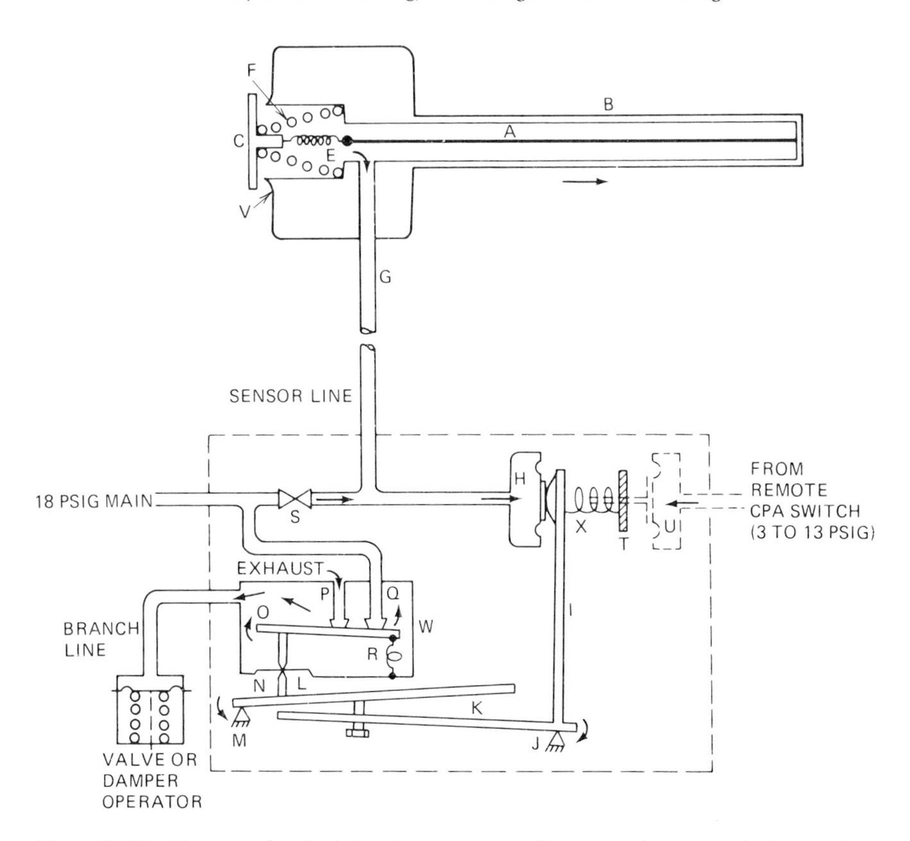

Figure 2-7(B) Diagram of a single-input, sensor-controller system increasing the branch-line pressure. *(Courtesy Honeywell, Inc.)*

Also, an additional input chamber A is added to allow a second force to be applied to the lever system. This force acts in the same direction as the input 1 chamber H force, regarding the main lever I.

In a typical application, compensation may be provided with this dual-input controller by using two remotely located sensors. To reset hot water, for example, the input 1 sensor is located in the supply water discharge and the input 2 sensor is located in the outdoor air. A drop in outdoor-air temperature reduces the input diaphragm A force. This has a similar effect on the main lever as increasing the set point adjustment spring X force. In other words, the drop in outdoor-air temperature raises the control point of the system. By setting the authority adjustment B, the relative effect of the input diaphragm A force may be varied, compared to the effect of the input diaphragm H force.

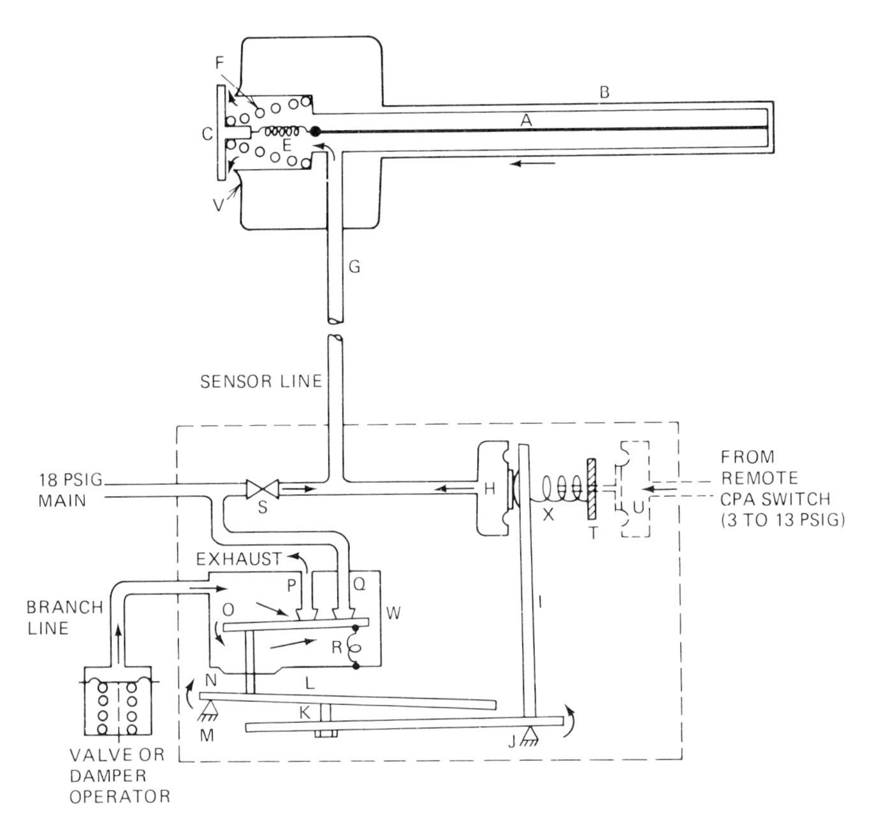

Figure 2-7(C) Diagram of a single-input, sensor-controller system decreasing the branch-line pressure. *(Courtesy Honeywell, Inc.)*

Considering the operation of the dual-input controller, an increase in the compensating medium temperature causes an increase in the input diaphragm A force. The resultant force acting on the main lever I through the authority lever C causes the main lever I to rotate clockwise about the pivot point J, effectively lowering the set point.

A decrease in the compensating medium temperature causes a decrease in the input diaphragm A force, reducing the force the authority lever C exerts on the main lever I, effectively raising the set point.

The system is shown direct-acting, but can be changed to reverse-acting as shown in detail "A" of Figure 2–7(A).

The CPA is a remote "control point adjustment," accomplished by varying the pressure at the CPA port using either a manual switch or a transducer controlled by a supervisory system.

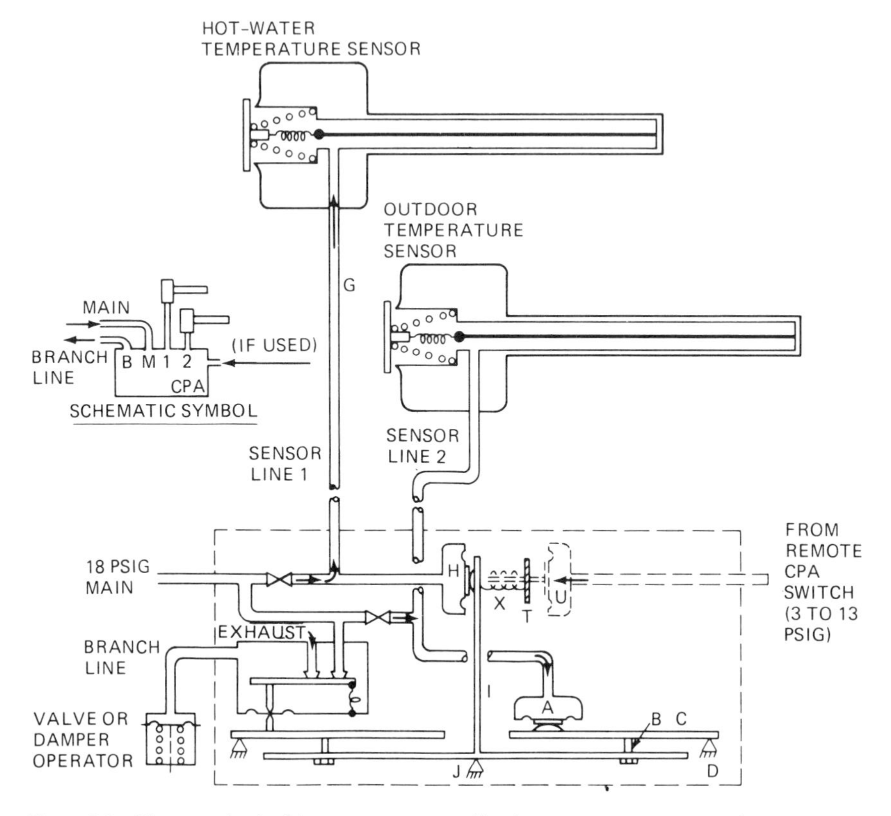

Figure 2-8 Diagram of a dual-input, sensor-controller, hot-water reset system with remote control point adjustment. *(Courtesy Honeywell, Inc.)*

These examples show temperature sensors but any kind of variable may be sensed—humidity, pressure, flow, etc.—with no change in the controller. Control differential and authority of primary and reset sensors may be adjusted over a fairly wide range at the controller.

2.2.5 Sensing Elements

Sensors have been referred to only in passing up to this point, noting that the controllers can be operated in response to various stimuli, such as temperature, pressure or humidity. A brief discussion of the more common sensing elements is in order.

2.2.5.1 Temperature-Sensing Elements include bimetal; vapor-filled bellows; and liquid-, gas- or refrigerant-filled bulb and capillary.

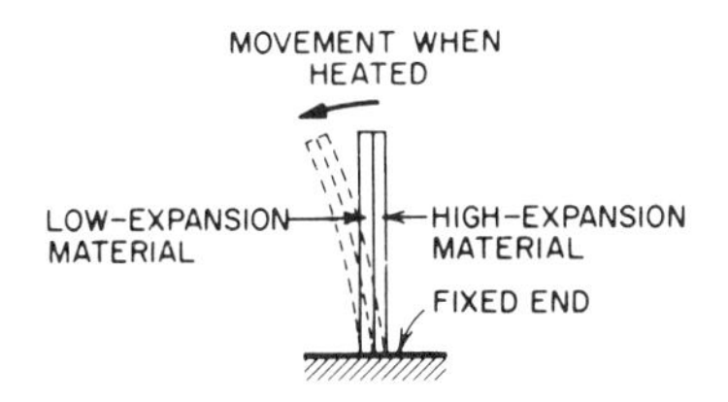

Figure 2-9 Bimetal sensor.

The bimetal element is the simplest and most often encountered sensor. It consists of two thin strips of dissimilar metals fused together. Since the two metals expand at different rates, the element bends as the temperature changes. The resulting motion can typically be used to vary the pressure at a pneumatic control port or to open and close an electrical contact. The device is most often made of brass (with a high coefficient of expansion) and invar metal which has a very low expansion coefficient. Figure 2–9 shows the movement of a bimetal element as it is heated. The bimetal may also be shaped in a spiral coil to provide a rotary motion output. The bimetal is often used in wall-mounted thermostats where it will sense ambient air temperature.

Another ambient temperature sensor is the vapor-filled bellows. The bellows is usually made of brass and filled with a thermally sensitive vapor, which will not condense at the temperatures encountered. Temperature changes will cause the bellows to expand and contract. An adjustable spring is used to control set point and limit expansion. The resulting movement can be used directly or through a linkage.

Bulb and capillary elements are used where temperatures must be measured in ducts, pipes, tanks or similar locations remote from the controller. There are three essential parts of this device: bulb, capillary and diaphragm operating head. The fill may be a liquid, gas or refrigerant, depending on the temperature range desired. Expansion of the fluid in the heated bulb exerts a pressure which is transmitted by the capillary to the diaphragm and there translated into movement (Figure 2–10).

The sensing bulb may be only a few inches long, as used in a pipe or tank, or it may be as long as 20 ft when used to sense average air temperature in a duct of large cross section. Special long bulbs are used for freeze protection. These will give a reaction if any one-foot section is exposed to freezing temperatures. Refrigerant is used in this type of bulb, since the refrigerant will condense at freezing temperatures, causing a sharp decrease in pressure.

Temperature-compensated capillary tubes are used to avoid side effects from the ambient temperature around the capillary. Capillaries may be as long as 25 or 30 ft.

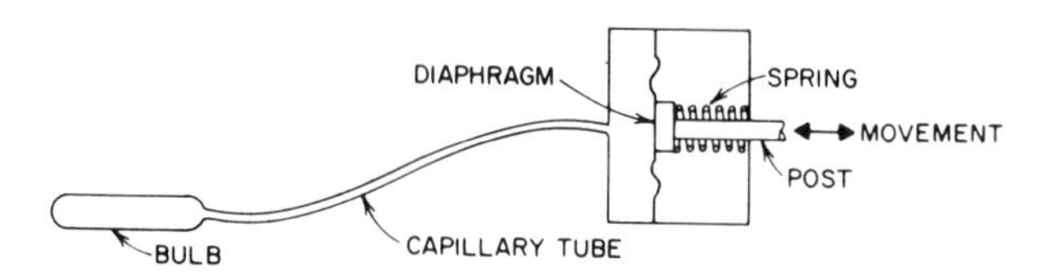

Figure 2-10 Bulb and capillary sensor.

Pressure on the diaphragm operates against an adjustable spring to move a post and/or lever. (See also 3.1.1.1 electric elements.)

2.2.5.2 Pressure-Sensing Elements include diaphragms, bellows and bourdon tubes.

The diaphragm is a flexible plate, sealed in a container so that fluid cannot leak past it. A force applied to one side will cause it to move or flex. A spring usually operates to control the movement and return the diaphragm to its initial position when the force is removed. Some diaphragm materials will spring back to the original shape without help. A wide variety of materials are used to cope with the various temperatures, pressures and fluids encountered.

A bellows is a diaphragm which is joined to the container by a series of convolutions (folds) so that a greater degree of movement may be obtained (Figure 2–11). The bellows may be completely sealed, as in a temperature-sensitive unit, or it may have a connection for sensing pressure, either internally or externally. The bellows acts like a spring, returning to its original shape when the external force is removed. Frequently a separate spring is added for adjustment and to increase reaction speed.

The bourdon tube (Figure 2–12) is widely used in pressure gages and other pressure instruments. It consists of a flattened tube bent into circular or spiral form. One end is connected to the pressure source and the other end is free to move. As pressure is increased the tube tends to straighten out, and this movement may be used, through an appropriate linkage, to position an indicator or actuate a controller.

2.2.5.3 Humidity-Sensing Elements are made of hygroscopic materials, which change size in response to changes in humidity. An ele-

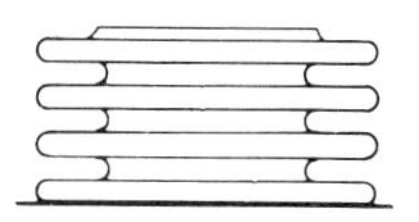

Figure 2-11 Bellows sensor.

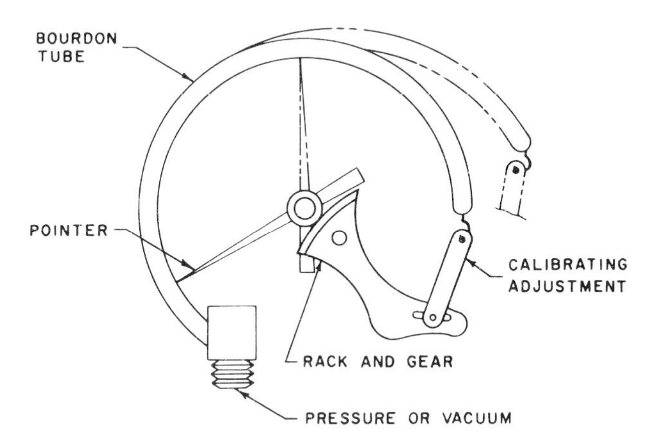

Figure 2-12 Pressure sensor, bourdon-tube type. *(Courtesy Johnson Service Company.)*

ment similar to a bimetal is made of two strips of unlike woods glued together. The different rates of hygroscopic expansion will cause the strip to bend as humidity changes. Yew and cedar woods are frequently used for this purpose.

Elements made of animal membrane, special fabrics or human hair will increase or shorten their length as humidity changes, with the resulting movement mechanically amplified. Current practice is to use nylon or similar synthetic hygroscopic fabrics.

2.2.6 Pneumatic Actuators

Sooner or later the output of a pneumatic controller activates a pneumatic actuator which positions a valve or damper. A pneumatic actuator is simply a piston and spring in a cylinder (Figure 2–13). When control air enters the cylinder it drives the piston to compress the spring until the spring pressure and the load on the connecting rod balance the air pressure. The stroke may be limited by adjustable stops. The connecting rod may drive a valve stem directly, or operate a damper by means of a linkage. Different spring ranges are available for sequenced control—for example, full 3 to 13 psi range is most often used but 3 to 8 psi, 8 to 13 psi and other ranges are available.

2.2.6.1 Positive Positioners A standard pneumatic actuator may not respond to small changes in control pressure, due to friction in the actuator or load, or to changing load conditions such as wind acting on a

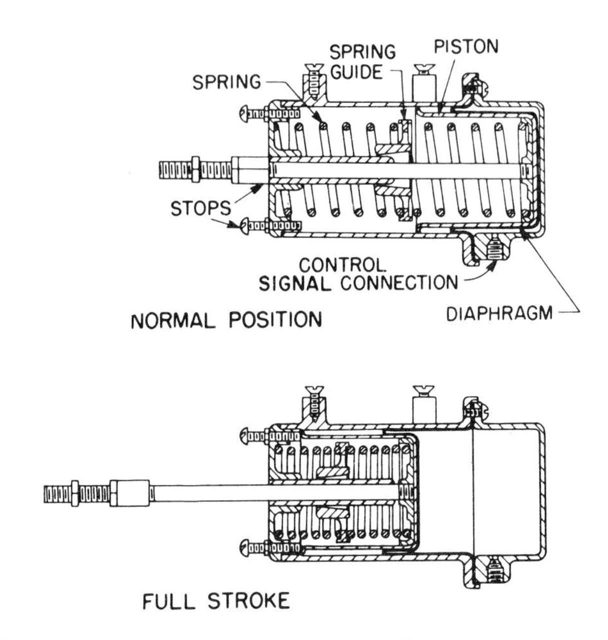

Figure 2-13 Pneumatic operator. *(Courtesy Johnson Service Company.)*

damper blade. Where accurate positioning of a modulating device in response to load is required, positive positioners are used.

A positive positioner (or positive-positioning relay) is designed to provide full main control air pressure to the actuator for any change in position. This is accomplished by the arrangement shown schematically in Figure 2–14. An increase in branch pressure from the controller (A) moves the relay lever (B), opening the supply valve (C). This allows main air to flow to the relay chamber and the actuator cylinder, moving the piston (not shown). The piston movement is transmitted through a linkage and spring (D) to the other end of the lever (B), and when the force due to movement balances out the control force the supply valve closes, leaving the actuator in the new position. A decrease in control pressure will allow the exhaust valve (E) to open until a new balance is obtained. Thus, full main air pressure is available, if needed, even though the control pressure may have changed only a fraction of a psi. The movement feedback linkage is sometimes mounted internally. Positioners may be connected for direct- or reverse-action. For large valves or dampers, main air to the actuator may be at a higher pressure than pilot air.

Some positioners have adjustments for start point and spring range, so that they may be used in sequencing or other special applications. *Start point* is the pressure at which the operator starts to move. *Spring range* is the pressure range required for full travel of the operator.

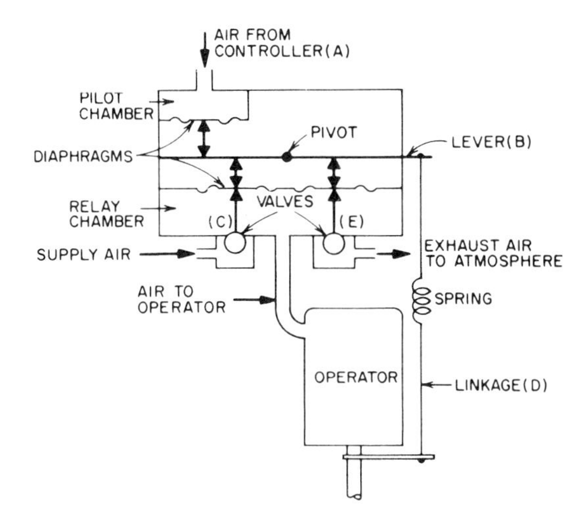

Figure 2-14 Positive positioner.

2.2.7 Relays

Many different types of pneumatic relays are manufactured. Mostly they use some variation of the non-bleed controller shown in Figure 2–2. An external force from another controller or relay replaces the force due to temperature or pressure change. Thus there may be a reversing relay (Figure 2–15) in which a direct-acting control input force may be changed to a reverse-acting output. The output of this and other relays may also be amplified or reduced with respect to input, so that one input to several relays may produce a sequence of varied outputs.

Another type of relay will produce an output proportional to the differ-

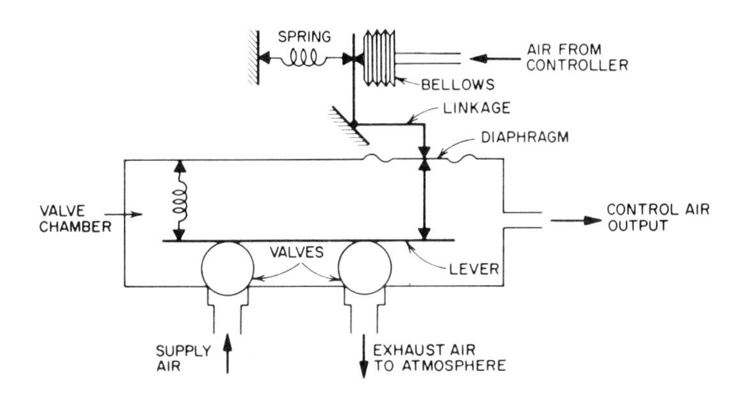

Figure 2-15 Reversing relay.

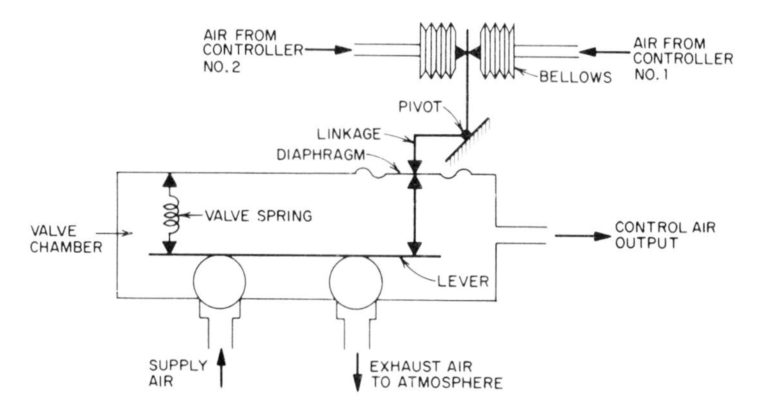

Figure 2-16 Relay: output proportional to difference between two signals.

ence between two inputs (Figure 2–16). Another produces an output equal to the higher (or lower) of two pressures (Figure 2–17). Two-position relays use the principles illustrated in Figures 2–5 and 2–6.

To illustrate sequencing, envision a controller with a 6 to 9 psi output over the desired control range. It is desired to operate three valves so that one goes from fully open to fully closed with a 3 to 8 psi signal, a second goes from closed to open with a 5 to 10 psi signal and a third has a control range of 8 to 13 psi from closed to open. These are to operate in sequence over the 6 to 9 psi control range. The first valve would utilize a relay which produced a 3 to 8 psi output from a 6 to 7 psi control input. The second relay would provide a 5 to 9 psi output over a 7 to 8 psi input and the third would provide an 8 to 13 psi output over the remaining 8 to 9 psi input change.

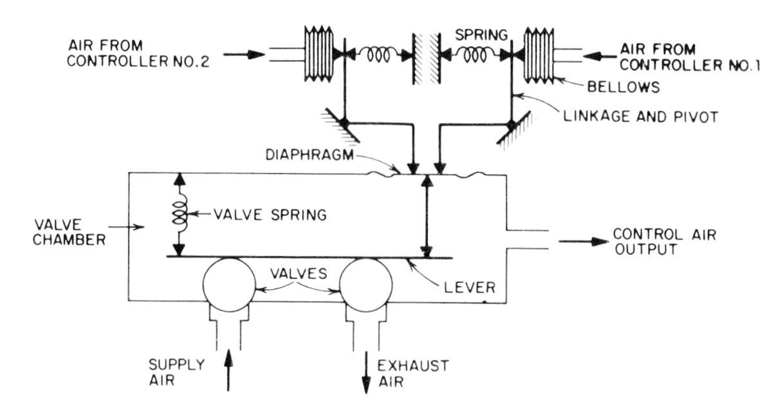

Figure 2-17 Relay: output proportional to higher of two pressures.

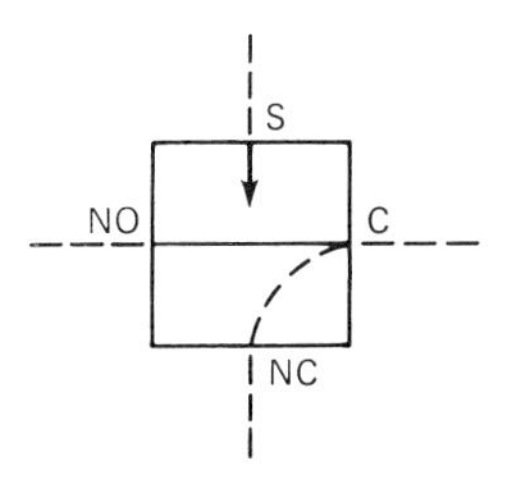

Figure 2-18 Switching relay.

The so-called "discriminator" relay will accept a large number of input signals—from six or seven to twenty, depending on the manufacturer—and select and output a signal output to the highest or lowest of the inputs. Some models will output *both* the highest and lowest. These relays are widely used in multiple zone HVAC systems for energy conservation. (See Chapter 7.)

An averaging relay will ouput the average of two to four input signals. It is thus similar in effect to the discriminator and may be more effective in some cases.

A switching relay is used to divert control signals in response to a secondary variable, typically outside air temperature. It is essentially a two-way valve. (See Figure 2–18.) With the switching signal above the set point the C port is connected to the NC port. With the switching signal below the set point the C port is connected to the NO port.

2.2.8 Master-Submaster Thermostats

The master-submaster arrangement was used extensively in control systems. It is seldom used in current practice, since the single controller with two sensors serves the same purpose. It consists of two thermostats, one of which (the master) senses some uncontrolled variable, such as outdoor temperature, and sends its output to the second (the submaster). This has the effect of resetting the set point of the submaster, which is sensing the controlled variable. Adjustments are provided for basic settings of both thermostats and for setting up the reset schedule, that is, the relationship between submaster set point and master output signal. Figure 2–19 shows the system schematically. Both instruments are basically of the non-bleed type. The master may also be a bleed-type controller.

This principle may be extended to utilize a single non-bleed controller with two remote bleed-type sensors as in Figure 2–20. In this case one sensor acts as a master to reset the set point of the other sensor. This arrangement is used extensively in current practice, both for master-submaster control

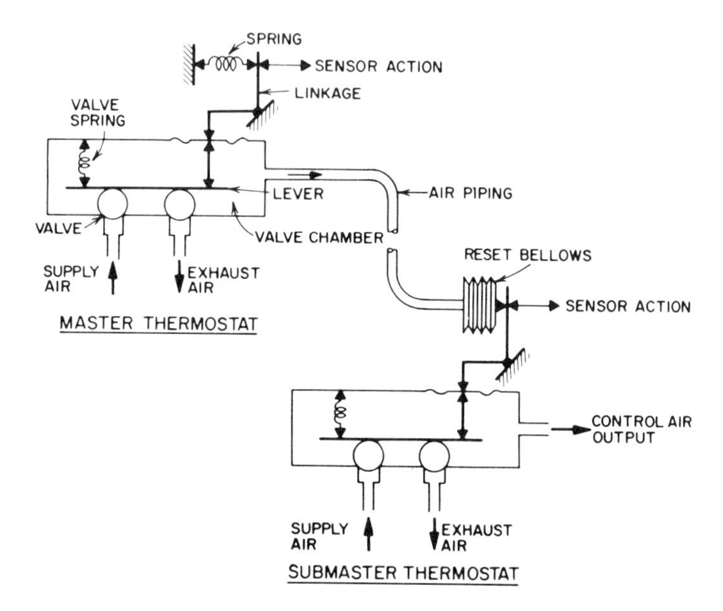

Figure 2-19 Master-submaster thermostat.

and, with one sensor, for single-point sensing and control without reset. It has the advantage that the output of the sensor is independent of the controller action, and can be used to transmit information to other devices, such as supervisory panels or pneumatic thermometers.

2.2.9 Dual-Temperature Thermostats

This classification includes day-night and summer-winter thermostats. Two related temperature settings may be made and the transfer from one to the other is made by changing the supply air pressure. The settings cannot overlap and there is a built-in differential, adjustable or fixed depending on the instrument.

A typical unit with two bimetal sensors is shown in Figure 2–21. The piloting relay is similar to those discussed previously. Two control ports are provided and the bimetals are adjusted to close one port sooner than the other. The port in use is determined by a diaphragm-operated transfer mechanism which is actuated by changing the main air supply pressure; typically from 15 to 20 psi.

This schematic illustrates a different way of providing negative feedback. The control air pressure moves a diaphragm which moves the levers supporting the bimetals.

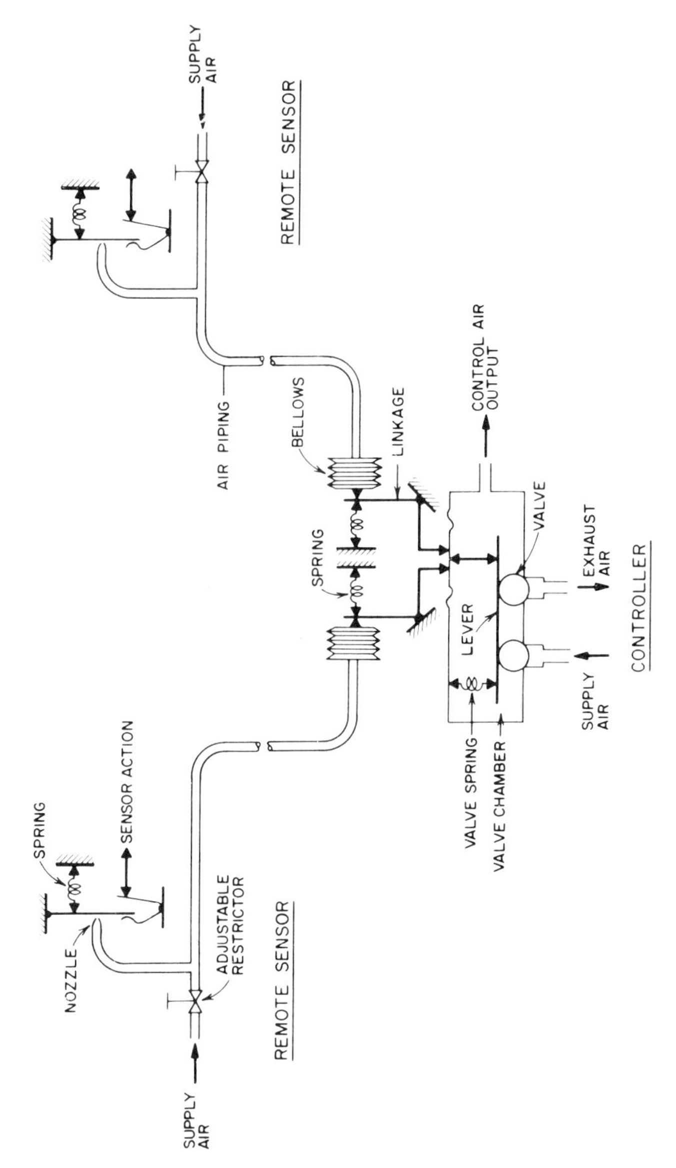

Figure 2-20 Controller with two remote sensors.

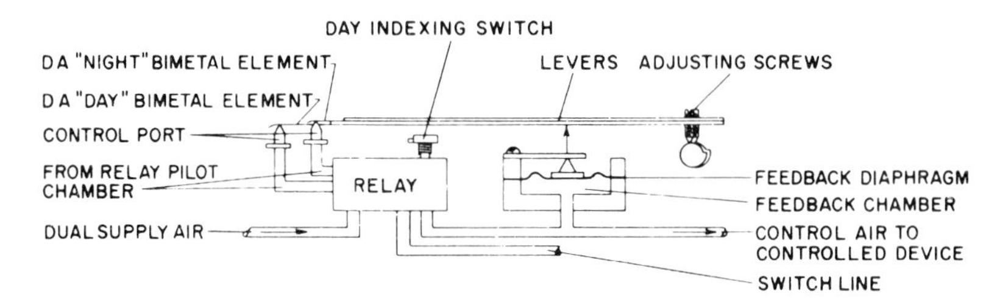

Figure 2-21 Dual-temperature thermostat. *(Courtesy Johnson Service Company.)*

If both bimetal elements are direct-acting, that is, they bend toward the control port on increase of temperature, then the thermostat functions as a day-night instrument. When used for heating, a lower temperature will be provided at night than during the day. If one of the bimetals is direct-acting and the other is reverse-acting, then the thermostat provides heating-cooling control, for year-round use.

Changeover from one supply pressure to the other may be accomplished in several ways: manually, by means of time clock or programmer, by means of an outdoor thermostat or by means of a sensor which detects supply water temperature in a two-pipe system. (Obviously, it does little good to call for heating when chilled water is being supplied!)

2.2.10 Dead-Band Thermostats

The *dead-band* or *zero-energy band* thermostat was developed to conserve energy. Its design is based on the assumption that the comfort range of temperature is fairly wide and no heating or cooling is required over that range. The dead-band thermostat therefore has a wide differential (dead-band) over which output remains constant as the temperature varies, with output changing only in response to temperature outside the differential range. (See Figure 2–22.) The set point is usually the upper limit of the dead-band range. Set point and dead-band are field adjustable.

2.2.11 Transducers

Broadly defined, a transducer is a device for transforming one form of energy to another. Since a pneumatic control system must frequently interface with electrical or electronic devices, transducers are necessary. These may be two-position or modulating.

The PE (Pneumatic-Electric) relay is simply a pressure switch which is ad-

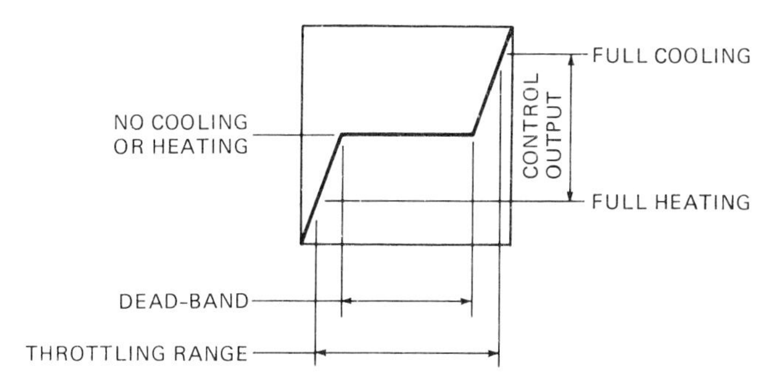

Figure 2-22 Dead-band thermostat performance.

justed to open or close an electrical contact at some value of the control pressure. Sequenced PE relays are frequently used for capacity control of refrigeration systems, for two-speed fan control and similar functions.

The EP (Electric-Pneumatic) relay is a solenoid air valve, usually three-way to supply or exhaust air to or from the pneumatic control circuit. A three-way valve may also be used to change supply pressure (by changing from one supply source to another).

Modulating transducers change a modulating air signal to a variable voltage output, in either the electric or electronic range; or a variable electric or electronic signal may produce a varying air pressure output. These will be discussed in the sections on electric and electronic controls.

2.2.12 Manually Operated Pneumatic Switches

Manual switches are used extensively in pneumatic control systems. A simple 2-position switch as shown in Figure 2–23 may be used to switch from one signal (A) to another (B). More commonly, control air is provided at port A, and port B is open to vent the line from AB when desired.

A more complex 2-position switch is shown in Figure 2–24. In one position ports 1 and 2 are connected as are ports 3 and 4. In the other position port 1 is connected to port 4 and port 2 to port 3.

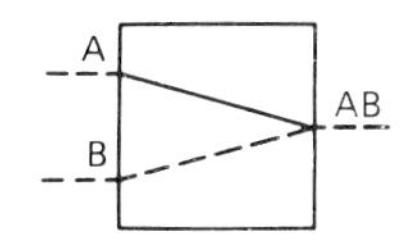

Figure 2-23 Manual switch; 2-position.

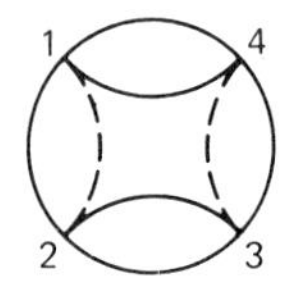

Figure 2-24 Manual switch; 2-position, 4 ports.

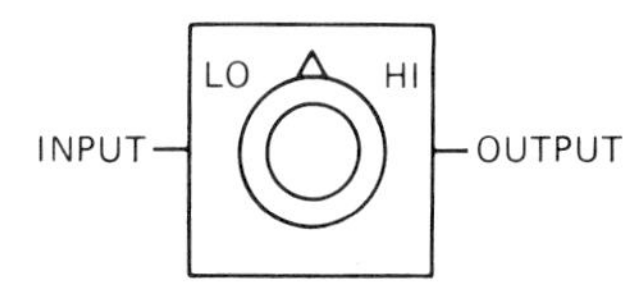

Figure 2-25 Hand "gradual" switch.

Many other switching arrangements are shown in manufacturer's catalogues.

Figure 2–25 shows a "gradual" switch which is used to provide a fixed output pressure (using control supply air as input). It is, therefore, an adjustable PRV (pressure reducing valve). Its primary function is to provide a minimum signal for a damper, valve, or speed controller. It is also used for set point (CPA) adjustment in sensor-controller systems.

2.3 CONTROL CABINETS

With any control system which includes more than 3 or 4 devices it is desirable to provide a control cabinet in which will be mounted all the controllers, relays, switches and indicating devices for the system. Control manufacturers supply standard cabinet sizes, up to about 36 by 48 inches. The cabinets are usually customized for the system. Indicating devices and switches are mounted on the door so that the cabinet may be closed and locked during normal operation.

Indicating devices may show status, temperature, humidity and control pressure at various points, as well as HVAC system static pressures, filter pressure drops and even air flow rates. Pilot lights may be added to show motor and safety control status.

While a control cabinet adds some cost, it is extremely useful for monitoring system operation and for trouble-shooting. Without some kind of status information it is very difficult to determine what is actually happening in the system.

Figure 2-26(A) Typical pneumatic control panel, door closed. Note temperature and static pressure indication, minimum outside air switch and selector switch. *(Courtesy Johnson Controls.)*

2.4 AIR SUPPLY

The air supply for a pneumatic control system must be carefully designed. It is of the utmost importance that the air be clean and dry, free from oil, dirt and moisture. Thus it is essential to use air dryers, oil separators and high-efficiency filters. Even small amounts of dirt, oil or water can plug the very small air passages in modern commercial pneumatic devices, rendering them useless.

Air consumption can be estimated from use factors for the components as shown by the control manufacturer. Good practice requires that the compressor have a capacity at least twice the estimated consumption. In com-

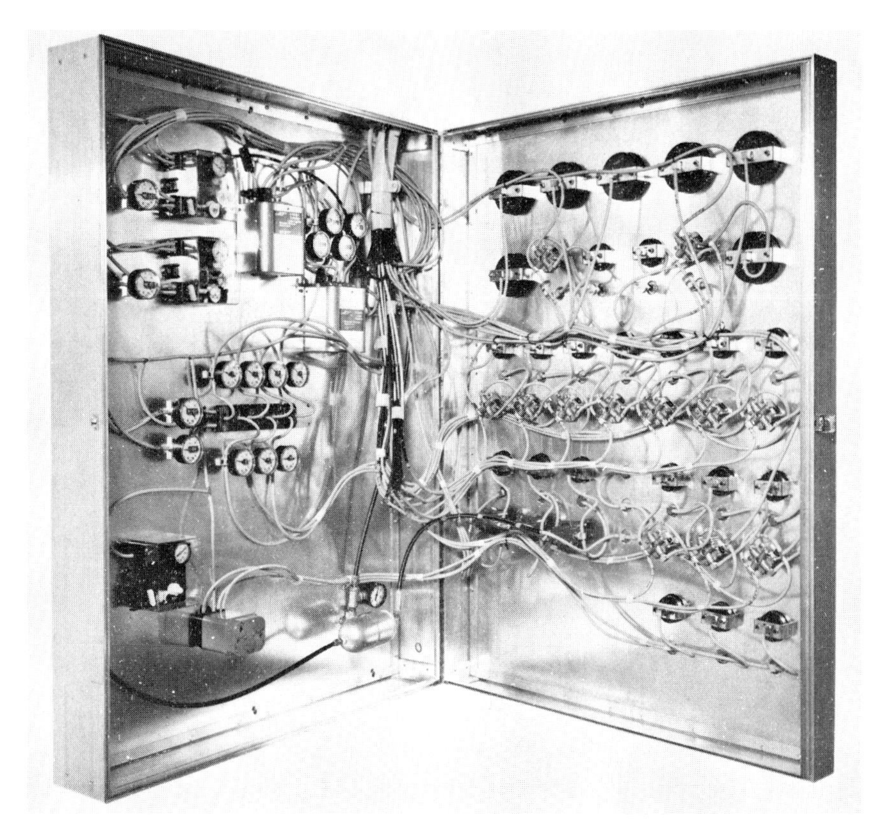

Figure 2-26(B) Typical pneumatic control panel with door open. Not the same panel as in the exterior view, this panel has a much larger group of display devices. Relays and receiver-controllers are mounted within the cabinet. Flexible tubing allows door to be opened and closed. *(Courtesy Johnson Controls.)*

paring compressor ratings it should be noted that *displacement* and *capacity* are not the same. Capacity will be from 60% to 80% of displacement.

If the system is very large, it is preferable to have two compressors to improve reliability. Generally air is compressed to 60 to 80 psig, and passes through driers, filters and a pressure-reducing valve. It is very important that the air be dry and clean, to minimize maintenance on the components.

2.4.1 Air Compressors

Compressors usually are reciprocating-type, single-stage, air-cooled. Typically they range in size from $\frac{1}{4}$ hp to 10 hp, and the smaller units are mounted directly on the receiver tank. Motors are started and stopped by a pressure switch on the receiver, set for 60 to 80 psig with about a 10 psi differential. When two compressors are used, the receivers are crossconnected and a single pressure switch with an alternator controls both motors. The alternator is a sequencing device which provides for running a different motor on each alternate start-stop cycle, thus equalizing motor running time.

2.4.2 Dryers and Filters

Two methods of drying are used. Refrigerated air dryers cool the air sufficiently to condense out the excess moisture which is then removed automatically by means of a trap, similar to a steam trap.

Chemical dryers are also used, with silica gel as the agent. For small systems an in-line unit may be used, and the chemical must be replaced at regular intervals. For large systems a double unit may be used, with one section being regenerated while the other is in use.

Oil may be removed in a coalescing filter, especially designed for oil mist entrapment and removal. The remaining oil and dirt may be removed in a three-micron high efficiency filter.

2.4.3 Pressure-Reducing Valves

The control manufacturer can and does supply the PRV as part of the pneumatic control system. If a dual-pressure system is used then two PRV's will be used, in series, as in Figure 2–27.

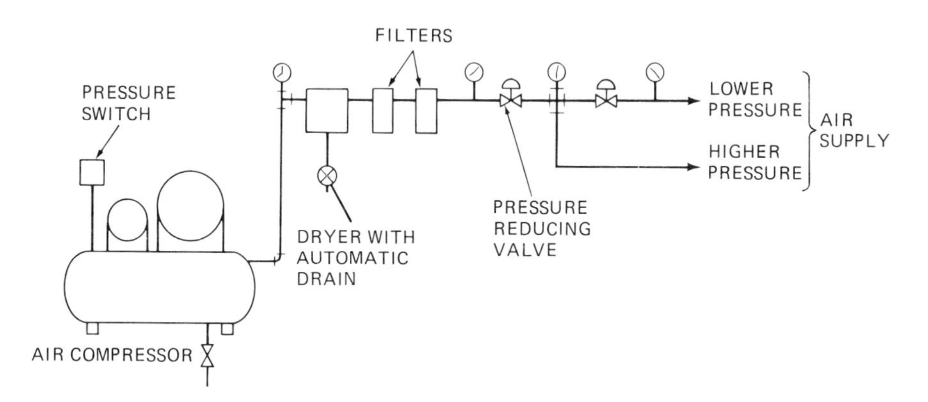

Figure 2-27 Compressed air supply.

2.4.4 Air Piping and Accessories

Air distribution piping systems almost invariably use soft copper tubing with soldered fittings. Compression fittings may be used at equipment connections. Branch piping is commonly $\frac{1}{4}$ in. outside diameter. Recent improvements in nylon-reinforced plastic tubing have led to acceptance of this material, with compression fittings, in accessible places, and in conduits. Specifications should call for a pressure test of the piping system at about 30 psi.

Air gages are a necessity for adjusting, calibrating and maintaining the control elements. Gages should be provided at main and branch connections to each controller and relay and at the branch connection to each controlled device. In this manner the status of the component is readily identifiable.

Linkages of various types are furnished for connecting operators to dampers. Many of these fit only a particular manufacturer's equipment, but many are universally adaptable. Manual valves of various kinds are available. Some of these are for shutoff only. Small needle valves may be used as adjustable restrictors which can be set in infinite steps to provide a desired maximum output pressure at a given flow.

3 Electric and Electronic Control Devices

3.1 ELECTRIC CONTROL DEVICES

Electrical controls are available in a wide variety of configurations, for every conceivable purpose in HVAC applications. All of these are based on one of four operating principles: the switch, the electromagnetic coil and solenoid, the two-position motor and the modulating motor.

It is highly desirable that the reader understand the basic principles of electrical circuits. This book will not discuss these principles, since they are readily available from many sources. The manufacturers' handbooks listed in the bibliography include some very good and simple presentations.

Any electrical circuit includes three elements: a power source, a switch and a load (Figure 3–1). The load represents resistance and consumes the power. The switch serves to turn the power on and off. In an HVAC control system the load will be an actuator or relay, the switch will be the sensor-controller. The power source is usually the building electrical power, which may be used at a normal 120 V or transformed to some lower voltage, typically 24 V. Some electrical devices use direct current (DC). This may be supplied by a battery or from an alternating current (AC) source by means of a transformer and rectifier.

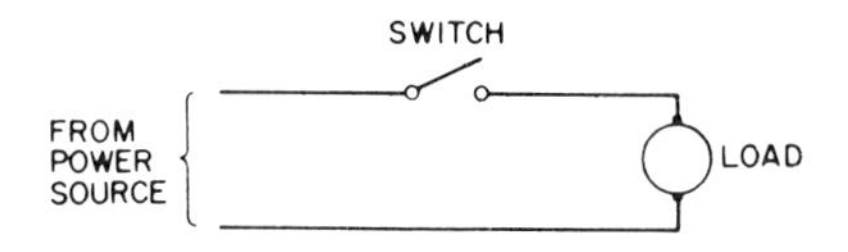

Figure 3-1 Electrical circuit.

3.1.1 Two-Position Controls

3.1.1.1 Sensors Any of the sensors described in the chapter on pneumatic controls may be used for two-position (on-off) electric control. Certain modifications may be necessary.

The bimetal is very commonly used in electric thermostats since it can serve to conduct electricity. Figure 3–2 illustrates a simple single-pole, single-throw (SPST) bimetal thermostat. When it is used for heating, a decrease in room temperature will cause the bimetal to bend toward the contact. When the contact is almost closed, a small permanent magnet affects the bimetal enough to cause a quick final closure and lock it in place. This magnet also causes a lag in the release, with resultant quick opening of the contact. This minimizes arcing and burning of the contacts, and eliminates chattering.

The bimetal may also be arranged in a spiral, fixed at one end and fastened to a mercury switch at the other (Figure 3–3). The mercury switch is simply a glass tube partially filled with mercury and with wiring connectors at one or both ends. It is loosely pivoted in the center so that when it turns past center the weight of the mercury running to the low end causes it to pivot farther. The mercury bubble acts as a conductor to connect the electrodes.

The mercury tube may also be used with a bourdon tube pressure sensor.

The diaphragm movement of bulb and capillary sensors can be used to trip an electric switch. Bellows sensors can be used in the same manner. The tripping action can be direct or through a linkage. A bistable spring or over-center mechanism is required to provide the snap action, as described above.

Humidity sensors may also be used to trip switches, through either bending action or expansion.

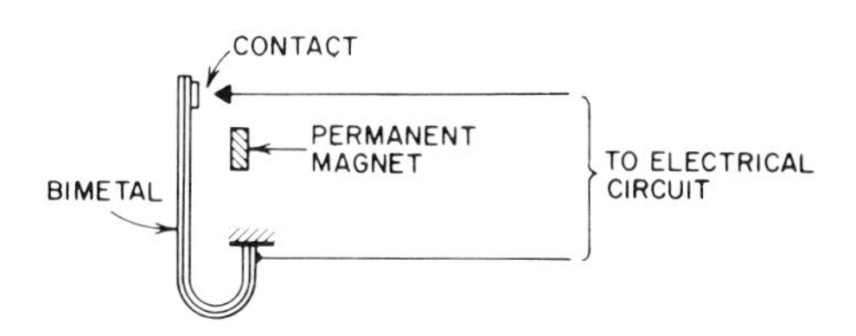

Figure 3-2 Bimetal sensor, electric.

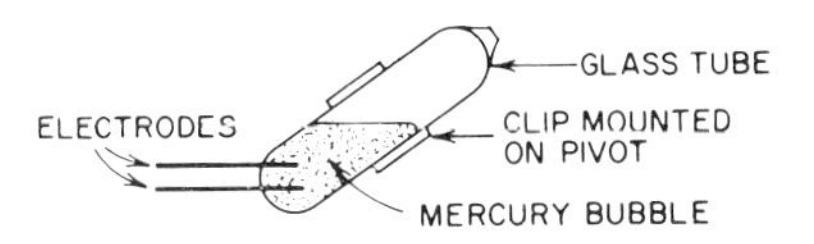

Figure 3-3 Mercury switch.

3.1.1.2 Safety Controls are used in HVAC systems for the detection of abnormally high or low temperatures and for smoke detection.

High temperature sensors or smoke detectors are required in most systems by NFPA and local codes. Low temperature sensors are used to prevent freezeup.

A high temperature sensor will usually have a bimetal or rod-and-tube element designed for insertion in the supply or return air duct. Factory temperature settings of 125°F to 135°F are provided. If the air temperature exceeds the control setting a switch will open and remain open until the device is manually reset. The control is commonly used to stop the air handling unit supply fan.

Smoke detectors for duct installation must be specifically designed for that use. The detector continuously samples the air stream in the duct and compares the sample with a standard. If products of combustion are detected a control contact is opened. Additional contacts may be provided for alarm service and reporting. The smoke detector may be used to stop the supply fan, but more often, in modern systems, is used to position dampers for smoke control and evacuation while the fan continues to run.

A low temperature duct sensor should have a long capillary so designed that a freezing temperature at any point will cause the sensor to open the relay contact. This prevents freezing due to stratified air streams. These devices may be automatically or manually reset.

3.1.1.3 Electromagnetic Devices Sometimes referred to as electromechanical devices, this class includes relays and solenoid valves, as well as motor starters.

All of these elements utilize the principle of electromagnetism. When an electric current flows through a wire a magnetic field is set up around the wire. If the wire is formed into a coil then the magnetic field may become very strong, and a soft iron plunger placed in proximity to the end of the coil may be drawn up inside it. This is the "solenoid" which can then be used to operate a valve or set of contacts.

Solenoid valves are made in many sizes and arrangements, for control of water, steam, refrigerants and gases. Figure 3–4 shows a typical two-way valve. This value is held in the normally closed position by fluid pressure.

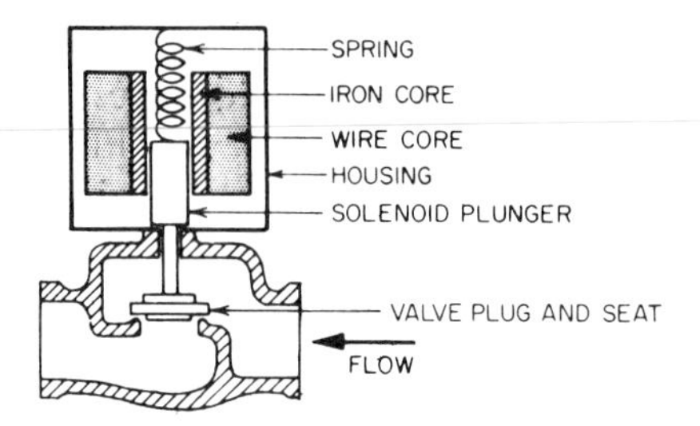

Figure 3-4 Solenoid valve.

When the coil is energized, the plunger is lifted and opens the valve. Some models are arranged with an internal pilot: a small port which is opened by the solenoid allowing fluid pressure to open the valve. Three-way arrangements are common and four or more ports are not unusual. The maximum size for a solenoid valve is about 4 in. pipe size. Large sizes lead to problems of pressure and water hammer due to quick opening and closing.

Control relays are designed to carry low-level control voltages and currents, up to about 15 A and 480 V. The contact rating of a control relay will vary with the voltage and with the type of load. The rating will be higher for a resistive load than for an inductive load. A relay can "make" a circuit with a much higher current than it can "break" without arcing and burning the contacts. The "breaking" capacity of the relay should therefore be the criterion.

One typical control relay configuration is the coil and solenoid as shown in Figure 3–5. The figure shows double-pole, double-throw (DPDT) contacts, but numerous arrangements from SPST to as many as eight poles are available. The armature is spring-loaded so that it will return to the "normal" position when the power to the coil is turned off. Typically, also, coils are available for most standard voltages, that is, 24, 48, 115, 120, 208, 240, 480.

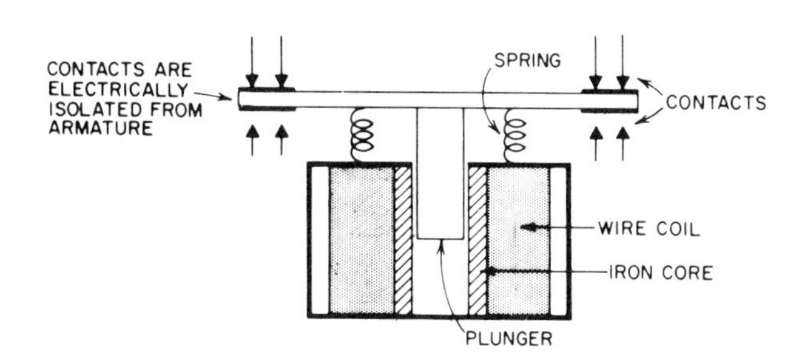

Figure 3-5 Relay coil and solenoid.

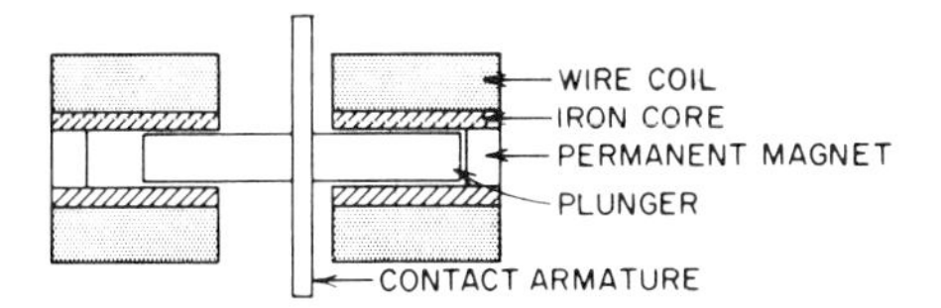

Figure 3-6 Permanent-magnet latching-type relay.

The solenoid-type relay is also available in "latching" arrangements, so that it may be driven to one position by a short-time energization of the coil and will stay in that position until returned by energizing a second coil. One latching system uses weak permanent magnets at each electromagnetic coil. These permanent magnets are not capable of displacing the solenoid, but will hold it in position (Figure 3–6). Another method uses a mechanical latch or detent, which is tripped when the electromagnetic coil is energized. The advantage of the latching relay is that no power is required to hold it in position. The disadvantage is that it does not "failsafe" when power is removed.

Electromagnetic coils are also used in "clapper-type" relays (Figure 3–7). The coil is mounted near a soft iron bar which is part of a pivoted, spring-loaded contact armature. When the coil is energized the arm is pulled over to close the contact. Several contact circuits may be mounted on a single armature. Double-throw contacts are also supplied. This type of relay usually, but not always, has a lower current and voltage rating than the solenoid type. The units are available as reed relays, and some miniature versions are made for electronics work.

Contactors are similar in all respects to solenoid-type control relays, but are made with much greater current-carrying capacity. They are usually used for electric heaters, or similar devices with high power requirements. A special kind of contactor uses mercury switch contacts, to allow for the frequent cycling required in electric heating applications.

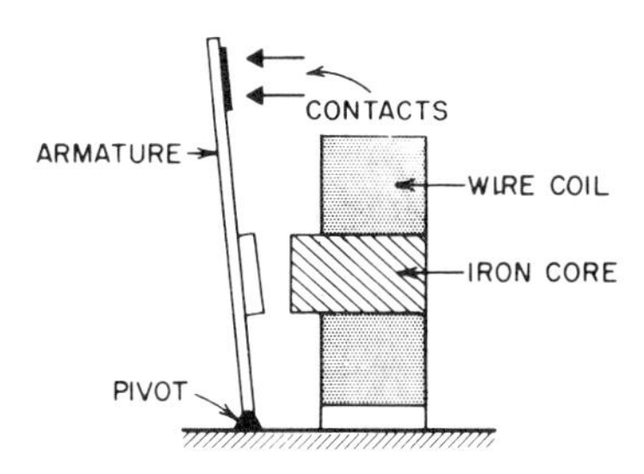

Figure 3-7 Clapper-type relay.

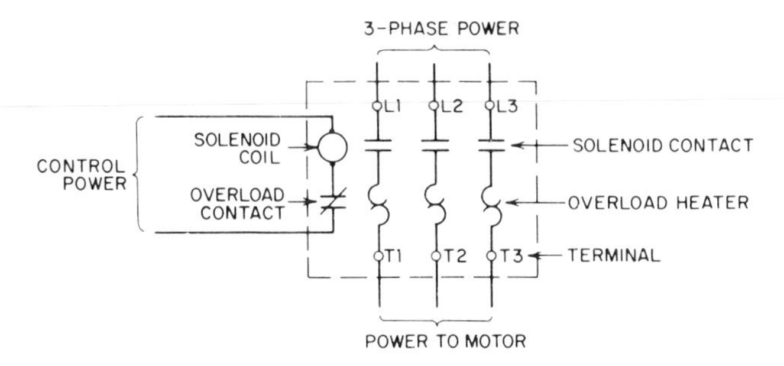

Figure 3-8 Motor starter.

Motor starters also use the solenoid coil actuator, and are similar to relays but with the addition of overload protection devices. These devices sense the heating effect of the current being used by the motor and break the control circuit to the coil if the current exceeds the starter rating. A typical across-the-line starter schematic is shown in Figure 3–8. This subject is discussed further in Chapter 8.

Time-delay relays, as the name implies, provide a delay between the time the coil is energized (on-delay) or deenergized (off-delay), and the time the contacts open and/or close. This delay may range from a small fraction of a second to several hours. Three general classes of time delay relays are available: solid-state, pneumatic and clock-driven.

Solid-state timers use electronic circuits to provide highly accurate very short delay periods. Timing ranges vary from 0.05 sec to 15 min or more. Pneumatic timers use the familiar solenoid coil principles but the movement of the solenoid is delayed by the diaphragm cover of a pneumatic chamber with a very small, adjustable leakport (Figure 3–9). In the on-delay sequence, when the coil is energized the solenoid pushes against the diaphragm, forcing the air out of the chamber. The leakport governs the rate of escape, and therefore the time delay. When the coil is deenergized, a check valve opens to allow the chamber to refill rapidly. Delays from about 0.1 sec to 60 min are available. Off-delay (after deenergization) is also available.

Clock timers utilize a synchronous clock motor which starts timing when the power it turned on. At the end of the timing period the control circuit contacts are opened or closed. When the power is turned off the device immediately resets to the initial position. Clock timers are available with ranges from a fraction of a second to about 60 h. Many special sequences and capabilities are used in process control.

Most of these time-delay relays may be provided with auxiliary contacts which open and/or close without delay, as in an ordinary relay.

Sequence timers include a synchronous motor which drives a cam shaft

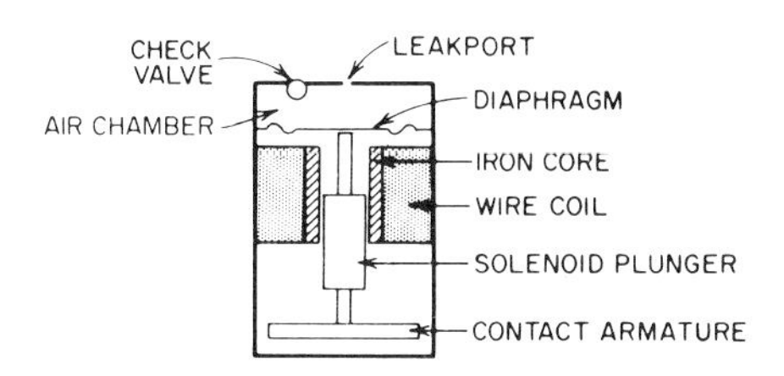

Figure 3-9 Time-delay relay, pneumatic type.

through a chain of reducing gears. Adjustable cams operate switches in any desired sequence. The timer may go through a single cycle, then stop; or it may run continuously, repeating the cycle over and over. Units with as many as 16 switches are available. Modulating or floating type motor control can also be used if needed. Then the sequence can be stopped at any point with the HVAC system stabilized at that condition. A typical use for this type of control is the sequencing of cooling tower fans.

3.1.2 Two-Position Motors

Two-position motors are used for operating dampers or for valves which need to open and close more slowly than a solenoid coil will allow. Motors may be unidirectional, spring-return; or unidirectional, three-wire.

A spring-return motor is shown in Figure 3–10. When the controller closes its SPST switch, the motor winding is energized from A to B. This starts the motor and it runs, driving (through a reduction gear) a crankshaft and linkage to open or close a valve or damper. A cam is mounted on the shaft and at the proper time (usually 180° of rotation but sometimes less) the cam throws the limit switch from B to C. This added coil resistance reduces the current to a "holding" level, which will hold the motor in this position but will not cause damage. When the controller switch opens, the spring returns the motor to its original position.

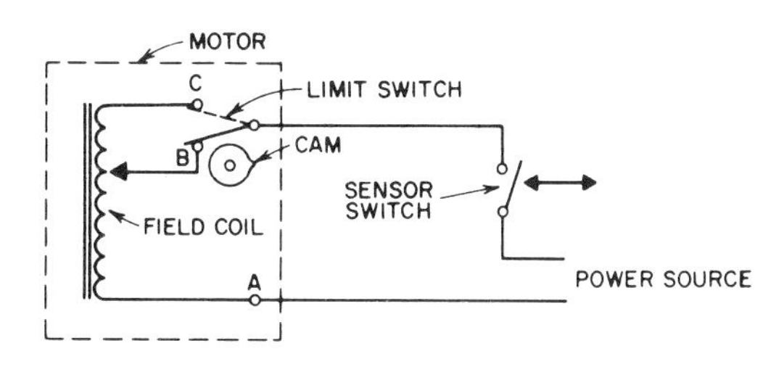

Figure 3-10 Two-position spring-return motor.

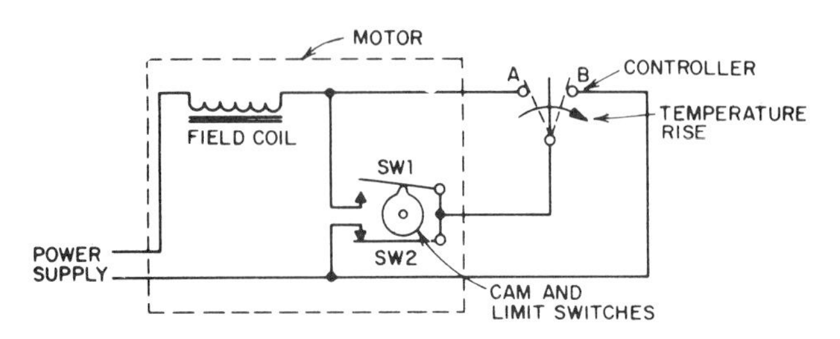

Figure 3-11 Two-position motor, three-wire arrangement.

Figure 3–11 shows a three-wire motor. For discussion purposes assume that this motor is operating a heating valve and the valve is closed in the position shown. A double-throw controller is required, and assume that this is a thermostat which closes contact B on temperature rise and contact A on temperature fall. The controller is shown in the satisfied position. The limit switches (marked SW1, SW2) are operated by a cam of the motor crankshaft. On a fall in room temperature the thermostat closes contact A establishing a circuit through SW2 to the motor field coil. The motor runs and almost immediately SW1 closes, establishing a maintaining circuit to the coil. Now the motor will run 180° regardless of what happens at the controller. When a 180° stroke has been completed SW2 is opened by the cam and breaks the circuit to the coil, stopping the motor. The valve is fully open. On a rise in temperature the controller breaks contact A and makes contact B, establishing a circuit through SW1 to the coil, and restarting the motor. As the valve starts to close SW2 closes and makes the maintaining circuit. When the valve is fully closed the cam opens SW1, stopping the motor.

In case of power failure the three-wire motor will stop and stay wherever it may be. The spring-return motor will return to normal position.

These motors may be provided with auxiliary contacts, cam-operated, which open and/or close at any desired point in the stroke.

3.1.3 Modulating Motors

Modulating motors are used for proportional and floating controls. They must, therefore, be reversible and capable of stopping and holding at any point in the cycle. The motors used in these devices will be either reversible two-phase induction motors or reversible shaded-pole motors.

3.1.3.1 Reversible Induction Motors Schematically, the two-phase induction motor (Figure 3–12) has two field windings directly connected at one end (C) and connected at the other by a capacitor. Power may be supplied at A or B on either side of the capacitor, with the other power

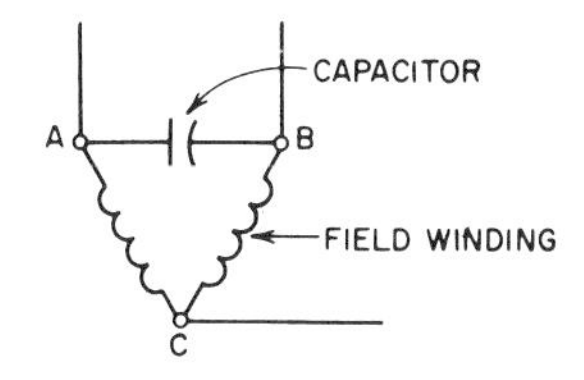

Figure 3-12 Reversible induction motor.

connection at C. If, for example, alternating current power is connected across A and C, then coil 1 is directly powered and coil 2 is indirectly powered through the capacitor. The effect of the capacitor is to introduce a phase shift between coils 1 and 2, thus a rotary motion is imparted to the motor armature. If the power is applied across B and C, the phase is reversed and the motor runs in a reverse direction. This unit may also be called a capacitor or condenser motor, although it is not the same as the capacitor-start, induction-run single-phase motor commonly found on large household appliances.

3.1.3.2 Shaded-Pole Motors The shaded-pole motor is constructed with a main field coil which is directly energized. However, by itself this will not start the motor. To provide a "biasing" effect shading coils are added as shown in Figure 3–13. These are powered by transformation effect

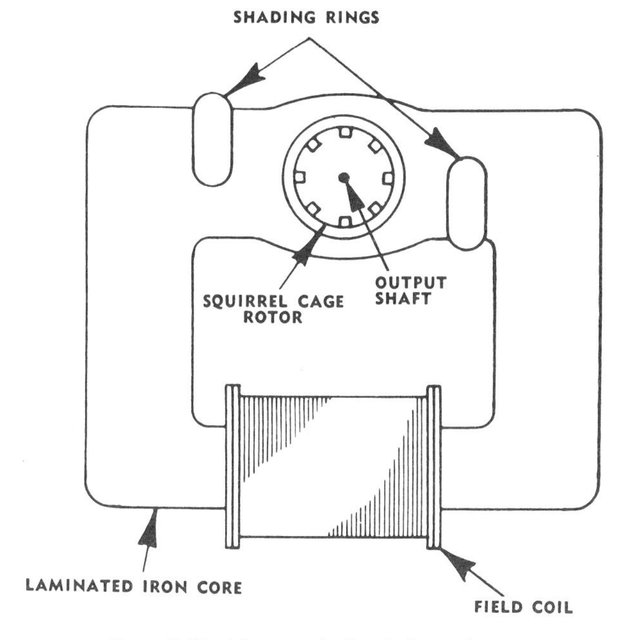

Figure 3-13 (*Courtesy Barber-Colman Company.*)

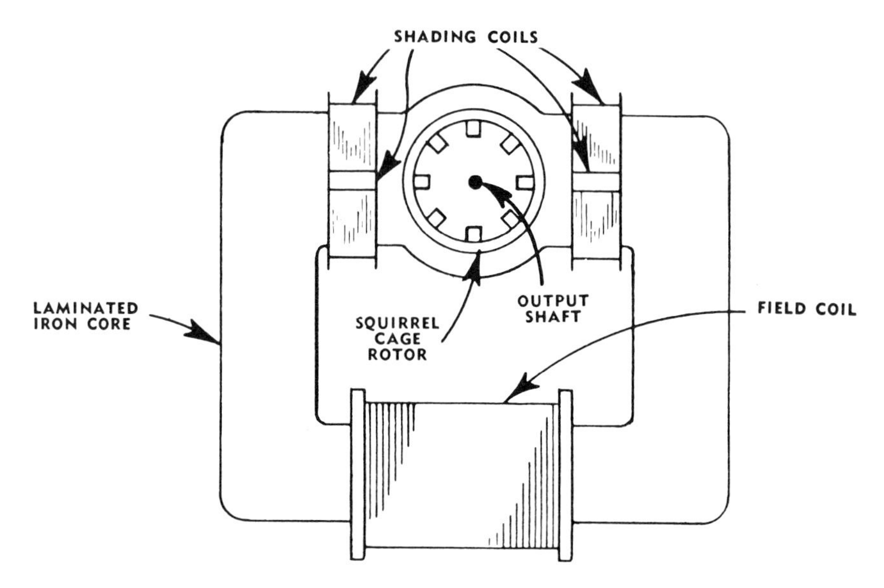

Figure 3-14 (*Courtesy Barber-Colman Company.*)

from the main field coil, and when both ends of the shading coil windings are shorted out, a phase lag is caused in part of the field. This causes a rotating field which starts the motor and improves its efficiency while running. This arrangement is unidirectional. To provide a reversing motor two additional shading coils are added (Figure 3–14) and wired as shown in Figure 3–15. Grounding one pair of shading coils causes the motor to run in one direction, grounding the other pair causes the motor to reverse rotation.

3.1.3.3 Modulating Motor Control Two general types of control configurations are used with modulating motors.

Floating control includes a three-wire controller with a center "dead spot" and no feedback. The motor is geared down to provide a slow change in the controlled device, so that sometimes system response can cause the sensor to return the switch to the center-off position before the motor has traveled the full stroke. Figure 3–16 shows a shaded-pole motor with this type of control. The limit switches are cam-operated at the end of the stroke (as described under two-position motors). The sensor controller is commonly called a "floating controller." This same controller may be connected in a similar manner to an induction motor.

Fully modulating control requires negative feedback at the motor as in Figure 3–17. Figure 3–18, which shows only the potentiometers and relay coils, will aid in understanding the following description. In the balanced

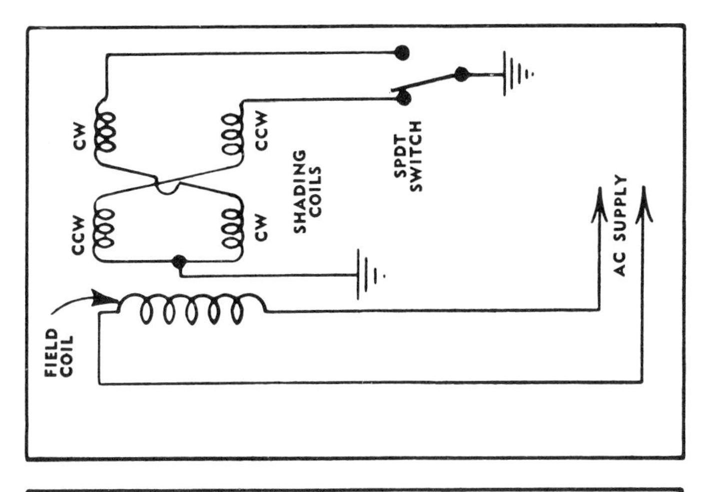

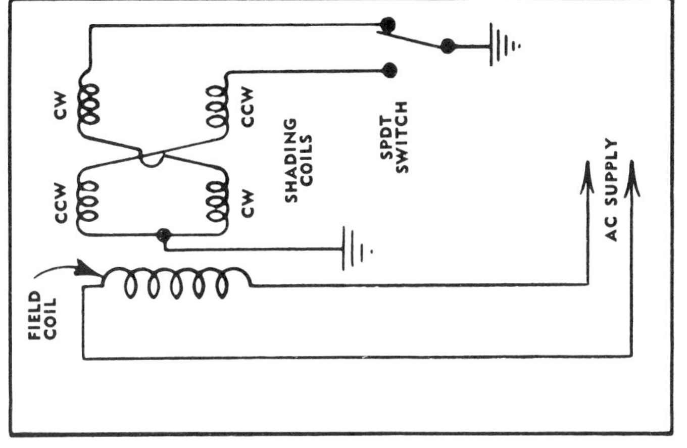

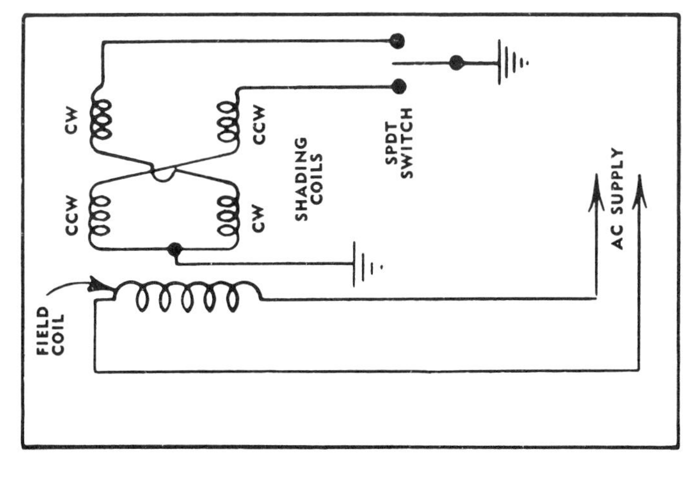

Figure 3-15 *(Courtesy Barber-Colman Company.)*

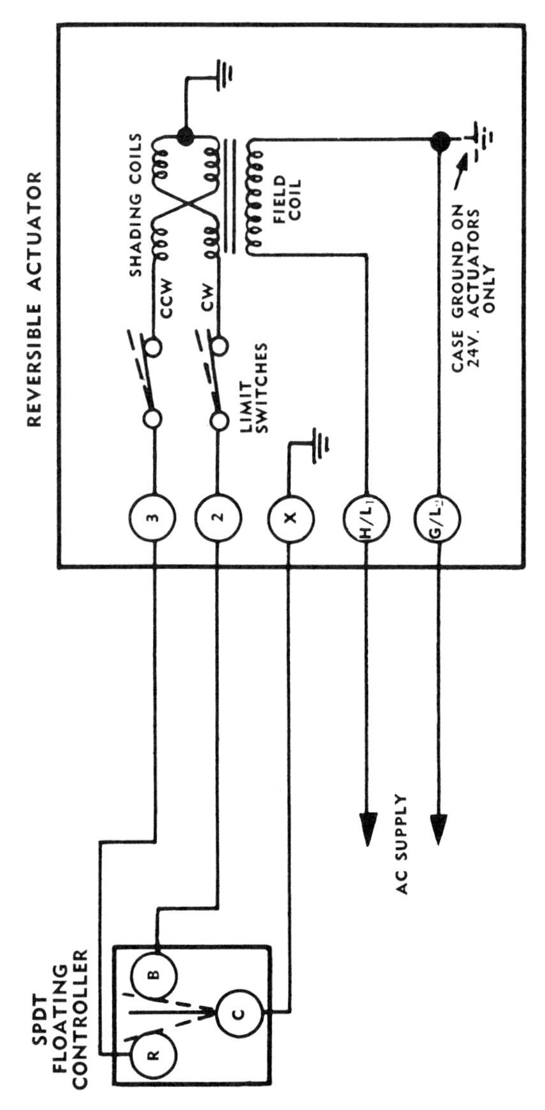

Figure 3-16 *(Courtesy Barber-Colman Company.)*

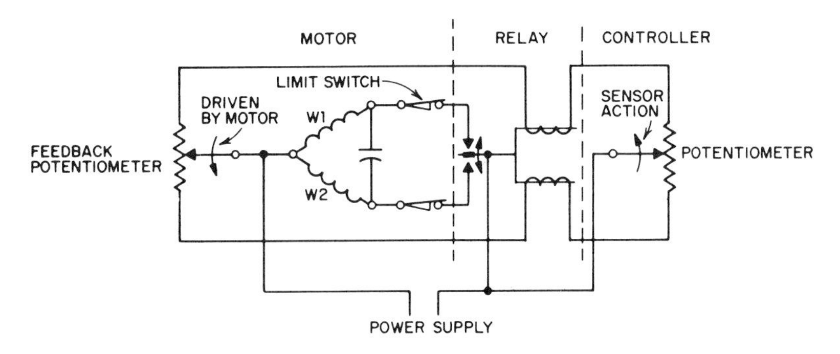

Figure 3-17 Modulating induction motor.

condition shown the wiper arm of the controller potentiometer is centered on its resistance winding, as is that of the feedback potentiometer. Under these conditions the currents through the two coils of the balancing relay are equal, and the relay arm is centered between contacts (Figure 3–18(A)). If the controller responds to temperature (or pressure) change and moves its potentiometer wiper arm, the circuits are unbalanced and the current differences in the two relay coils cause the relay arm to swing to one contact (Figure 3–18(B)). This starts the motor and causes it to actuate a valve or damper to offset the variation from set point. The motor operation also moves the feedback potentiometer wiper, which offsets the effect of the controller.

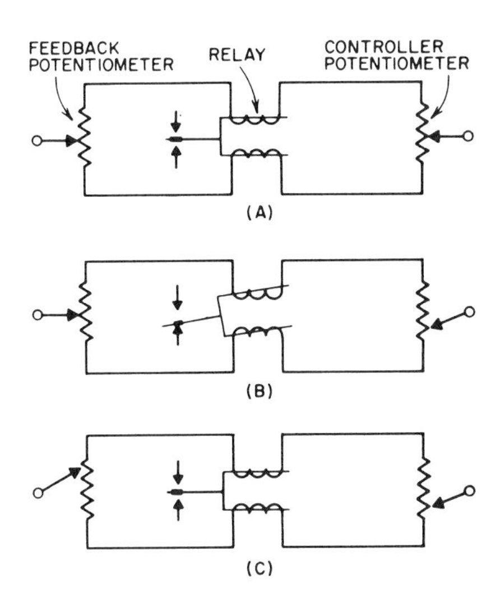

Figure 3-18 (A) Balanced position. (B) Controller moved, relay unbalanced. (C) Feedback moved, relay rebalanced.

When balance is restored (with the motor in a new position) the relay coil currents are again equal, the relay arm is centered and the motor stops (Figure 3–18(C)). The limit switches are needed to stop the motor at the end of the stroke if the sensed conditions are beyond the throttling range of the controller.

The modulating motor may be equipped with cam-operated auxiliary switches which open or close at any point in the stroke. It may also drive an auxiliary potentiometer which provides input to make another modulating motor "follow" the action of the first motor driving another valve or damper.

3.2 ELECTRONIC CONTROL DEVICES

Electronic controls are distinguished from electrical controls by the use of low voltages and solid-state devices. Power supply voltage is typically 24 V AC or DC but signal level voltage ranges are commonly 0–5 or 0–10 V. *Current* is also used as a signal, the usual standards being 4–20 Ma (milliamperes) or 10–50 Ma ranges. Simplified discussions of operating principles are included here but without details of construction and theory.

The increasing sophistication and decreasing cost of electronic devices, as well as their ease of interface with computer-based controls, are resulting in their more frequent use in preference to pneumatic devices.

3.2.1 Bridge Circuits

The original and most commonly used bridge circuit is the "Wheatstone bridge" (Figure 3–19). The bridge is formed by four resistances connected as shown. Power is connected to two "corners" of the bridge and output to the two opposite corners. One or more of the resistances may be variable (R4 is shown here as variable by the arrow across it.) When all resistances are equal the output is zero.

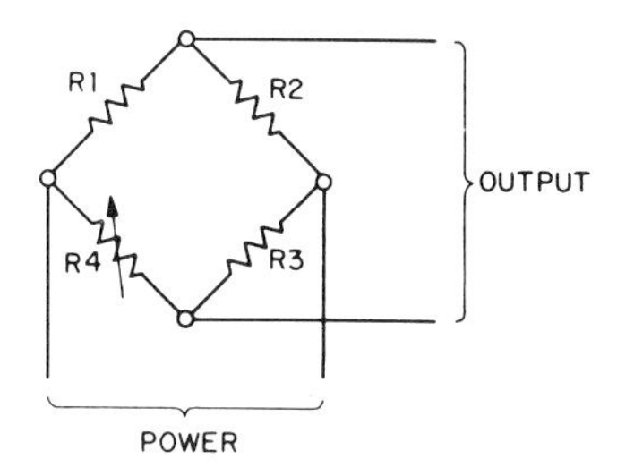

Figure 3-19 Wheatstone bridge.

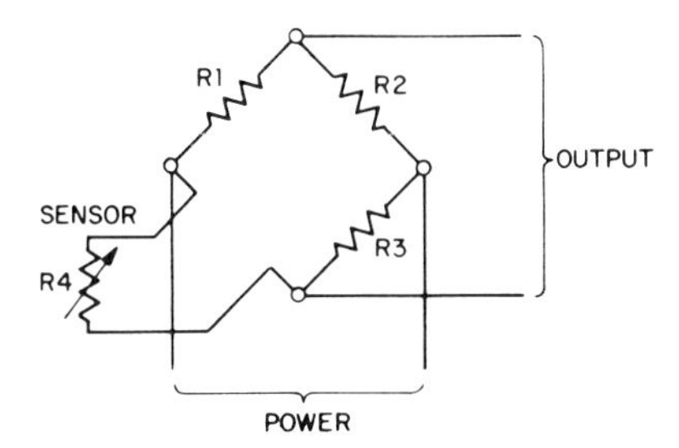

Figure 3-20 Wheatstone bridge with remote sensor.

Then if one or more of the resistances is changed the bridge becomes "unbalanced" and an output signal results which is approximately proportional to the resistance change.

In the ordinary electronic sensor, the variable resistor is the sensing element and is often mounted remote from the rest of the bridge and the amplifier (Figure 3–20). When remote mounting is required, some method must be used to compensate for the length of the conductors, and each control manufacturer has his own method of compensation.

This very simple arrangement does not allow for adjustment of set point, or calibration. For these functions it is necessary to add two more resistances (Figure 3–21). The set point adjustment is in series (or parallel) with the sensor resistor and the calibration adjustment is a potentiometer with an adjustable wiper arm.

To provide negative feedback for modulating controls a "throttling range bridge" must be added to the circuit. This is wired in series with the main bridge (Figure 3–22). It includes a variable potentiometer with a wiper arm which is driven by the controlled device motor. When the main bridge is

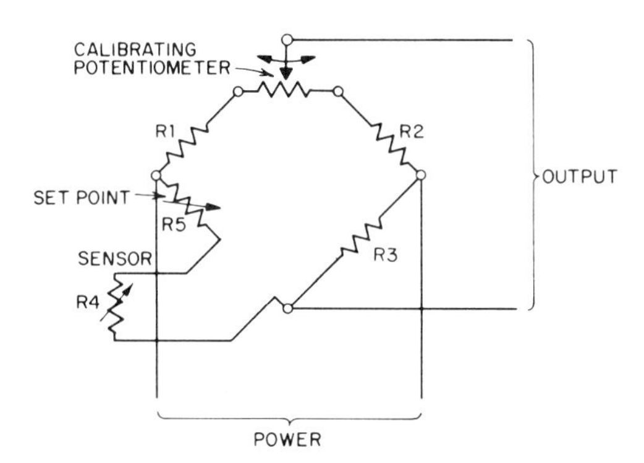

Figure 3-21 Bridge with calibration and set point.

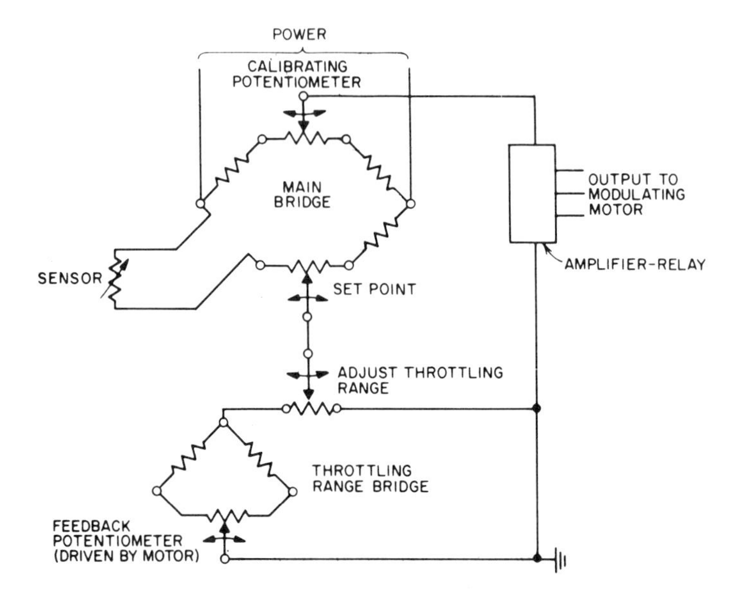

Figure 3-22 Throttling range bridge.

unbalanced and causes the motor to run, this potentiometer adjusts the throttling range bridge to offset the effect of the sensor and rebalance the system. In a simpler arrangement, the motor-driven potentiometer may be placed in series with the sensor, but then throttling range is not adjustable.

Sensing and relating two control points, such as room temperature and discharge temperature, may be done with a single bridge circuit (Figure 3–23). Here the two sensors are shown on opposite sides of the bridge. The room sensor has the same basic resistance as the other legs of the bridge. The discharge sensor however usually has only 10% to 20% of the basic resistance, with the balance being furnished by fixed resistors. This gives the discharge sensor less "authority," the amount being expressed as the ratio of discharge sensor resistance to room sensor resistance. Typical values are 1000 ohms for room sensor and 100 ohms for discharge sensor. Then the authority of the discharge sensor is 100:1000 or 10% which means that a 10° rise in discharge temperature is necessary to rebalance the bridge after a 1° fall in room temperature.

Thus, when room temperature decreases, the resistance of the room sensor R1 is decreased. This unbalances the bridge and causes an output to the amplifier. This may cause a hot water valve to open, which increases the discharge air temperature. This is sensed by the discharge sensor R2 and results in an increased resistance at the sensor until the bridge is rebalanced.

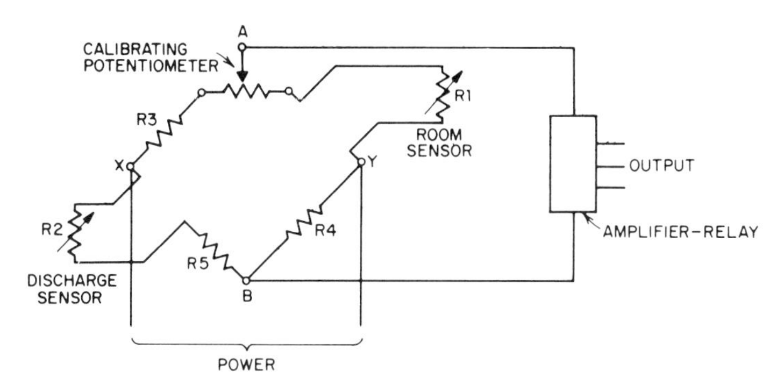

Figure 3-23 Bridge with two sensors.

Figure 3–24 will serve to clarify the situation. Figure 3–24(A) shows the bridge in the balanced condition, with a 12 V drop being shared equally by each of the two resistors in each side. Then there is no difference in potential across A–B; therefore no output. When the resistance of R1 decreases the voltage drop across it will also decrease, say to 5 V. But to maintain the total 12 V drop, the drop across R3 must increase to 7 V (Figure 3–24(B)). Now there is a difference in potential from A to B, which the amplifier will sense and utilize to operate a valve motor. As the discharge temperature increases, resistance R2 will increase until the voltage drop across R2 + R5 equals 7 V

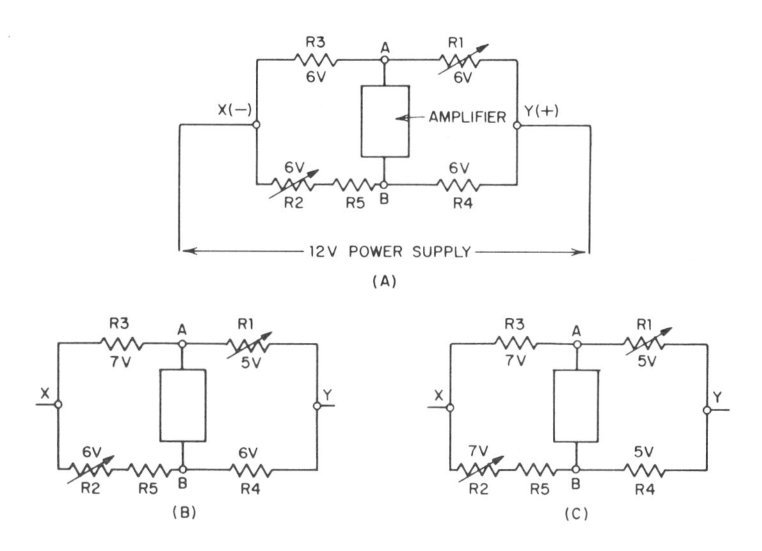

Figure 3-24 (A) Bridge balanced. (B) Room temperature decrease, bridge unbalanced. (C) Discharge temperature increase, bridge rebalanced.

(leaving 5 V across R4) and the bridge is rebalanced (Figure 3–24(C)). Output ceases and the valve motor is stopped at this position.

The system as shown in set up for heating. For cooling it is necessary to interchange the location of R4 with that of R2–R5.

3.2.2 Electronic Sensors

3.2.2.1 Temperature Sensors Electronic sensors for temperature sensing in modern HVAC systems are usually resistance temperature detectors (RTD), in which resistivity varies as a function of temperature. The least expensive and most common RTD is the *thermistor*, which uses a solid-state device in which resistance varies with temperature. A thermistor has a high reference resistance, typically 1000 ohms at 0° C. Thermistors tend to drift—get out of calibration—and need recalibration yearly or oftener. Unless special manufacturing techniques are employed a replacement thermistor may not have the same operating characteristics as the thermistor being replaced. Despite these maintenance disadvantages thermistors are popular and widely used.

Platinum RTD's, using pure-platinum wound-wire resistors, will retain their calibration indefinitely. Typically, the platinum RTD has a reference resistance of 100 ohms. This low resistance makes the resistance of the leads significant and three- or four-wire leads are used to compensate.

A fairly new development is the solid-state platinum RTD. This is a temperature-sensitive resistor made by deposition of a thin platinum film on a silicon chip. The device is very small and has a high reference resistance of 1000 ohms or more.

Thermocouples are seldom used in electronic control for HVAC. However there is one thermocouple application which is common in small residential gas-fired heating systems. A special form of thermocouple called a thermopile is inserted in the gas burner pilot flame. The heat of the flame generates enough electric current to power a special gas valve, through a bimetal thermostat. Since the small available current will not operate a conventional solenoid valve, the valve used is a balanced diaphragm type. A small solenoid pilot valve opens to allow gas pressure from the supply line to move the diaphragm and open the valve.

3.2.2.2 Humidity Sensors There are several types of electronic sensors for humidity. Synthetic fabrics which change dimensions with humidity changes are still used, but they have poor accuracy and need frequent recalibrations. A good device for measuring dew point uses a tape impregnated with lithium chloride and wound with two wires which are connected to a power supply. As the lithium chloride absorbs moisture from the atmosphere it creates an electrical circuit which heats the system until it is in balance with the ambient moisture. The resulting temperature is measured. The device is quite accurate but requires frequent maintenance to retain accuracy. Solid-state humidity sensors (Figure 3–25(A)) use thin or thick polymer film elements so that resistance or capacitance varies with relative humidity. The most accurate dew-point sensor presently available is the chilled mirror type. (Figure 3–25(B)) A polished stainless-steel mirror is provided with a small thermoelectric cooling system. A light beam is reflected from the mirror to a photo-cell. When the mirror is cooled to the ambient dew-point temperature, moisture condenses on it. The resulting change in light reflectivity is noted and the surface temperature is measured by a platinum RTD. This is the dew-point, usually accurate to less than 1°F. The only maintenance required is periodic cleaning of the mirror. The relative humidity is obtained by means of a transmitter/signal conditioner or, if the sensor is connected to a computer, by suitable software.

3.2.2.3 Pressure Sensors Electronic pressure sensors can use the same sensing elements as pneumatic and electric devices while providing an "electronic-level" signal. Peculiar to electronic circuitry is the *strain gage*,

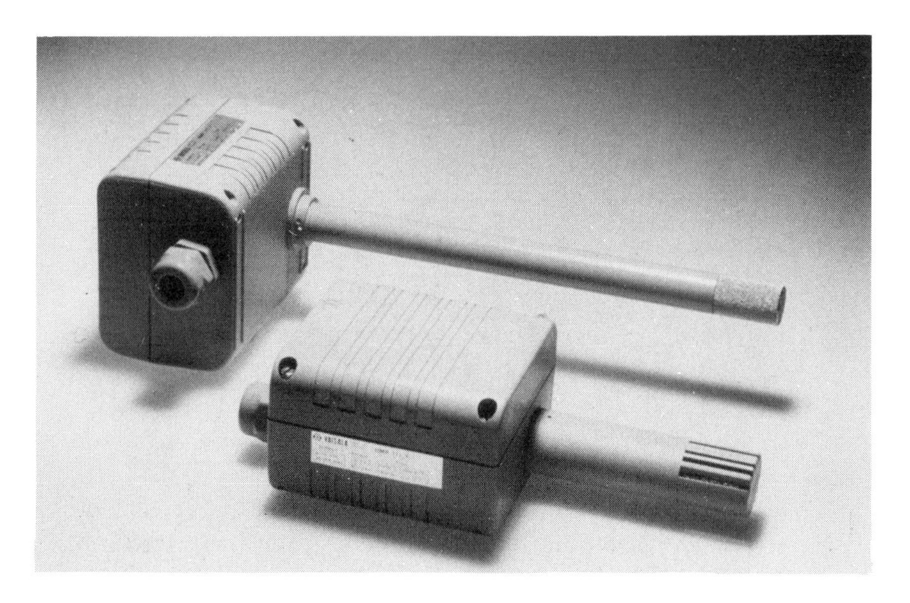

Figure 3-25 (A) Solid state humidity sensor. (*Courtesy Vaisala, Inc.*)

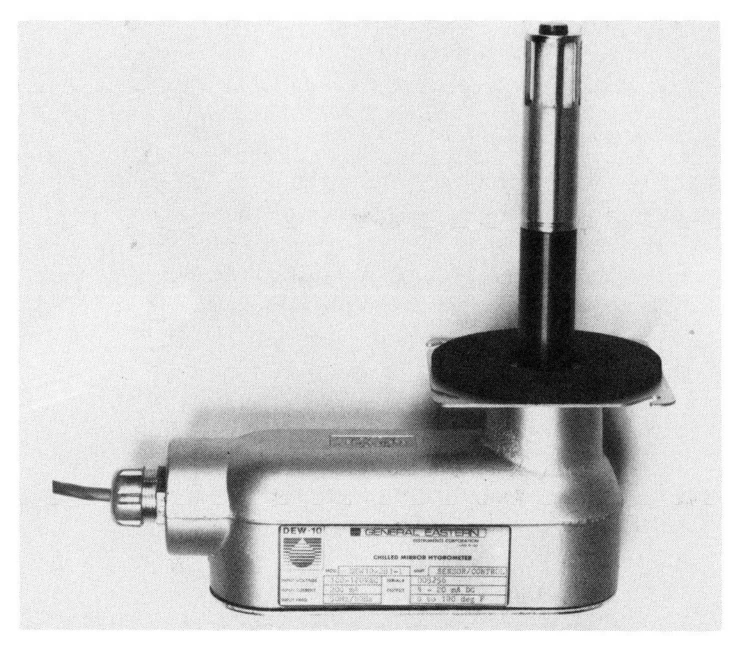

Figure 3-25 (B) Chilled mirror dewpoint sensor. (*Courtesy General Eastern Instrument Corp.*)

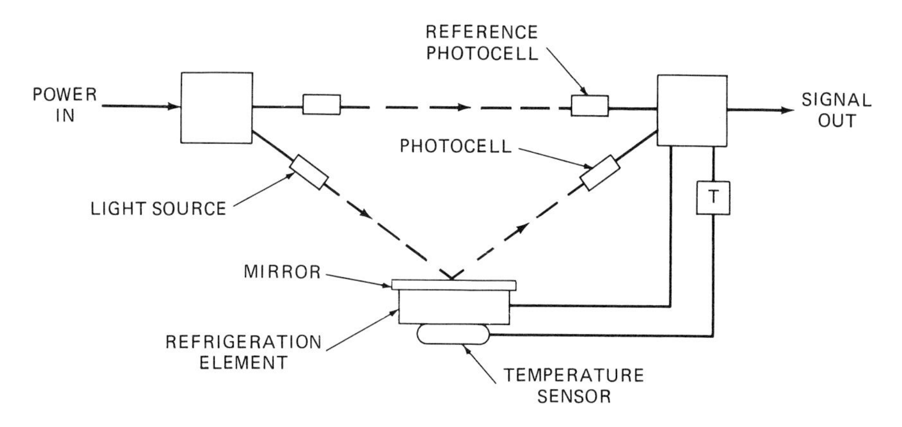

Figure 3-25 (C) Principle of chilled mirror dewpoint sensor.

which makes use of the piezo-electric effect. A small solid-state device is connected to the diaphragm of a pressure sensor. When distorted by pressure changes the resistivity of the device varies. A small distortion produces a significant change so that very small pressure changes can be measured, as low as a few hundredths of an inch of water column.

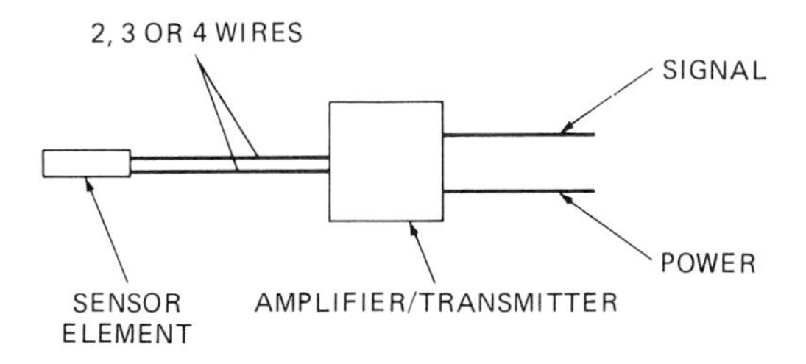

Figure 3-26 Sensor-transmitter.

3.2.3 Amplifiers and Transducers

The output of an electronic sensor usually is of such a low level that an amplifier is required. Amplifiers use bridge circuits and other techniques to "condition" the signal—including linearization if needed—and raise it to a level adequate for transmission and use by controllers. (See Figure 3–26.)

An electronic to pneumatic transducer can be current to pneumatic (I/P) or voltage to pneumatic (E/P). One type consists of a bleed-type relay controller as described in paragraph 2.2.3 together with what might be called a "variable solenoid" which positions a bleed valve as shown in Figure 3–27. As the controller output varies the power to the solenoid coil, the spring-loaded plunger varies the output of the bleed valve, changing the output of the pneumatic relay controller. The electronic power may vary infinitely or in small discrete increments (pulse-width modulation).

A pneumatic to electric or electronic transducer (P/I) uses a potentiometer, with the wiper arm driven by a bellows which senses the output of the pneumatic controller. Then the potentiometer resistance varies with branch line

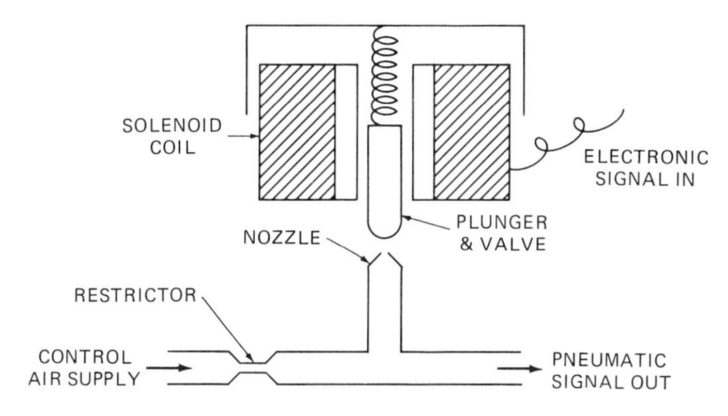

Figure 3-27 E/P (or I/P) transducer, variable bleed.

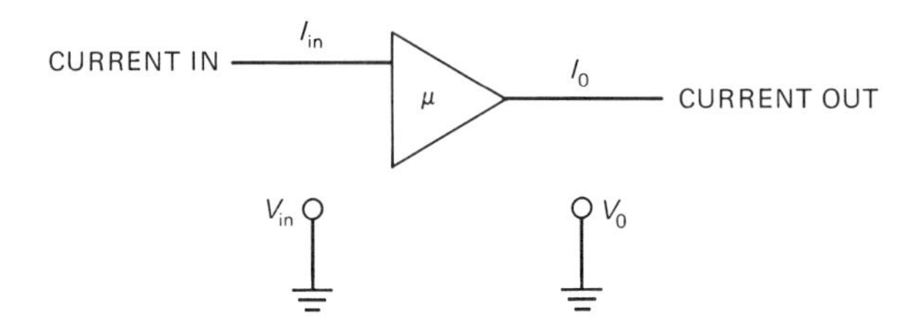

Figure 3-28 Ideal operational amplifier (OP Amp).

pressure. Solid state P/I transducers are also available, utilizing the principle that current flow in solid-state devices will vary as physical pressure is applied to the device.

3.2.4 Electronic Controllers

The basic element in an electronic controller is a device called an *operational amplifier* (OP Amp). The OP Amp is a multistage solid-state amplifier capable of providing a large gain while handling signals which vary with time, and over a wide frequency range. The *gain* of the OP Amp is the negative of the ratio of voltage out to voltage in, thus:

$$u = -\frac{V_o}{V_i} \tag{3–1}$$

(See Figure 3–28.)

To make the OP Amp useful, input and feedback impedance circuits must be added. *Impedance* refers to resistors or capacitors or combinations thereof.

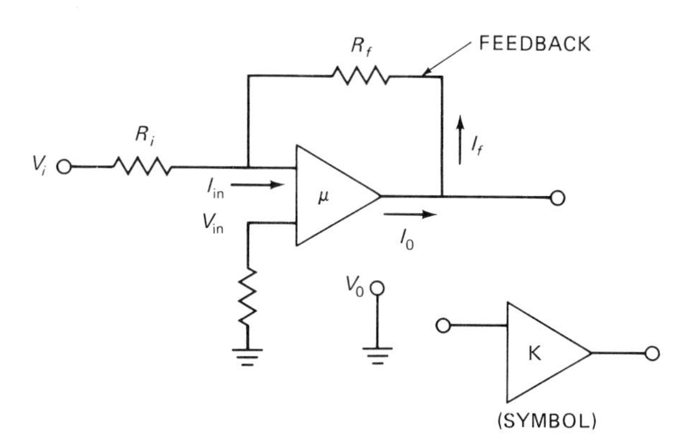

Figure 3-29 Basic proportional OP Amp.

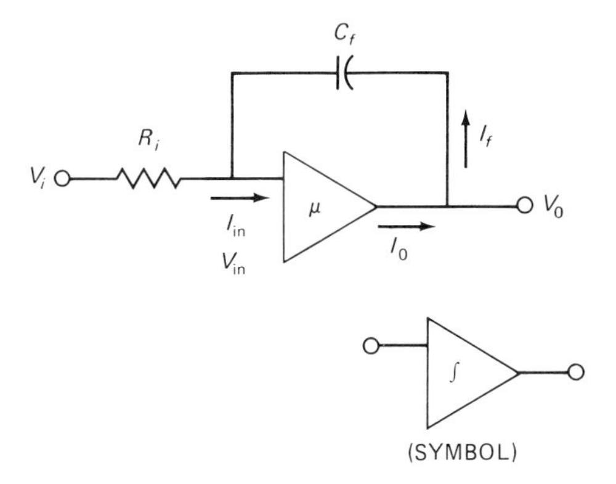

Figure 3-30 Integral OP Amp.

Figure 3–29 shows a proportional amplifier, which provides a proportional mode output as described in paragraph 1.5.5. In this case the gain is a function of the ratio of feedback and input resistors. The symbol used in control diagrams is also shown in Figure 3–29.

For integral and derivative modes (paragraphs 1.5.6 and 1.5.7) a combination of capacitance and resistance must be used. Figure 3–30 shows the integral mode arrangement with an input resistor and feedback capacitor. The gain becomes a function of the charging time of the capacitor. Figure 3–31 shows a derivative mode arrangement with an input capacitor and feedback resistor.

To combine the various modes as desired by the designer other OP Amps are used as *summers*. The simple summer circuitry is shown in Figure 3–32. Subtraction is shown in Figure 3–33. Two input voltages are added algebra-

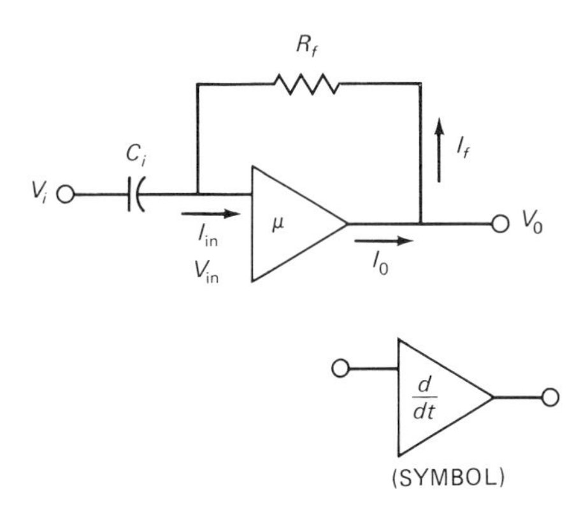

Figure 3-31 Derivative OP Amp.

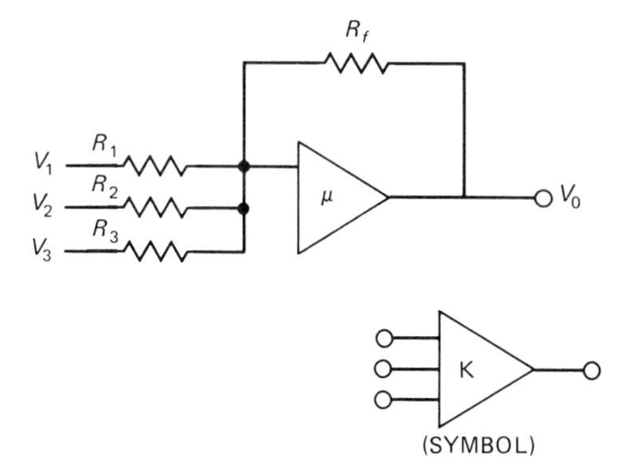

Figure 3-32 Summing OP Amp.

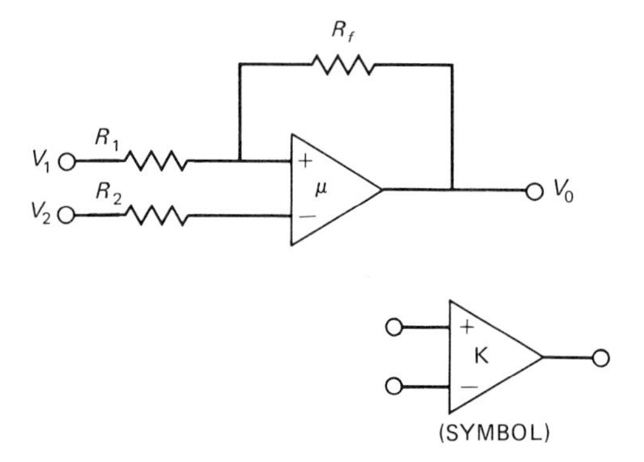

Figure 3-33 Subtraction with an OP Amp.

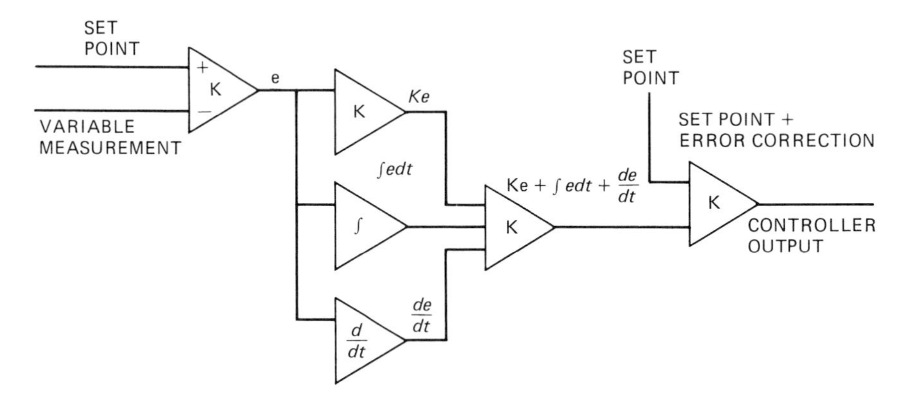

Figure 3-34 Ideal OP Amp controller.

ically (with due regard for negative and positive signs) to compare setpoint and measured value of the variable. Then a controller can be created with one, two or three modes (any combination of P, I and D) as shown in Figure 3–34.

All of the circuits shown are in simplified form, showing only the essential elements. In practice, additional circuitry is required to control voltage drift and provide stable operation. Adjustable resistors and capacitors are needed to provide adjustment of times and gains.

4 Fluidic Control Devices

4.1 INTRODUCTION

The first use of fluidic principles in control applications occurred about 1960. Fluidic devices are now used extensively in process control systems but their use in HVAC controls is somewhat limited. Fluidic amplifiers were used because of the high ratios available. So-called "self-powered" control systems using duct air supply pressures are essentially fluidic in nature. Although they use lower pressures than pneumatic devices, more air is consumed, so fluidic systems do require more power to operate.

Fluidic devices utilize the dynamic properties of fluids, in contrast to pneumatic and hydraulic devices which depend primarily on static properties. Almost any gas or liquid can be used, though compressed air is the preferred medium. Logically, fluidic devices resemble electronic devices, and the same logic terms are frequently applied.

The principles most commonly used in fluidic devices are wall attachment, turbulence amplification and vortex amplification.

4.2 WALL ATTACHMENT DEVICES

A wall attachment device functions because of the *Coanda effect:* the property of a jet to attach itself to the surface of an adjacent plate (or

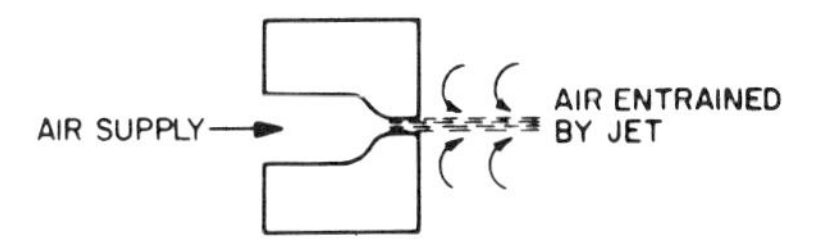

Figure 4-1 Elementary air jet.

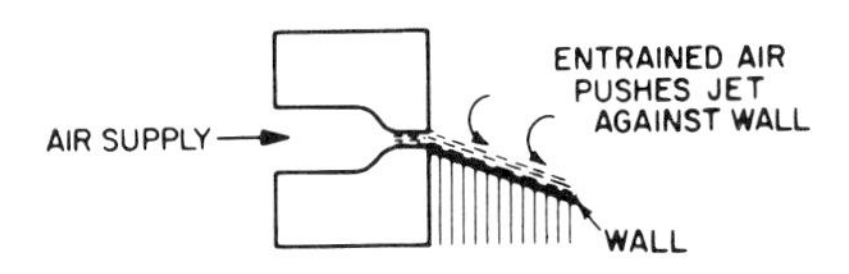

Figure 4-2 Wall attachment principle.

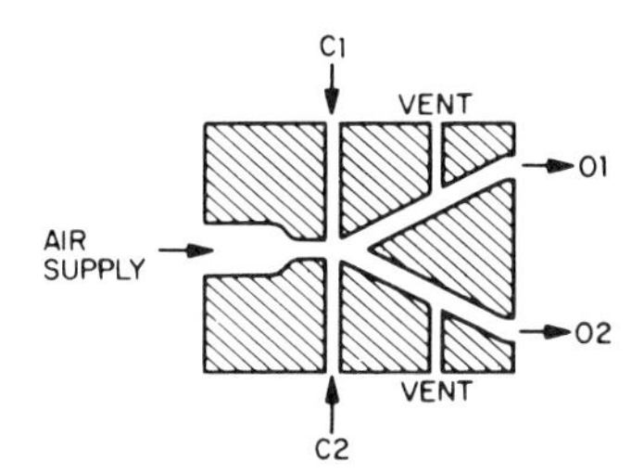

Figure 4-3 Bistable amplifier.

wall). A jet of air flowing from a nozzle entrains air with it (Figure 4–1). If the jet is confined between two parallel surfaces (assumed to be in the plane of the page), then the entrainment takes place only in that plane. If a plate or wall surface is added at one side of the jet (Figure 4–2) air entrainment on that side is reduced. This reduces the air pressure near the wall and the jet is "bent" until it is attached to the wall. If a similar and symmetrical wall is provided on the other side of the jet, then it may become attached to either side with only a slight nudge. Although in theory the shape of the jet is not important, in practice a rectangular jet is found to work best.

This effect can be used to create a logic relay-amplifier with a memory (Figure 4–3). If air is supplied at control port C1 the jet is deflected away from the port and attaches itself to the opposite wall, providing an output signal at O2. If the control signal is removed, the jet will still continue to supply O2. If now a signal is applied at C2 the jet will switch to the opposite wall and the output signal will appear at O1. Amplifications on the order of 3 or 4 to 1 (output to control signal) are possible. The vents are provided to allow continuing airflow if the output is blocked, since interruption of airflow would cause the device to operate improperly. This device is called a "bistable amplifier with memory."

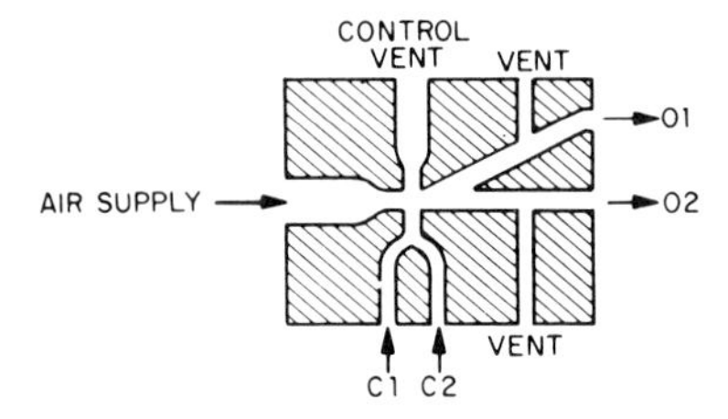

Figure 4-4 OR-NOR element (monostable amplifier).

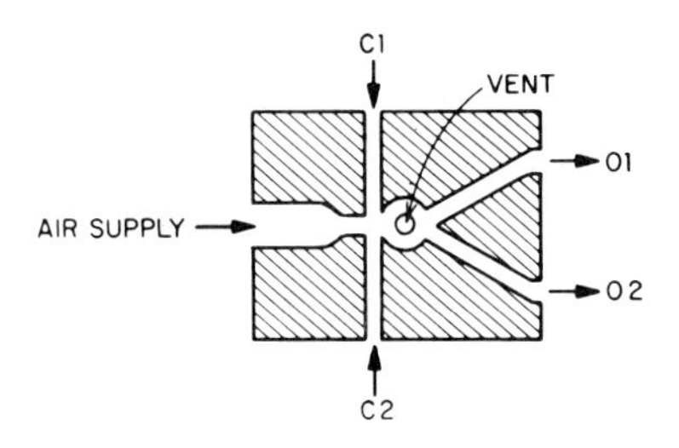

Figure 4-5 Proportional jet amplifier.

A similar but different device is the "OR-NOR" amplifier of Figure 4–4. With no signal at C1 or C2 the output is at O2 (NOR). With a signal at C1 or C2 or both, the output is at O1 (OR). When the signal or signals are removed the output returns to O2.

Figure 4–5 is similar to the bistable amplifier of Figure 4–3 but with the walls cut away just beyond the control ports. With no walls and no control signal, the supply jet goes straight out and produces two equal outputs at O1 and O2. If a control signal is applied at C1 or C2 the jet is deflected in proportion to the strength of the signal, and the outputs at O1 and O2 become unequal. This is a "proportional beam deflection amplifier."

All of the above devices operate best at low pressures: 1 to 5 psig supply and 0.1 psig or less control. Output pressures are usually less than 1 psig.

4.3 TURBULENCE AMPLIFIERS

A jet issuing freely from a nozzle can be adjusted to a fairly long laminar flow pattern (in the order of 1 in. or more). A receiving nozzle in the path of this pattern will then receive some flow and pressure (Figure 4–6). A small control jet directed at the laminar jet will cause it to break up, thus decreasing the pressure felt at the receiver. This change in output is proportional to the control pressure change, with considerable amplification (as

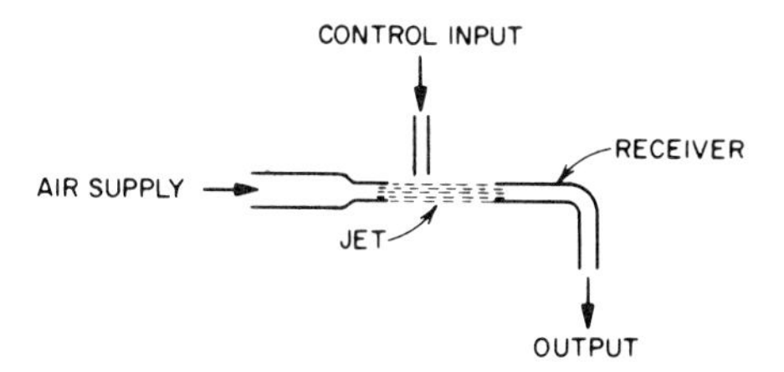

Figure 4-6 Turbulence amplifier.

much as 10:1). Also, more than one control jet can be used. The turbulence amplifier is a very low pressure device, typically utilizing supply pressures of 10 in. water and control pressures of 0.4 in. water.

4.4 VORTEX AMPLIFIERS

A vortex amplifier has a cylindrical body with supply air introduced at the side and output at one end (Figure 4–7). With no control flow, the air flows directly to the outlet, and flow is not restricted. When a control signal is applied at a tangent to the cylinder wall and at right angles to the supply flow, a swirling action is created, forming a vortex with high resistance and, therefore, lower flow. The control pressure must be higher than the supply flow, and supply output can be reduced, in a proportional manner, to as little as 10% of supply input. The device is most often used as a valve or variable restrictor.

4.5 RADIAL JET AMPLIFIER

A variation of the turbulence amplifier is the radial jet amplifier (Figure 4–8). Two jets are arranged to oppose one another. At the point where they meet a radial pressure area is created, and when this is confined in a chamber an output signal can be produced. The diagram shows a reference jet (B) which can be adjusted to a desired pressure (set point), and a

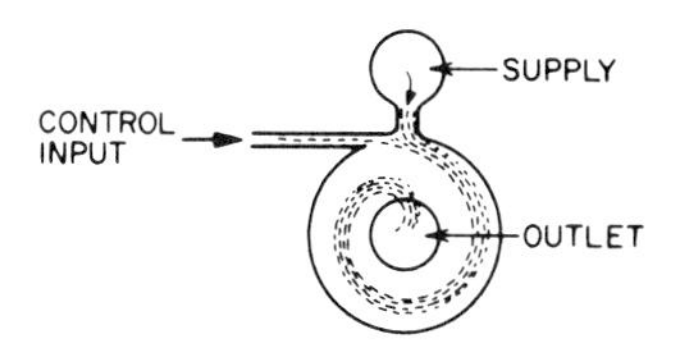

Figure 4-7 Vortex amplifier.

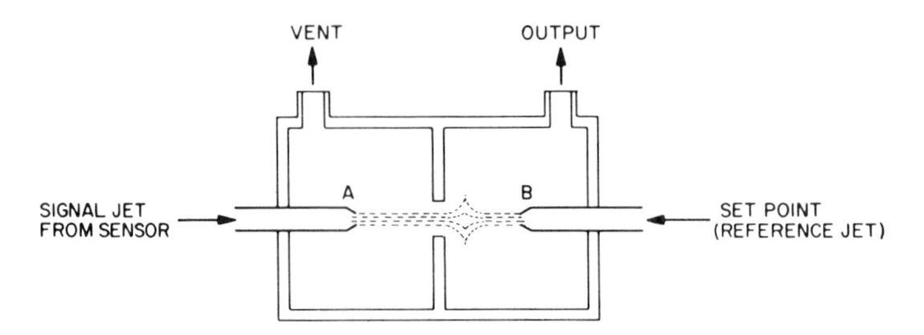

Figure 4-8 Radial jet amplifier.

signal input jet (A) from the sensor. When A is less than B the radial jet forms in the left-hand chamber and is vented to atmosphere. When A is greater than B then the pressure is developed in the right-hand chamber and an output signal is generated which can power an actuator or reset a relay.

4.6 FLUIDIC TRANSDUCERS

You will have noticed that fluidic control pressures are very small. To operate controlled devices such as valves or motors, it is necessary to provide transducers. Fluidic to pneumatic transducers are most common, since both use air, though at greatly different pressures. All fluidic transducers operate on one of two basic principles: direct force or assisted direct force.

In Figure 4–9 the fluidic signal operates against a diaphragm to deflect it. This *direct force* action could close an electrical switch or move a pilot valve in a pneumatic relay. Even with a large diaphragm the available force is small.

In Figure 4–10 a high-pressure air supply is provided to *assist* or amplify the fluidic signal. With a low fluidic signal the high-pressure air is vented. When the fluidic signal is increased the diaphragm 1 moves to restrict or close the vent nozzle, increasing the pressure on diaphragm 2. This higher pressure is adequate for positioning valves or dampers directly. The action

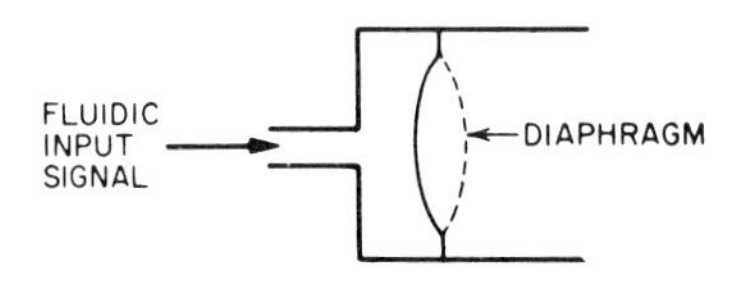

Figure 4-9 Direct force transducer.

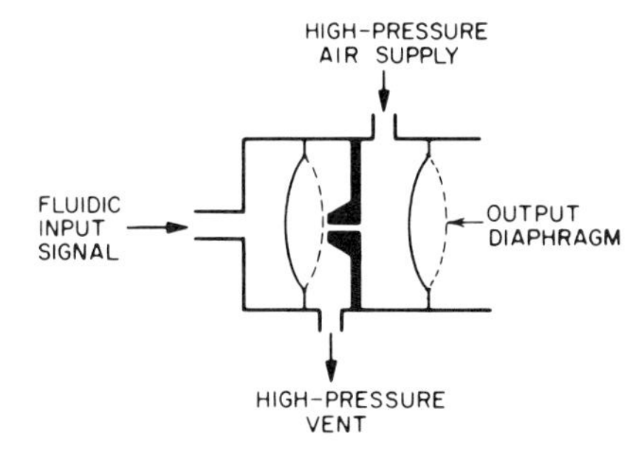

Figure 4-10 Assisted force transducer.

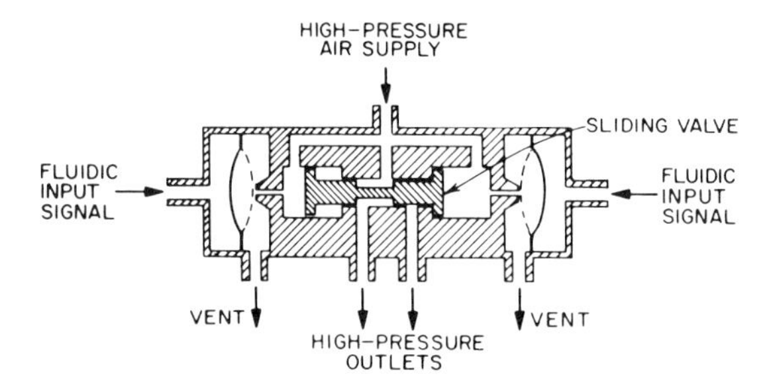

Figure 4-11 Three-way valve (with assisted force operation).

can be proportional or two-position, depending on the construction of the diaphragm.

Figure 4–11 shows a three-way two-position pneumatic valve, driven both ways by two assisted fluidic signals operating on a piston. This could also be constructed to use a single signal with spring return. Such a valve has a very fast response time.

Fluidic-electronic transducers are also available. These generally use strain gages or pressure-sensitive transistors to modify the low-power electronic signal. They are not generally used in HVAC systems at this time.

4.7 MANUAL SWITCHES

A simple fluidic pushbutton switch operates on the bleed principle (Figure 4–12). The air supply is vented through an orifice and chamber under the button. When the button is depressed the vent is closed and the output signal increases.

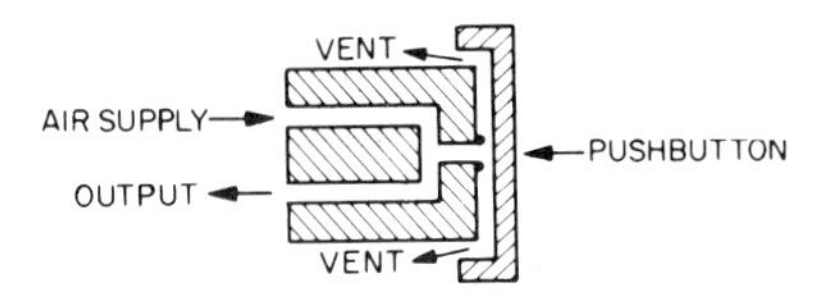

Figure 4-12 Fluidic pushbutton.

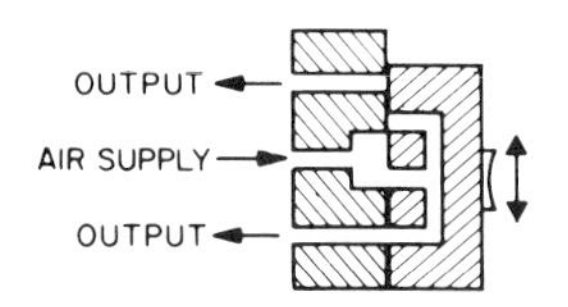

Figure 4-13 Selector switch.

Figure 4–13 shows a selector switch which provides a choice of two outputs. This can be built in a rotary or slide arrangement, with a choice of many outlets or combinations thereof.

Any of the manual or automatic air valves commonly used for pneumatic systems can also be used with fluidic controls.

5 Flow Control Devices

5.1 DAMPERS

Dampers are used for the control of airflow to maintain temperatures and/or pressures. Some special types of dampers are used in HVAC equipment, such as mixing boxes and induction units, but the control dampers in air ducts and plenums are almost invariably the "multi-leaf" type. Two arrangements are available: parallel-blade and opposed-blade. As Figure 5–1 shows, in parallel-blade operation all the blades move in the same (or parallel) way. In opposed-blade operation adjacent blades move in opposite directions. Without going into the physics of the situation it would appear that the opposed-blade damper is superior to the parallel-blade damper for modulating control. This is, in fact, true. Parallel-blade dampers, like flat-seated valves, should be used only for two-position control except for special applications such as air mixing to avoid stratification. (See paragraph 6.3.)

5.1.1 Pressure Drop

As with control valves, there is an optimum relation between control linearity and the initial pressure drop in the system. For a typical parallel-blade damper to achieve essentially linear modulating control it is

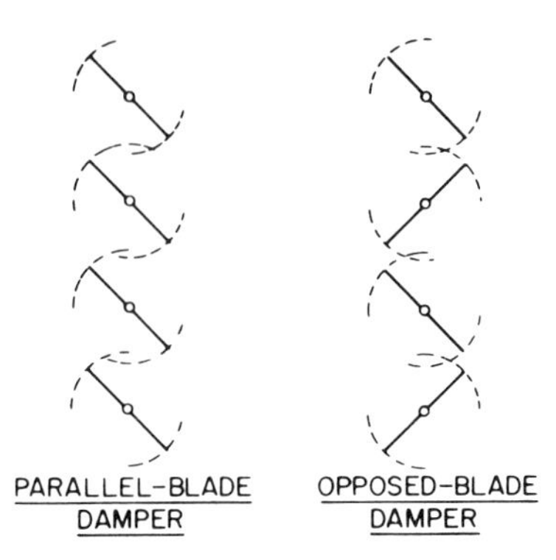

Figure 5-1 Airflow dampers.

necessary that the pressure drop through the damper in the full open position be about 50% of the total system pressure. This is shown in Figure 5–2 (curve K) and Table 5–1 as a typical flow-vs-stroke curve for these conditions. To accomplish an equivalent linearity with a typical opposed-blade damper, the initial pressure drop through the damper need be only about 10% of the system pressure drop. (See Figure 5–3 (curve F) and Table 5–1.)

Since most of the energy loss creates noise, the opposed blade damper is preferable for modulating service from both noise and energy loss standpoints.

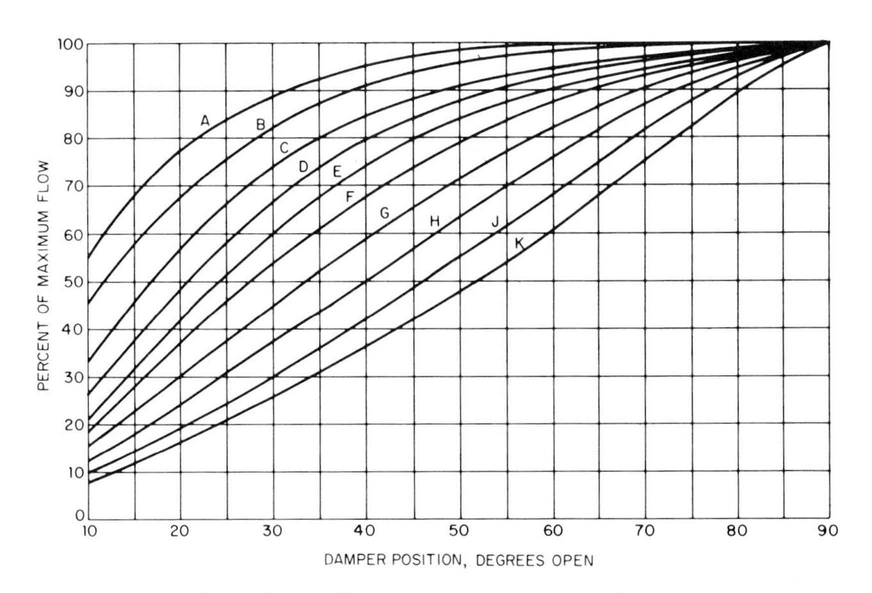

Figure 5-2 Parallel-blade damper flow characteristics. (Redrawn from E. J. Brown, *Heating, Piping and Air Conditioning,* April 1960, p. 171.)

TABLE 5–1 Ratios of Damper Resistance to System Resistance Which Apply to Flow Characteristic Curves of Figure 5-2 and 5-3[a]

PARALLEL-LEAF DAMPERS		OPPOSED-LEAF DAMPERS	
Open Damper Resistance. Percent of System Resistance	Flow Characteristic Curve	Open Damper Resistance. Percent of System Resistance	Flow Characteristic Curve
0.5–1.0	A	0.3–0.5	A
1.0–1.5	B	0.5–0.8	B
1.5–2.5	C	0.8–1.5	C
2.5–3.5	D	1.5–2.5	D
3.5–5.5	E	2.5–5.5	E
5.5–9.0	F	5.5–13.5	F
9.0–15.0	G	13.5–25.5	G
15.0–20.0	H	25.5–37.5	H
20.0–30.0	J		
30.0–50.0	K		

[a]From E. J. Brown, *Heating, Piping and Air Conditioning,* April 1960, p. 171.

5.1.2 Leakage

Unlike single-seated control valves, dampers are inherently leaky. The amount of leakage at close-off is a function of the damper design. Minimizing leakage increases cost, so we should always look at system require-

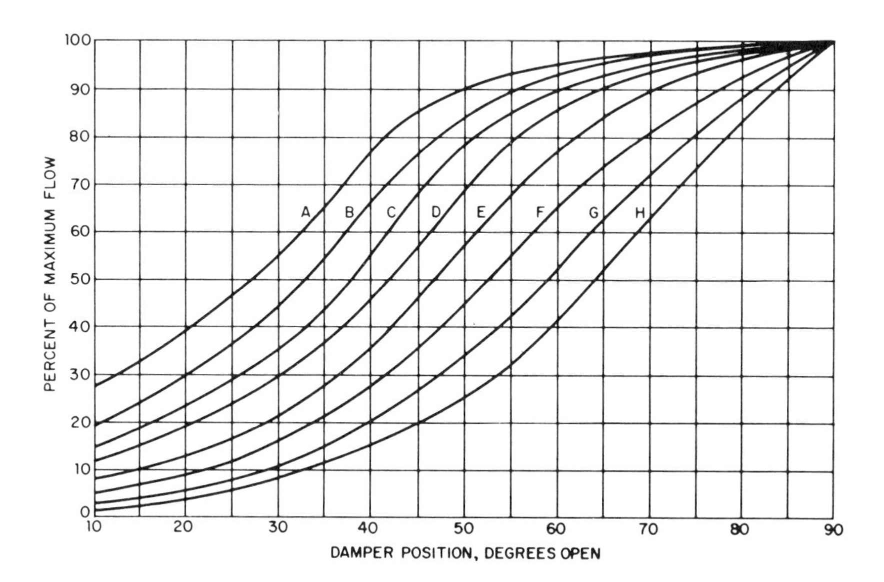

Figure 5-3 Opposed-blade damper flow characteristics. (Redrawn from E. J. Brown, *Heating, Piping and Air Conditioning,* April 1960, p. 171.)

ments and select the damper accordingly. Most manufacturers publish leakage ratings. A simple inexpensive damper may have a leakage of 50 cfm per square foot of face area at 1.5 in. of water pressure, while a very carefully designed and constructed damper may leak as little as 10 cfm per square foot at 4 in. of water pressure. Damper leakage can become a serious problem in some applications. For example, outside air dampers must have minimal leakage to prevent equipment damage due to freezing air.

5.1.3 Operators

Damper motors may be pneumatic, electric or hydraulic. They must have adequate power to overcome bearing friction and air resistance, and in the case of tight-fitting low-leakage dampers, to overcome the "binding" friction of the fully closed damper.

5.1.4 Face and Bypass Dampers

Face and bypass dampers are frequently used with preheat coils, and less often with direct-expansion cooling coils. Figure 5–4 shows a typical preheat installation. The preheat coil is sized with a face velocity based on the manufacturer's recommendations; and with all the air flowing through the coil. The open face damper will then have this face velocity. The face damper and coil pressure drops can then be determined from manufacturer's information (see also reference 5 in the bibliography). Assume a bypass damper size and determine its pressure drop when full open and with all the air flowing through it. This drop should be approximately equal to the sum

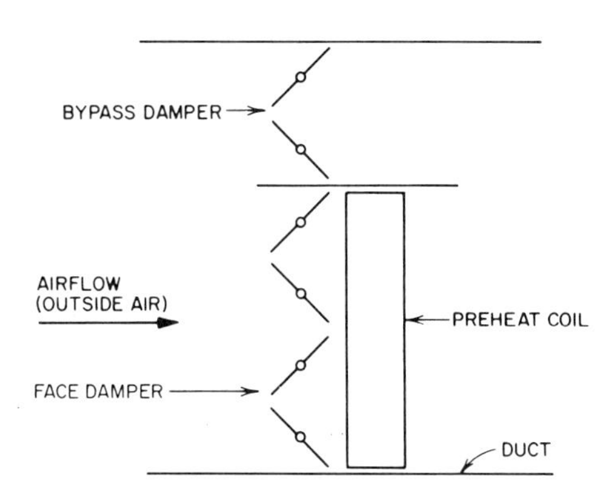

Figure 5-4 Face and bypass dampers.

of the face damper and coil pressure drops. Adjust the bypass size until equality is achieved. For the sake of economy, standard damper sizes should be used if possible.

5.2 STEAM AND WATER FLOW CONTROL VALVES

The proper selection and sizing of valves for the control of steam and water flow requires an understanding of both the valve characteristics and the system in which it is to be used. This section will therefore discuss valve types and sizing methods.

The size of a heat exchanger or coil and the fluid flow rate through it must be based on some maximum design load. But, in practice, the equipment usually operates at part load, so the valve must control satisfactorily over the whole range of load conditions.

The rate of flow through a valve is a function of the pressure drop or head across the valve, in accordance with the hydraulic formula $V = K\sqrt{2gh}$

where: V = fluid velocity in the valve, feet per second.
K = a constant which is a function of valve design.
g = acceleration due to gravity, feet per second, per second.
h = pressure drop across the valve, feet.

The change in pressure drop and flow in relation to stroke, lift or travel of the valve stem is a function of the valve plug design. Different types of valve plugs are required to accommodate different control methods and fluids.

5.2.1 Two-Position Valves

These should be of the flat-seat or quick-opening type shown in Figure 5–5. The accompanying graph of percent flow vs percent lift (Figure 5–6, curve A) shows that nearly full flow occurs at about 20% lift. Two-position valves should be selected for a pressure drop of 10% to 20% of the piping system pressure differential, leaving the other 80% to 90% for the heat exchanger and its piping connections.

5.2.2 Modulating Valves

The stroke or "lift" of a modulating valve actuator will usually vary directly with the change in the load. If the valve is designed with a linear characteristic, as in Figure 5–7 and curve B in Figure 5–6, then flow rate will

Figure 5-5 Quick-opening valve. *(Courtesy Barber-Colman Company.)*

vary directly with lift or nearly so. By varying the angle of this V-port arrangement, curves above and below the linear relation can be obtained. There is a minimum flow which is obtained immediately upon "cracking" the valve open. This is due to the clearances required to prevent sticking of the valve. This minimum flow is generally on the order of 3% to 5% of maximum flow and is defined as the "turndown ratio" of the valve. At 5% the turndown ratio would be 100:5 or 20:1. This is a typical ratio for "commercial" quality valves. "Industrial" valves with turndown ratios of 50 or 100 to 1 or even higher are available. Higher ratios are needed only for very close control and are not justified in most HVAC applications.

A linear characteristic valve is excellent for proportional control of steam flow since the heat output of a steam heat exchanger is directly proportional to the steam flow rate. This is because the steam is always at the same temperature, and the latent heat of condensation varies only slightly with pressure change.

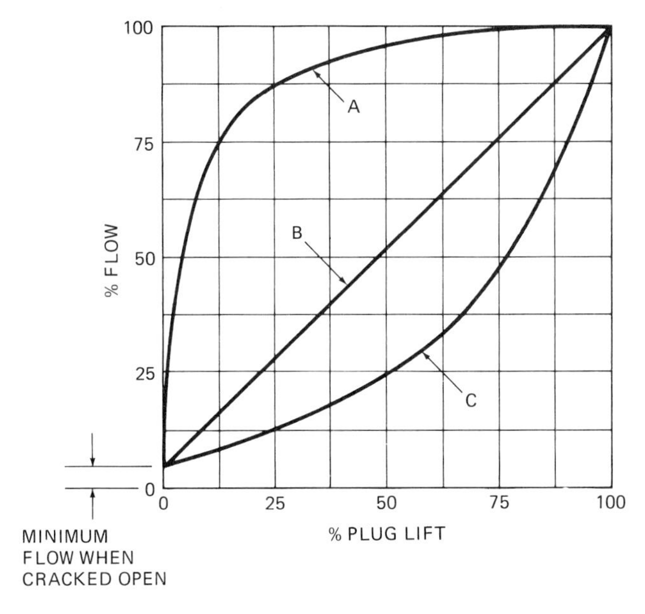

Figure 5-6 Valve characteristics *(Courtesy Barber-Colman Company.)*

Figure 5-7 Linear or V-port valve. *(Courtesy Barber-Colman Company.)*

Hot water, however, creates a different requirement, since reduction in flow will be accompanied by an increase in the temperature change of the water. The net result may be only a small reduction in heat exchange for a large reduction in flow. Figure 5–8 shows capacity vs flow for a typical hot water coil with 200° entering water and 50° entering air and a design water temperature drop of 20° at full load. This shows that a 50% reduction in water flow will cause only about a 10% reduction in heat output, with an increase to a 40° drop in the water temperature. In order to cut the capacity by 50% the water flow must be reduced by about 90%.

In order to get a better relationship of lift vs output for this case, an "equal percentage" valve is used (Figure 5–9 and curve C of Figure 5–6). The equal percentage valve has a plug shaped so that flow varies as the square root of lift. In order to take full advantage of this configuration a long stroke or "high lift" must be used (greater than ½ in.). The graph shows a plug with a square root curve. This type of valve should be used for all hot water heating control.

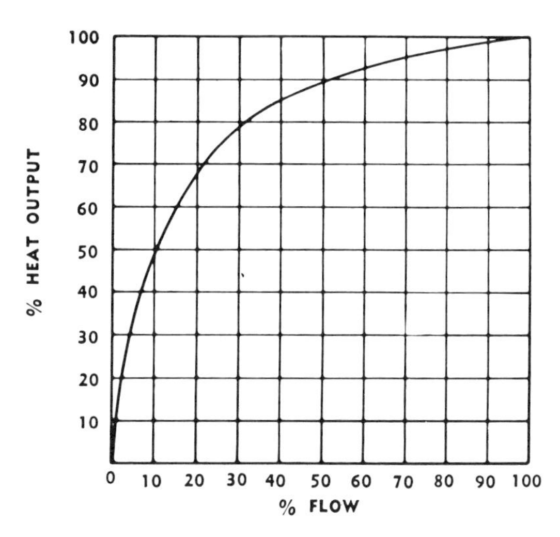

Figure 5-8 Flow vs heating capacity in a hot water heat exchanger. *(Courtesy Barber-Colman Company.)*

Figure 5-9 Equal percentage valve. (*Courtesy Barber-Colman Company.*)

5.2.3 The Flow Coefficient

Most manufacturers publish valve capacity tables based on the flow coefficient C_v. This is defined as the "flow rate in gallons of 60°F water that will pass through the valve in one minute at a one-pound pressure drop." The flow rate at pressure drops other than one pound is found by the formula:

$$GPM = C_v\sqrt{PD}$$

where *PD* is the desired pressure drop across the valve.

Valve capacity tables usually show C_v and then flow rate at various pressure drops.

The rated C_v is established with the valve fully open. As the valve partially closes to some intermediate position the C_v will decrease. The rate at which it decreases determines the shape of the curve in Figure 5–6. Some industrial valve manufacturers publish C_v data for partially open valves.

5.2.4 Valve Pressure Drop

To determine the required pressure drop for a modulating water valve at full design load it is necessary to look at the entire system. Three conditions are possible:

1. Throttling or closing off any individual control valve has little or no effect on the pressure differential from supply main to return main.
2. There is only one control valve, so that throttling or closing off this valve changes the flow correspondingly in the entire piping system.
3. Other conditions between these two extremes.

In the first case the heat exchanger and related piping between supply and return mains can be considered as a subsystem. (Figure 5–10(A)). First consider the subsystem without a control valve. Now if somehow the pressure differential between supply to return mains is varied, the flow rate in the subsystem will vary exponentially as shown in Figure 5–10(B). But a constant differential pressure between supply and return mains has been assumed, with a control valve as shown in Figure 5–11(A). This valve pressure

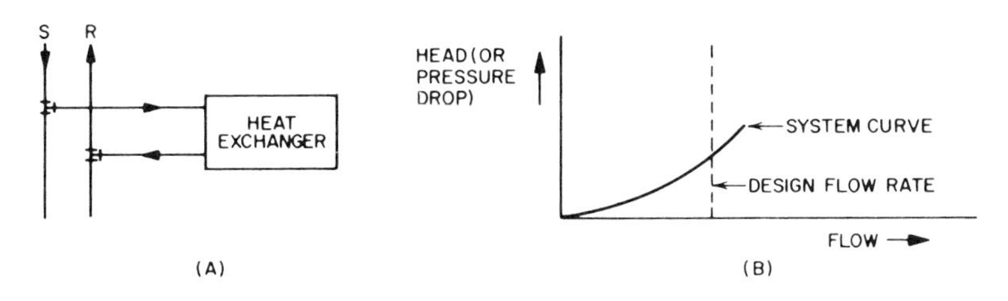

Figure 5-10 A heat exchanger subsystem.

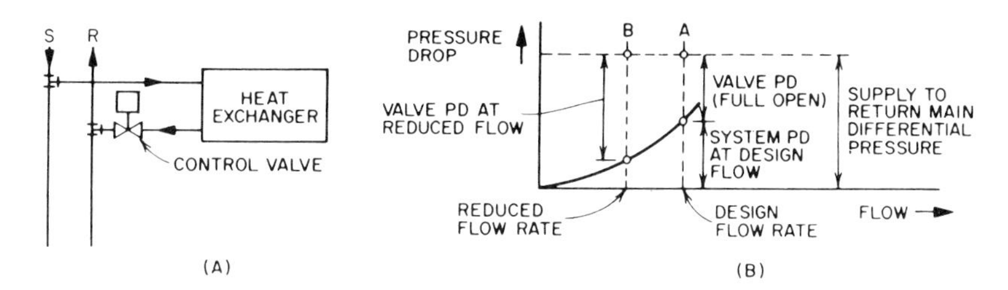

Figure 5-11 Subsystem with control valve.

drop at design flow should be at least 50% of the total available head. In order to provide adequate control at all conditions the major portion of the pressure drop must be taken by the valve. Otherwise the subsystem pressure drop would govern during part of the stroke.

If the valve pressure drop is 50% of the total head then Figure 5–11(B) shows the result. This is the same curve as before for the subsystem without the valve, but the supply to return main differential has been doubled to allow for the valve loss at design flow. Now, as the valve throttles to reduce flow the subsystem pressure drop decreases, and at some flow rate *B* the valve pressure drop must have increased to compensate.

In the second case exactly the opposite condition prevails. With only one control valve our system would look schematically like Figure 5–12. For this system, flow is again related to pressure drop by an exponential curve as in Figure 5–10(B). A typical pump performance curve of head vs flow will look like Figure 5–13. With the control valve full open these two curves will establish some operating flow rate *A* as in Figure 5–14. As the control valve

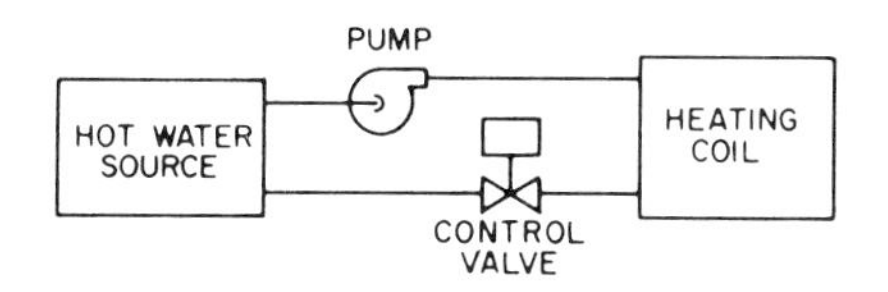

Figure 5-12 A simple system with one control valve.

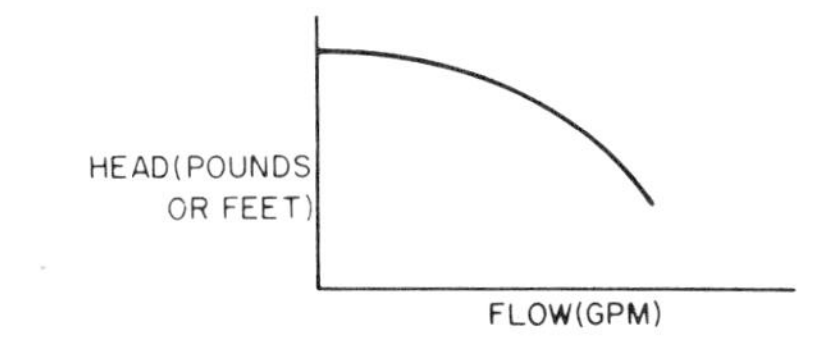

Figure 5-13 Typical pump curve.

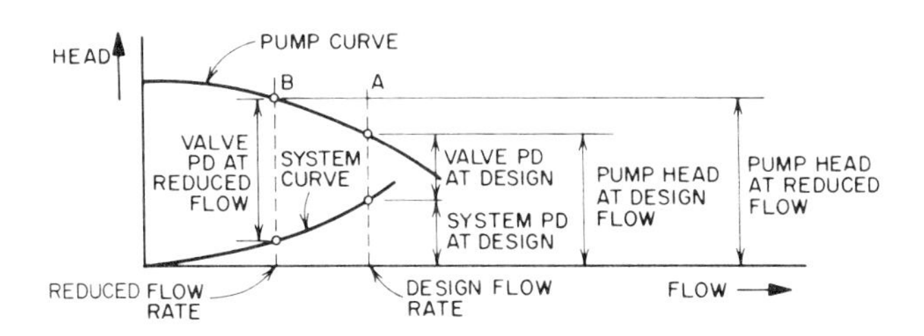

Figure 5-14 Simple system at design and reduced flow rates.

modulates toward the closed position, flow decreases to *B*. But the pump head must increase as flow decreases. This means simply that the control valve pressure drop has had to increase more for the same flow reduction than in the first case we considered. That is, the valve will need a longer stroke or a different plug characteristic at the same stroke in order to obtain the same flow reduction as in the first case.

Cases which fall between these two extremes result from small systems with only a few control valves or from systems in which a few control valves handle most of the water flow. Then any one of these control valves will have a noticeable but not completely governing effect on the pump pressure and flow.

These last two cases can be avoided by the use of three-way valves or pressure-controlled bypass valves, which will maintain the total flow rate (and therefore the supply to return main differential pressure) essentially constant. The use of three-way valves should be minimized in the interest of energy conservation, as noted in some of the discussions in Chapters 6, 7, and 12.

5.2.5 Shutoff Head and Static Head

Any flow control valve will have a flow direction symbol on the outside of the body. For a modulating valve the direction of flow should always be such that the flow and pressure tend to hold the valve open (Figure

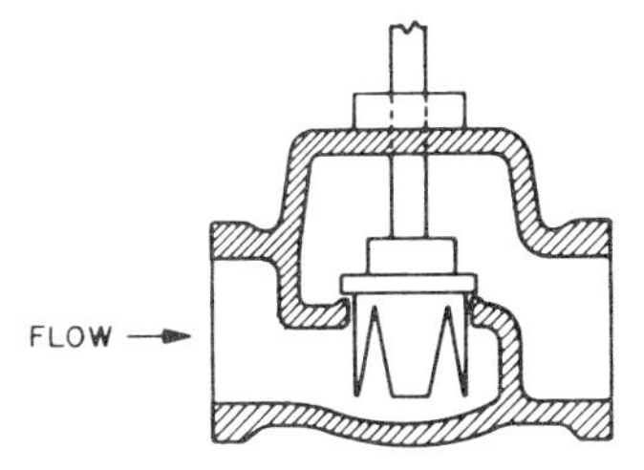

Figure 5-15 V-port valve.

5–15). As the valve modulates to near the closed position the flow velocity around the plug becomes very high. If the flow and pressure were arranged to help close the valve, then, at some point near closing, the velocity pressure would overcome the spring resistance and force the valve closed. Flow would then stop, with velocity pressure going to zero. But the static pressure differential alone would not be enough to hold the valve closed, so it would open and the cycle would be repeated. This causes "chatter" and is usually a result of installing the valve backward.

When the valve is properly installed, the closing force must overcome both velocity and static pressure. When the sum of these two pressures reaches a maximum, it is known as the close-off pressure and is one criterion for valve selection.

The static pressure rating of a valve is the highest pressure the valve body will stand without damage or leaking.

5.2.6 Three-Way Valves

Three-way valves are generally used to provide a constant flow rate through the piping system while varying the flow rate through the heat exchanger. They may also be used to provide constant flow through the heat exchanger while the primary flow rate varies (as in Figure 6–14).

"Heat exchanger" in this context may include an air-to-fluid heat exchange coil, a chiller, a heater, a condenser, a cooling tower or a fluid-to-fluid heat exchanger.

Three-way valves are made in two types: mixing and diverting. The internal designs are different (see Figures 5–16 and 5–17), the difference being necessary so that the valve will seat against flow, as discussed in the section on shutoff head above. The "mixing" valve uses a typical linear V-port plug with an added taper on top to seat in the second inlet port. The "diverting" valve uses two V-port plugs which seat in opposite directions and against

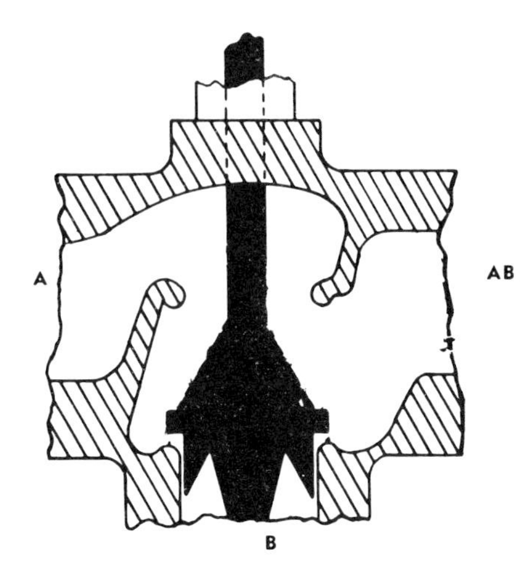

Figure 5-16 Three-way mixing valve. *(Courtesy Barber-Colman Company.)*

the common inlet flow. It can be seen that using either design for the wrong service would tend to cause chatter.

Although the valve connections will be marked "A," "B" and "AB," for logic purposes they are usually designated "normally closed," "normally open" and "common." Either the "A" or the "B" connection may become normally open, depending on the arrangement of the motor operator. Typically, however, the "B" connection is normally open.

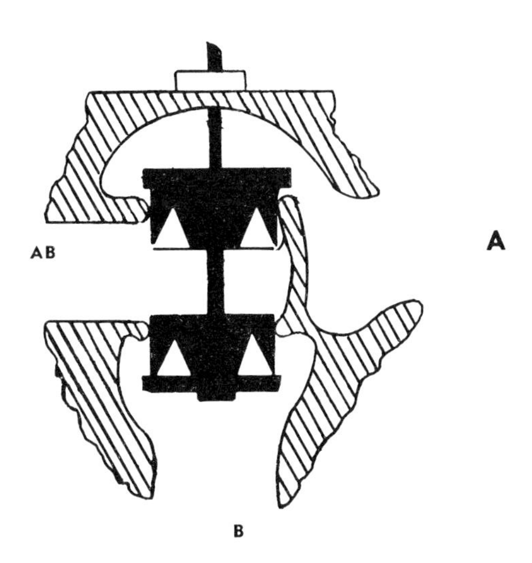

Figure 5-17 Three-way diverting valve. *(Courtesy Barber-Colman Company.)*

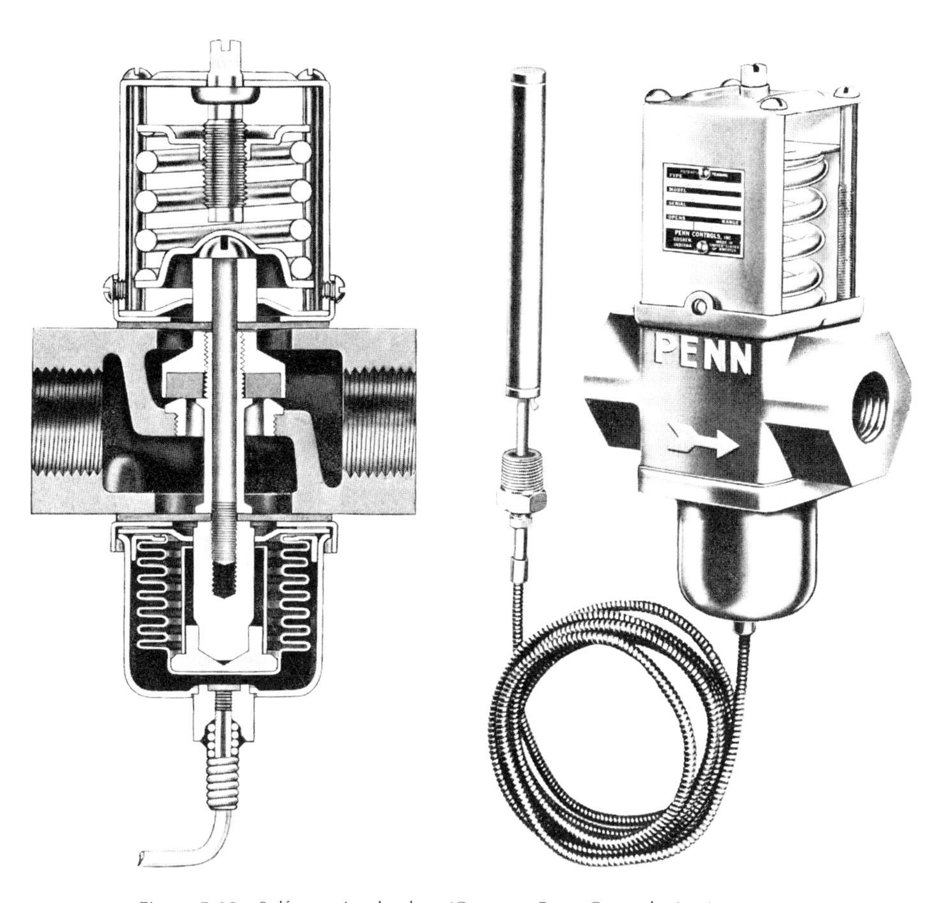

Figure 5-18 Self-contained valve. *(Courtesy Penn Controls, Inc.)*

5.2.7 Valve Operators

Valve operators may be pneumatic, electric or electronic motors. Such motors have been described in Chapters 2 and 3. Valves may also be "self-contained"; that is, the operator-motor will derive its power from the expansive force of a thermally sensitive fluid in a bulb and capillary tube arrangement (Figure 5–18). Another operator, which is technically electronic, is actually hydraulic, and contains an oil reservoir and a tiny pump which pumps oil from one side of a diaphragm to the other.

5.2.8 Criteria for Valve Selection

Flow control valves should be carefully selected to match the characteristics of the system which they are to control. Necessary criteria include flow rate, pressure drop, close-off pressure, static pressure, and type of action. The type of operator used should be compatible with the rest of the control system. Close-off pressure is most often overlooked, especially with large valves (3 inch and larger). It is often necessary to use oversize operators or higher operating air pressures to provide adequate close-off pressures.

5.3 SYSTEM GAINS

In Chapter 2 the *gain* of a controller was defined and discussed. It was noted that a high gain may cause a system to be unstable while a low gain will cause a slow sluggish response. The design of the HVAC system and its control components can also affect the system response. The factors involved are referred to as *system gains.*

The gains due to an improperly designed system may make the system uncontrollable or, at least, make it impossible to control to the desired parameters. Consider specifically how the selection of a control valve affects system gain.

Typically, a control valve selection is based on a calculated design fluid flow rate—gpm or pounds per hour—at an arbitrarily selected pressure drop. As previously noted, this design pressure drop should be substantially equal to the pressure drop through the heat exchanger and its related piping. The valve is selected from a manufacturer's catalogue based on the wide-open C_v at the design pressure drop. Since the design C_v will seldom match the catalogue C_v the next larger valve is usually selected, meaning that the valve is somewhat oversized. It is recognized that the design load is seldom, if ever, obtained, which means that the valve will always operate at less than fully open, making it even more oversized.

Looking again at Figure 5–8 reveals that in a water to air coil capacity does not change in a linear way with respect to flow rate. The shape of the curve is a function of the difference between entering air and entering water temperatures. If this delta T is large the curve is more convex. If the delta T is small the curve is more nearly linear. Thus, the traditional use of 180° F entering water temperature, to provide a 90° to 100° F temperature, creates further control valve difficulties. That is, while the controller response is essentially linear, the valve response may be exponential and the system response is even more exponential. If this heating or cooling coil is installed in a variable air volume (VAV) system, where the air flow rate is commonly 60% to 70% of design, the coil also becomes oversized and a source of system gains. Now the control system has serious problems!

These problems are not insoluble if the designer is aware of them during the design process. Usually, it is more appropriate to undersize control valves slightly in preference to oversizing.

Dampers are analogous to valves. This common practice of making dampers the same size as the connecting ducts can only lead to oversizing and lack of accurate control.

5.4 SUMMARY

The preceding discussion of control devices, while brief, is comprehensive enough to provide an understanding of the variety of elements available and their principles of operation.

The following chapters show how these elements are combined to make systems and subsystems for control of HVAC functions. In most cases any type of energy may be used, the criteria generally being convenience, economics, reliability, and accuracy.

6 Elementary Control Systems

6.1 INTRODUCTION

The preceding chapters discussed the various control elements or units, and how they function with different types of energy. The next four chapters will consider the application of these units to form combinations or "control systems" for the control of HVAC. (The word "system" becomes somewhat overworked here, applying as it does to both "control systems" and "HVAC systems." The difference is always clear in context.) This chapter will consider only small segments of the larger overall systems. Each of these is a complete control system within itself, and all large systems are made up of complexes of these elementary systems.

The following discussions will deal with function only, and except in a few cases, assume that the function can be accomplished by means of any type of energy--electric, pneumatic, electronic or fluidic.

6.2 OUTSIDE AIR CONTROLS

Before deciding on how to control the amount of outside air it is necessary to determine how much is required by the HVAC system and why. For example, certain areas such as laboratories and special manufacturing processes may require 100% exhaust and make-up. Clean rooms require that a positive internal pressure be maintained to prevent infiltration from surrounding areas, while spaces such as chemical labs and plating shops require a negative pressure to prevent exfiltration.

When there are no special requirements, such as those just described, the minimum amount of outside air required is that needed to meet the code requirements for ventilation rates. Typically 5 to 10 cfm per person is required, though more may be necessary in certain applications and where smoking is allowed. However, where the outdoor air quality is suitable (not too humid or dirty) "economy cycle" control is often used, as described below.

Once the criteria have been determined, one of the following methods of control can be selected.

6.2.1 Minimum Outside Air

By far the simplest method of outdoor air control is to open a "minimum outside air" damper whenever the supply fan is running (Figure 6–1). This provides the air required for ventilation or exhaust makeup with no complications and no frills. It was one of the first ideas developed and is still used extensively.

6.2.2 Economy Cycle Outside Air

It became apparent, however, that when nominal and fixed amounts (10% to 40%) of outside air are used, there are many times when it is necessary to operate the cooling coil even when outdoor air temperatures are near or below the freezing mark. This gave rise to the so-called "economy cycle" (Figure 6–2), with outside, return and relief dampers con-

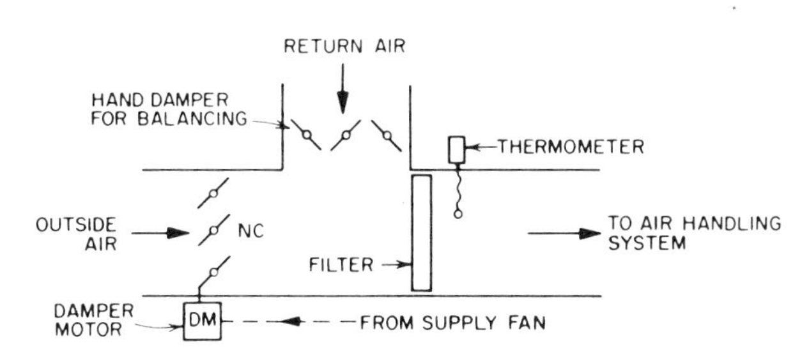

Figure 6-1 Outside air; two-position control.

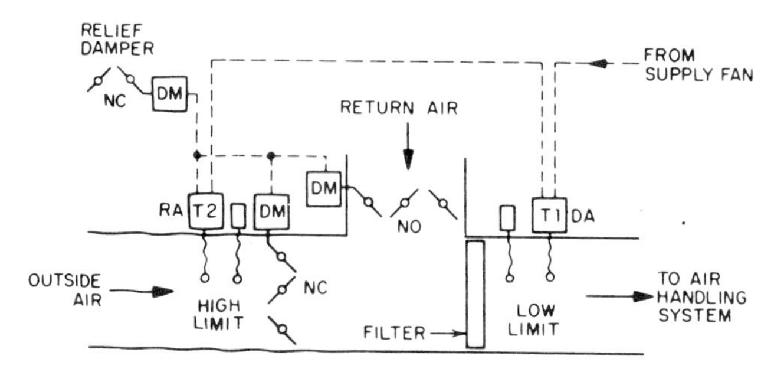

Figure 6-2 Outside air; economy cycle, no minimum.

trolled by air temperature. With outside air at design winter temperature the outside air damper and relief dampers are in minimum open positions (as determined by ventilation and exhaust requirements) and the return air damper is correspondingly in maximum open position. As outside air temperature increases, the mixed air thermostat (T1) gradually opens the outside air damper to maintain a constant low-limit mixed air temperature. Return and relief dampers modulate correspondingly. At some outside temperature, usually between 50° and 60° F, 100% outside air will be provided and used for cooling. As the outside air temperature continues to increase, at 70° to 75° F an outdoor air high-limit thermostat (T2) is used to cut the system back to minimum outside air, thus decreasing the cooling load. This system is extensively used today. Notice the interlock from the supply fan. It is provided in all outside air control systems so that the outside air damper will close when the fan is off.

Figures 6–3 and 6–4 show two alternative methods of economy cycle control. Figure 6–3 is the older arrangement, with a fixed minimum outside air

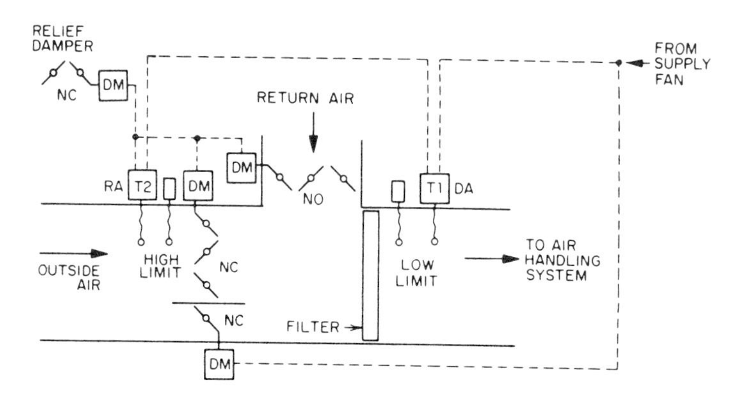

Figure 6-3 Outside air; economy cycle, fixed minimum.

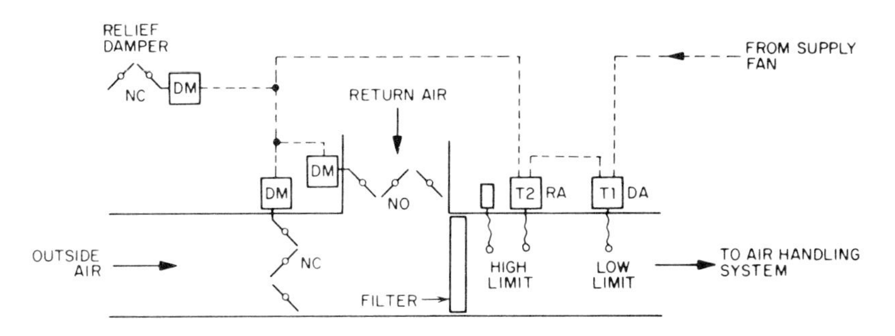

Figure 6-4 Outside air; alternate economy cycle.

damper which opens whenever the fan runs and is not affected by the temperature controls. The rest of the system operates exactly as described previously.

Figure 6–4 utilizes the high-limit controller in a different way. As before, with outside air at minimum design temperature, the low-limit mixed air thermostat (T1) is in control and outside air is at a minimum. As outdoor temperature increases, the outside air damper gradually opens until, when outside air temperature reaches the low-limit temperature set point, 100% outside air is provided. This condition is maintained until the outside air reaches the high-limit temperature setting. Then the high-limit controller takes over, the outside air damper starts closing and continues to modulate toward the closed position as outdoor temperature increases. This control system differs from the other two in two ways: It is not as economical with higher outdoor temperatures; and if the building itself at startup is warmer than the high-limit set point and the outside air, then the high-limit control will have the opposite effect to that intended. Admittedly, this last is a rare case, but it has occurred!

The above discussion supposes a fixed low-limit set point. It can be shown that this is not the best approach for energy conservation and that it will actually increase heating requirements as compared to a fixed minimum outside air. To solve this problem it is necessary to reset the low-limit controller set point as a function of building heating and cooling load. Some examples are shown in Chapter 7.

6.2.3 Enthalpy Control

It can be shown that outside air "economy cycle" control based on dry bulb temperatures (as described in paragraph 6.2.2) is not always the most economical. That is, in very humid climates the total heat (or enthalpy)

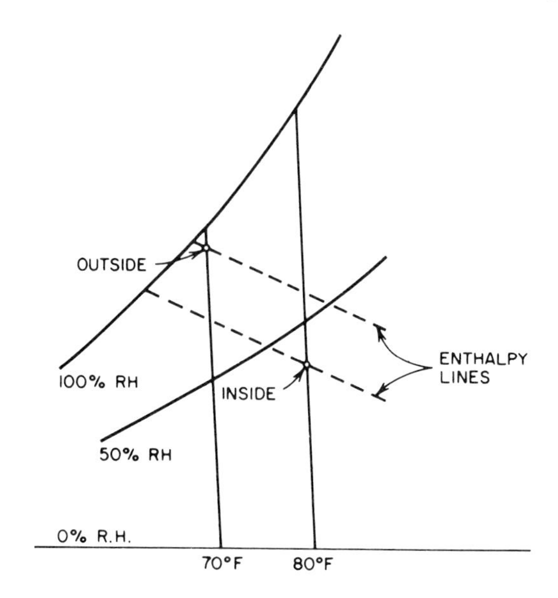

Figure 6-5 Partial psychrometric chart.

of the outside air may be greater than that of the return air even though the dry bulb temperature is lower. For example, in Figure 6–5, the psychrometric chart shows outside air nearly saturated at 70° DB while return air at 80° DB but much drier, has a lower enthalpy. Since the cooling coil must remove the total heat from the air to maintain the desired condition, it is more economical in this case to hold outside air to a minimum. To measure enthalpy it is necessary to sense two characteristics of the air: dry bulb plus either wet bulb, relative humidity, or dew point. Several manufacturers now have instruments which will sense dry bulb plus dew point, determine enthalpy, and do this simultaneously for return and outside air, providing an output to control the dampers. (See Figure 6–6.)

While enthalpy control is ideal in theory, it is questionable in practice. The accuracy of commercial humidity sensors is difficult to maintain without frequent calibration. The actual seasonal energy savings due to enthalpy control is too small, in most climates, to offset the cost of the equipment and maintenance.

6.2.4. Static Pressure Control

For those spaces requiring a constant positive or negative pressure with respect to their surroundings, the outside, return and relief air

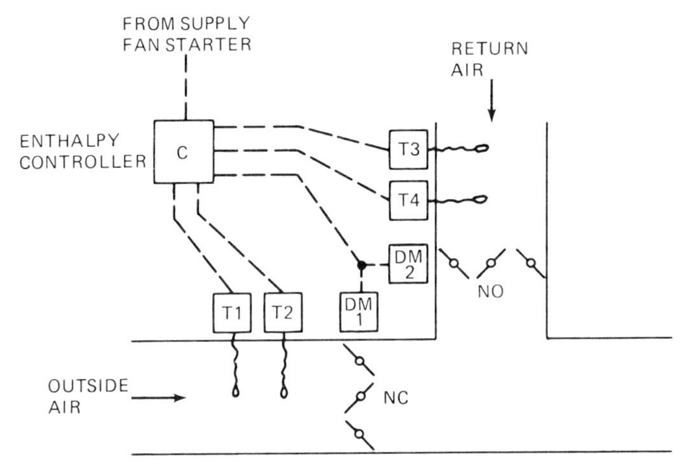

Figure 6-6 Differential enthalpy control of mixed air.

dampers will be controlled by static pressure controllers. This will usually be part of a larger system as described later in this chapter. In its simplest form (Figure 6–7), the static pressure controller senses the difference in pressure between the controlled space and a reference location (either adjacent to the controlled space or outdoors) and adjusts the dampers to maintain that pressure differential. The amount of outside air provided must be sufficient to make up any exhaust and to pressurize the space. Proportional plus integral controls are required. A delaying device or "snubber" is desirable to prevent surges due to sudden changes, such as caused by opening a door. In Chapter 9 some other pressure balance control systems are described.

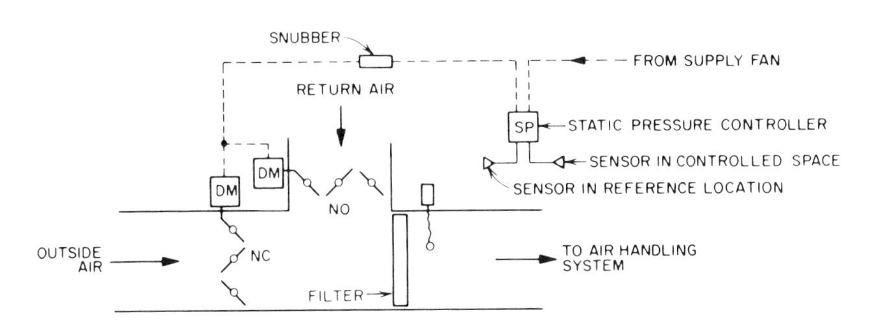

Figure 6-7 Outside air; static pressure control.

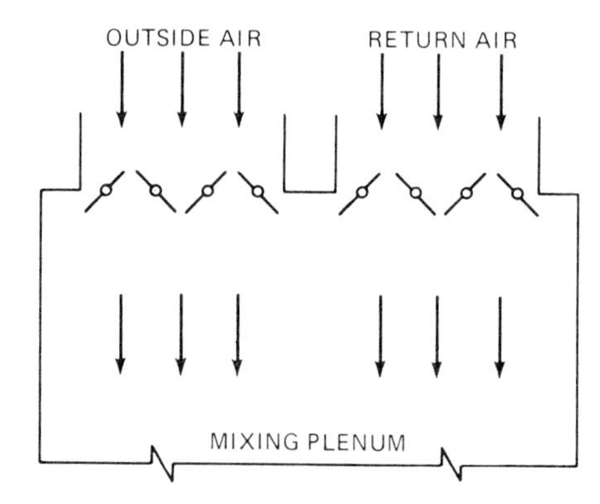

Figure 6-8 Air streams side by side; no mixing.

6.3 AIR STRATIFICATION

Stratification of return air and outside air streams in "mixing" plenums can be a serious problem. In a "worst condition" case, such as Figure 6–8, the two air streams will not mix and will remain separate for a long distance through filters, coils and even double-inlet centrifugal fans. If the outside air temperature is below freezing, the separate air stream can cause localized freezing in heating and cooling coils. Or, if a good low temperature safety control is provided, the HVAC system will cycle off and on, or shut down completely. In many other configurations, mixing will be minimal and problems will result.

The mixing plenum and its dampers should therefore be designed to pro-

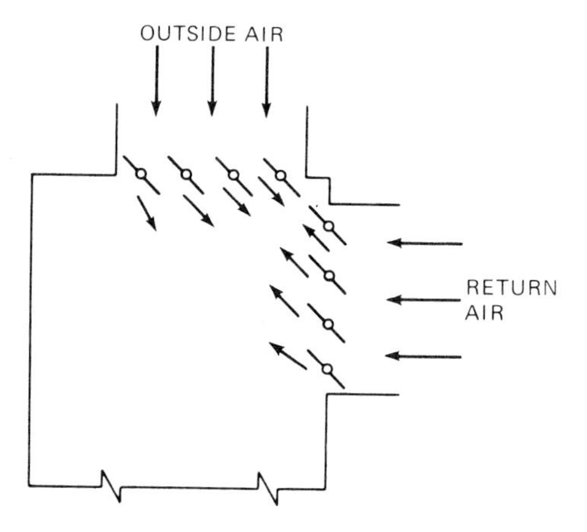

Figure 6-9 Air streams at 90° angle; parallel blade dampers, good mixing.

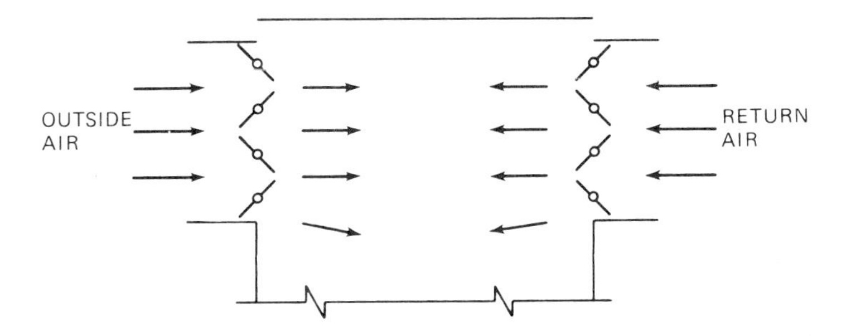

Figure 6-10 Opposed air streams; good mixing.

mote good mixing of the air streams. One of the simpler methods is shown in Figure 6–9. Parallel blade dampers are so arranged that the air streams meet head-on. If the air streams enter opposite sides of the mixing plenum, as in Figure 6–10, then opposed blade dampes will be preferred.

Static mixers can also be used. These impart a whirling, mixing motion to the air. A distance of a few feet is required. Another approach, sometimes used in retrofit situations, is to add propeller fans, set at right angles to the air streams, to provide mechanical mixing (Figure 6–11).

6.4 Heating

Heating in HVAC systems is usually provided by steam or hot water coils with remote boilers. Electric heating coils, heat pumps and direct gas-fired duct heaters are also used, and are discussed in other sections of this book.

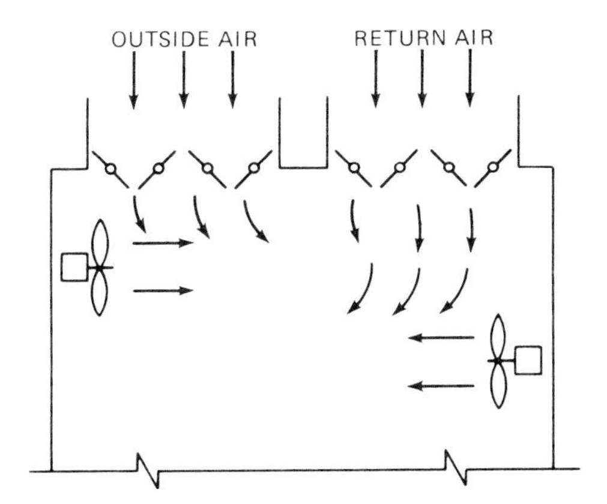

Figure 6-11 Dynamic mixing using propellor fans.

Heating may be done as preheat of outside air or mixed air, reheat for humidity control, individual zone control or what we might call "normal" heating. Each of these has its own special control requirements.

6.4.1 Preheat

Preheating is used with large percentages of outside air to prevent freezing of downstream heating and cooling coils and to provide a usable mixed air temperature for dual-duct or multizone systems (if the cold-duct temperature is too low the room control may be unstable). The principal problem in preheating is freezeup of the preheat coil itself. Several methods are used to prevent this.

Figure 6–12 shows the simplest. This is a two-position valve in the steam or hot water supply with an outdoor thermostat which opens the valve whenever the outdoor temperature is below 35° or 40° F. (This, incidentally, is an open-loop control.) The filter is downstream of the coil to prevent snow loading in severe winter storm weather. Since no control of leaving air temperature is provided, the preheat coil must be carefully selected to prevent overheating at, say, 30° outside, while still providing adequate capacity at perhaps −10° or −20° F outside design conditions. This is obviously difficult, if not impossible, so a compromise is worked out.

Face and bypass dampers are added at the coil and controlled by means of a downstream thermostat (T2, Figure 6–13) to provide a usable mixture temperature. The difficulty here is stratification of the two air streams. There have been cases where a downstream cooling coil has been frozen by a bypass air stream while the preheat coil was in full operation. The preheat coil should always be located in the bottom of the duct, and, even so, it is desirable to provide mixing baffles. Given sufficient distance in which to provide adequate mixing, this system works well.

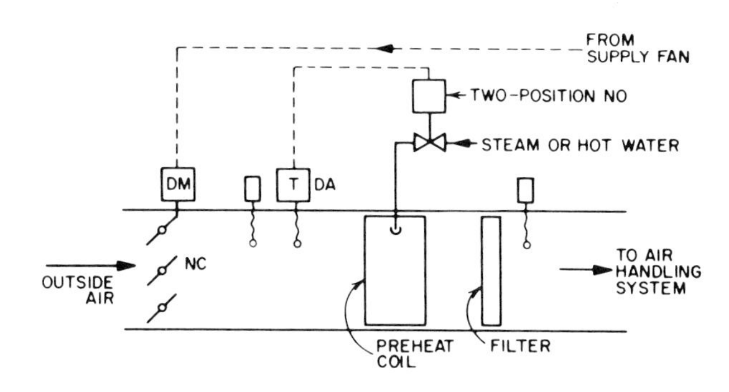

Figure 6-12 Preheat; outside air thermostat.

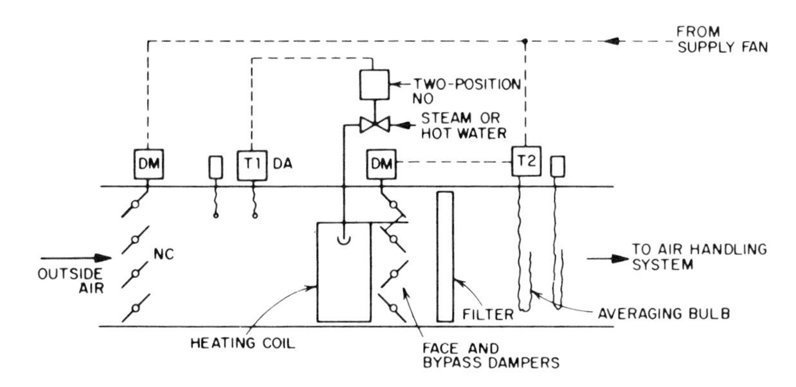

Figure 6-13 Preheat; face and bypass dampers.

In many cases such distance is not available. The best solution in this case is to use hot water with a recirculating pump (Figure 6-14). Now there can be full flow through the coil at all times with the temperature of the water varied to suit requirements. No air is bypassed so there are no mixing problems. Very accurate control of air temperature is possible. Notice the "opposed" flow arrangement with the hot water supply entering the "air leaving" side of the coil.

When dealing with freezing air certain precautions are necessary. For hot water, it has been shown experimentally that water velocities of 2½ to 3 ft/sec in the coil tubes are sufficient to prevent freezing at outdoor temperatures down to − 30° F or somewhat lower provided some hot water is being added. If it is necessary to contend with − 40° F or below the use of direct-fired systems, gas, oil or electric is recommended. Glycol solutions may also be used.

Steam coils in freezing air should be the double-tube distributing type with a good slope or vertical arrangement to drain condensate, adequate trap

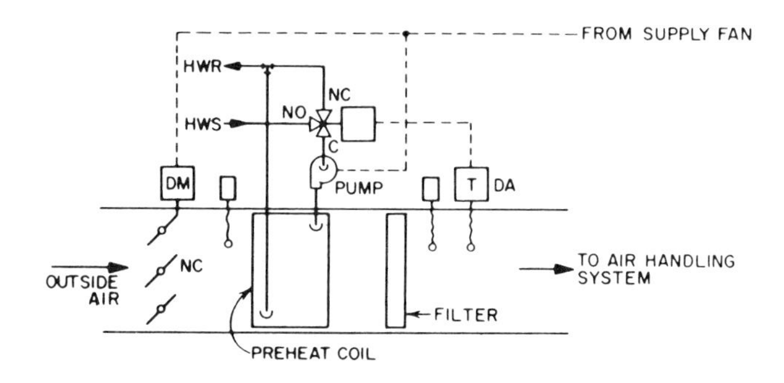

Figure 6-14 Preheat; secondary pump and three-way valve.

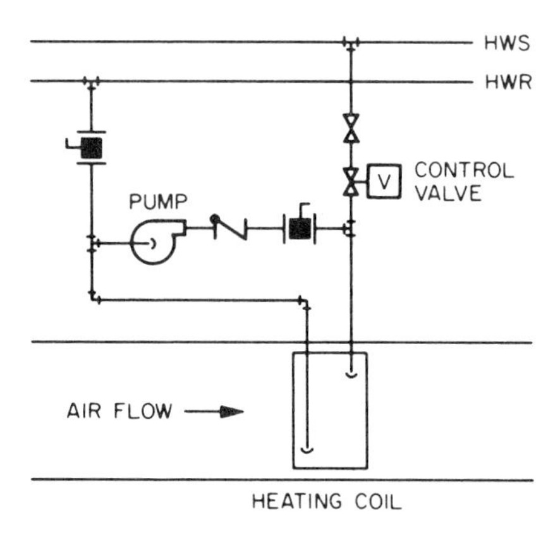

Figure 6-15 Preheat coil with circulating pump.

capacity and vacuum breakers. Even then, problems may occur if steam flow is modulated. Traps and drain lines must be insulated if exposed to freezing air.

An alternative pumping arrangement is shown in Figure 6–15. This allows the use of a straight-thru valve. The pump head and horsepower may be somewhat less than in the three-way valve arrangement. (See bibliography reference 22 for a full discussion of comparative performances of these two arrangements.) It should be noted here and remembered in the discussions which follow that in some situations the straight-thru valve may be preferred over a three-way valve. This is particularly true in large systems with multiple chillers and for boilers and circulating pumps. Three-way valves would require that full pumping capacity always be available, even at light loads. With straight-thru valves the flow is reduced at part loads and some of the pumps and chillers can be shut off, with a savings in energy consumption.

See Chapter 12 for a full discussion of this topic.

6.4.2 Normal Heating

"Normal" heating refers to the coil in a single-zone, multi-zone or dual-duct air system which handles all or a major portion of the system air at entering temperatures of 45° to 50° F or higher. In the case of a single-zone unit (Figure 6-16) the supply valve is controlled by a room thermostat (T1), frequently with a high-limit discharge thermostat (T2) added. The discharge thermostat acts as a feedback control, and decreases the operating differential.

Alternatively the supply valve may be controlled to provide a variable discharge air temperature with reset from the zone temperature (Figure

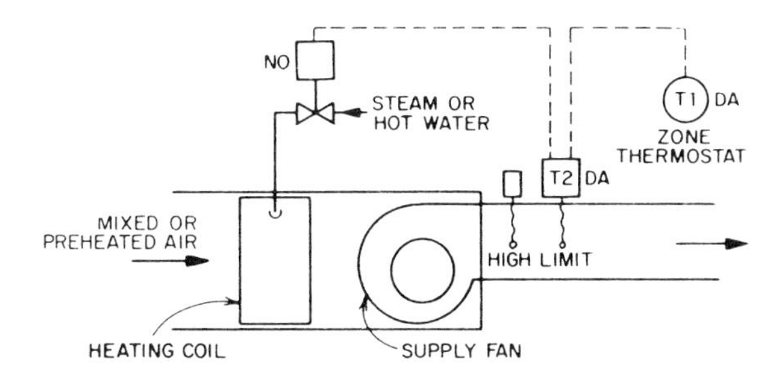

Figure 6-16 Heating, single-zone.

6–17). Either of these systems can be used for cooling, heating, or a combination of the two, with heating and cooling coils in series (See Chapter 7).

In dual-duct or multizone systems the supply valve is controlled by a hot plenum thermostat (Figure 6–18). To improve overall controllability, it is desirable to add outdoor reset, decreasing hot plenum temperature as outdoor temperature increases. "Discriminator" control is also used for reset, as described in Chapter 7.

If a recirculating pump system is used as in Figures 6–14 or 6–15, one coil may sometimes serve for both preheat and normal heating in a single-zone air handler. The choice will depend on the degree of control required, and the economics.

6.4.3 Reheat

Reheat is used for humidity control or individual zone control. In either case, control of steam or hot water supply valves is usually by room thermostat, sometimes in series with a supply duct high-limit thermostat.

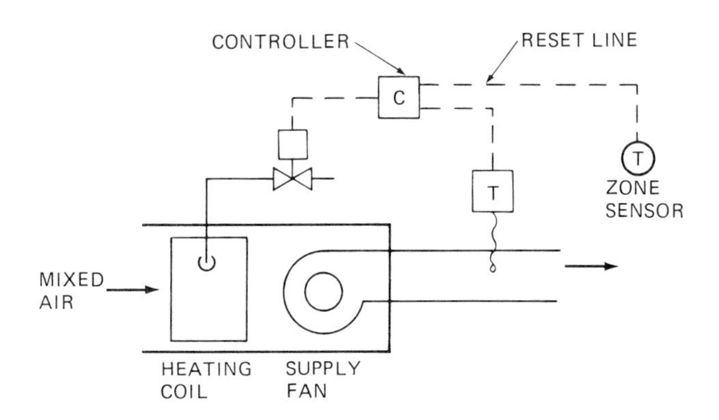

Figure 6-17 Heating, single-zone, discharge control.

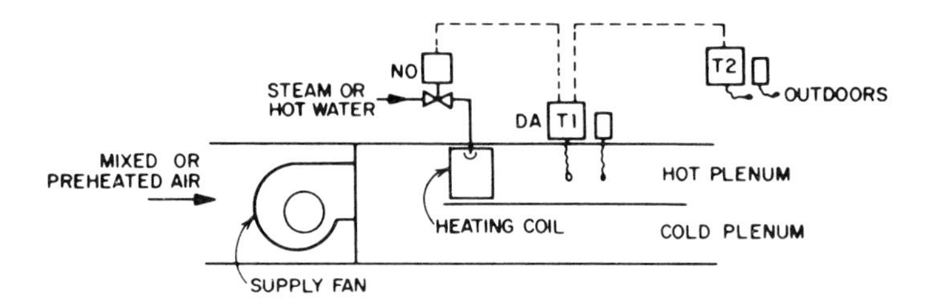

Figure 6-18 Heating, multizone or dual-duct.

6.5 COOLING COILS

Cooling coils are generally confined to the air handling unit, although occasionally recooling coils are required, as, for example, with chemical dehumidifiers. There are two types: direct-expansion (DX) coils and those using chilled water or brine.

6.5.1 Direct-Expansion Coils

DX coils must, by their nature, use two-position control with its inherently wide operating differential. Nonetheless, this system is often used, particularly in small units and where close control is not required. Figure 6–19 shows a typical DX coil control. The room thermostat opens the solenoid valve, allowing refrigerant liquid to flow through the expansion valve to the coil. The expansion valve modulates in accordance with its setting to try to maintain a minimum refrigerant suction temperature. A low-limit discharge thermostat, T2, keeps the supply air temperature from becoming too cold.

Controllability can be improved by providing face and bypass dampers (Figure 6–20), but this may lead to such complications as lack of humidity control and coil icing at high bypass rates. It follows that maximum bypass

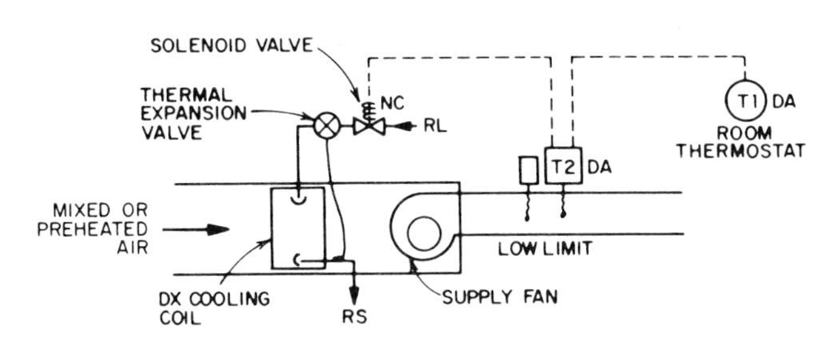

Figure 6-19 Direct-expansion cooling; two-position control.

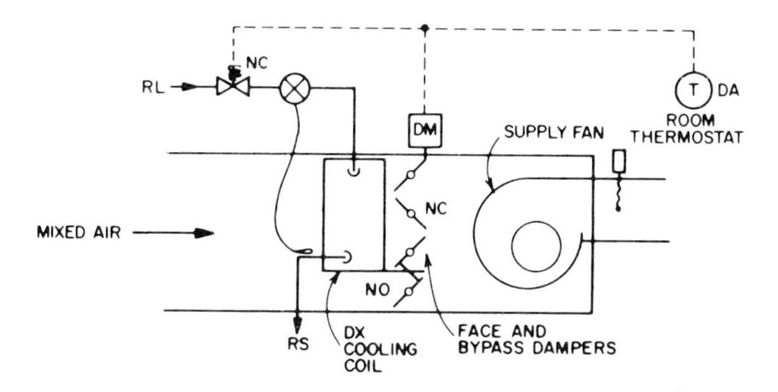

Figure 6-20 Direct-expansion cooling; face and bypass control.

rates must be established and the system may not provide adequate control at very light loads.

A different approach adds a variable back-pressure valve in the refrigerant suction line, controlled by the room thermostat (Figure 6–21). As the room temperature decreases the valve is throttled, increasing the suction temperature at the coil and decreasing the coil capacity. A reversing relay allows the back-pressure valve to be normally open, a necessary condition when the solenoid valve is first opened.

This scheme can lead to problems in the refrigerant circuit and should be used only by an expert in the refrigerant piping design.

Hot gas bypass may also be used for capacity control, as shown in Figure 6–22. A constant pressure expansion valve is used to maintain the evaporator pressure (and temperature) at a constant level, regardless of load. There are limitations on the percentage of total refrigeration flow which may be bypassed, and on pressure drops in the piping system. Consult a good manual on refrigeration practice.

Two-stage direct expansion will often provide adequate capacity control. The stages should be made by rows of coil rather than by sectioning the coil.

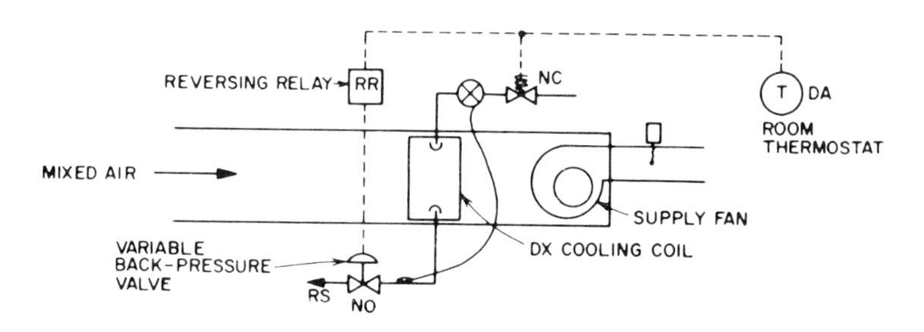

Figure 6-21 Direct-expansion cooling; suction pressure control.

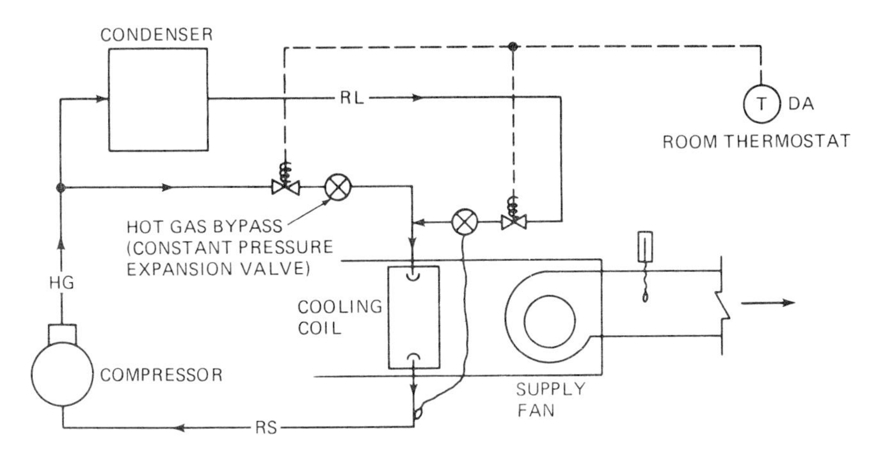

Figure 6-22 Hot gas bypass capacity control.

In the latter case, the active section may ice up forcing most of the air flow through the inactive section and reducing the coil capacity. In a multi-row coil the first stage should be the first row in the direction of air flow and the second stage the rest of the rows, since the first row of a 3- or 4-row coil does at least half the cooling. A two-stage thermostat is used (Figure 6–23).

6.5.2 Chilled Water Coils

Chilled water or brine coils are controlled in much the same way as heating, with a three-way or straight-through valves, modulating or two-

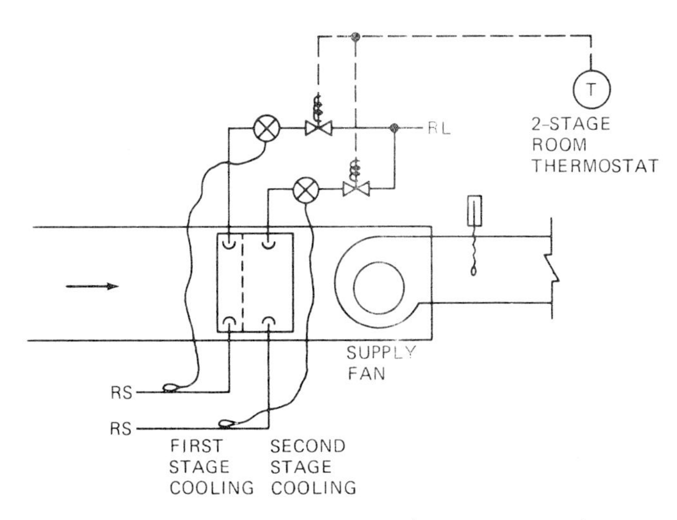

Figure 6-23 Direct expansion cooling; 2-stage control.

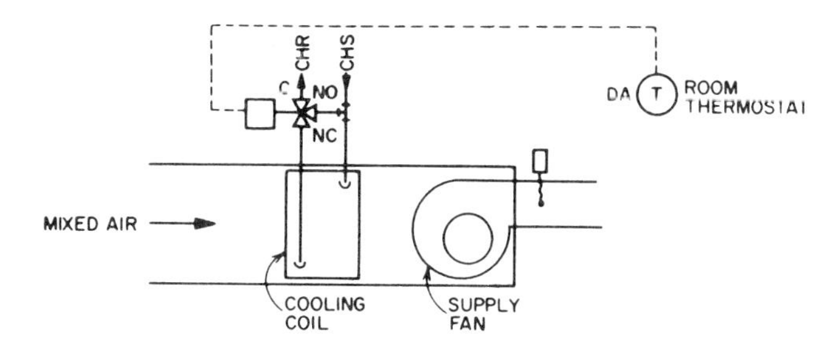

Figure 6-24 Cooling; chilled water, three-way valve.

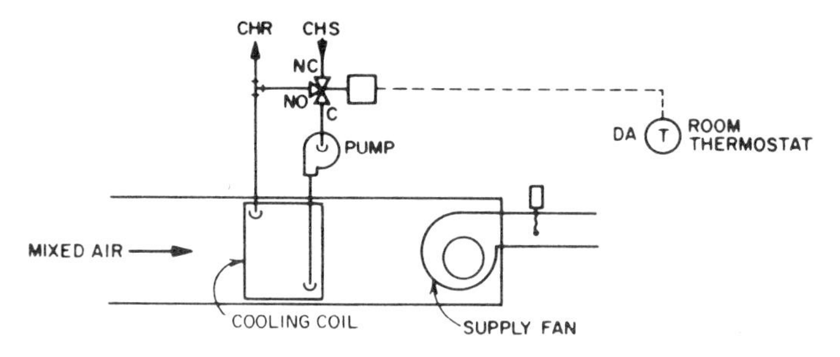

Figure 6-25 Cooling; chilled water, three-way valve and circulating pump.

position. Generally, however, it is desired that cooling coil control valves fail to the closed position since this allows the use of direct-acting controllers. The three-way valve arrangement would then appear as in Figure 6–24 or, if a recirculating pump is used, as in Figure 6–25 or Figure 6–15.

The recirculating pump arrangement is very useful in two cases: (1) for extremely accurate temperature control, and (2) to avoid freezing in those situations where system geometry may make it impossible to avoid stratification of mixed or partially preheated air.

6.5.3 Parallel and Counter Flow

You will notice in both the preceding figures that water flow is shown counter to airflow. This was mentioned before in connection with hot water heating coils, and it is very important in terms of heat transfer efficiency. Consider the cooling coil with air flowing through it, decreasing in temperature from 80° to 55° and with water flowing parallel to the air and increasing in temperature from 42° to 52°. On a graph of temperature vs distance this process would appear as in Figure 6–26(A). For counter flow Figure 6–26(B) applies. From heat transfer theory we learn that the heat

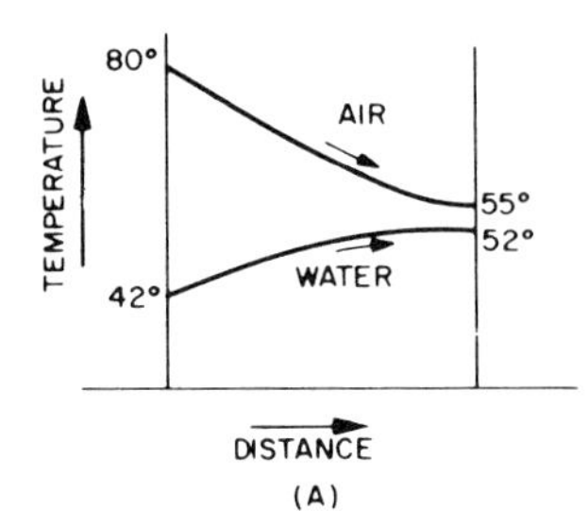

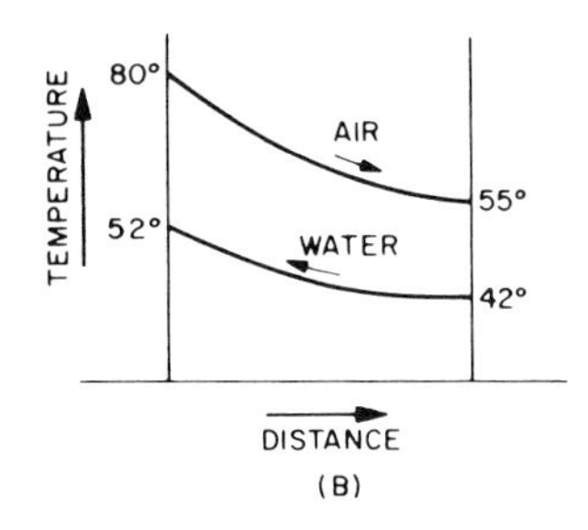

Figure 6-26 Chilled water flow in a cooling coil. (A) Parallel flow. (B) Counter flow.

transfer from air to water in a coil is a function of the *K* factor through the tube wall and air and water films, the total internal tube surface area and the MED (mean equivalent temperature difference). Given a fixed coil size and fixed rates of air and water flow, the only difference between parallel and counter flow arrangements is the MED. The MED can be calculated mathematically by means of equation (6–1).

$$MED = \frac{GTD - LTD}{\ln \frac{GTD}{LTD}} \tag{6–1}$$

where: GTD = greatest temperature difference between water and air.
LTD = least temperature difference between water and air.
MED = mean equivalent temperature difference.
ln = natural log (log to base e).

If we calculate the MED for each of the two flow arrangements, we get:

1. Parallel flow: $GTD = 80 - 42 = 38$
   $LTD = 55 - 52 = 3$

$$MED = \frac{38 - 3}{\ln \frac{38}{3}} = 13.8$$

2. Counterflow: $GTD = 80 - 52 = 26$
   $LTD = 55 - 42 = 13$

$$MED = \frac{26 - 13}{\ln \frac{26}{13}} = 18.7$$

This is an increase of nearly one-third in the heat transfer capacity, which of course cannot occur without a change in temperatures or flow rates. In practice, it can be shown that, in this example, to keep the same flow rate, capacity and air temperatures, the water temperature can be increased by 5° in going from parallel flow to counter flow, without any decrease in capacity. This would allow water temperatures of 47° in and 57° out, and would actually make the design condition (55° leaving air) impossible with parallel flow. The higher water temperature increases chiller efficiency and capacity.

The MED equation also holds when used with a condensing or evaporating fluid. But now counter flow does not apply, since condensing steam or evaporating refrigerant provides essentially a constant temperature in the coil.

6.6 HUMIDITY CONTROL

At different times and for different reasons it may be necessary to raise or lower the humidity of the supply air in order to maintain selected humidity conditions in the air conditioned space.

6.6.1 Air Washer

Consider first the air washer (Figure 6–27). Often used for its sensible cooling capability, it is then known as an evaporative cooler. Whether an inexpensive wetted-pad residential-type unit or a large industrial unit with an elaborate system of sprays and eliminators, any air washer operates on the adiabatic cooling principle. That is, the cooling is accomplished by using the sensible heat of the air to evaporate water. Thus, the air passing through the washer changes conditions along a constant web bulb line, with the final state being dependent on the initial state and the saturation efficiency of the washer (generally 70% to 90%). There is no control of humidity. This is shown in the psychrometric chart of Figure 6–28.

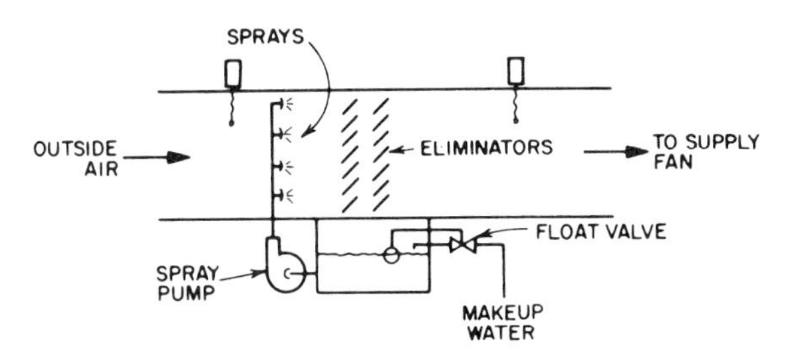

Figure 6-27 Evaporative cooling (air washer).

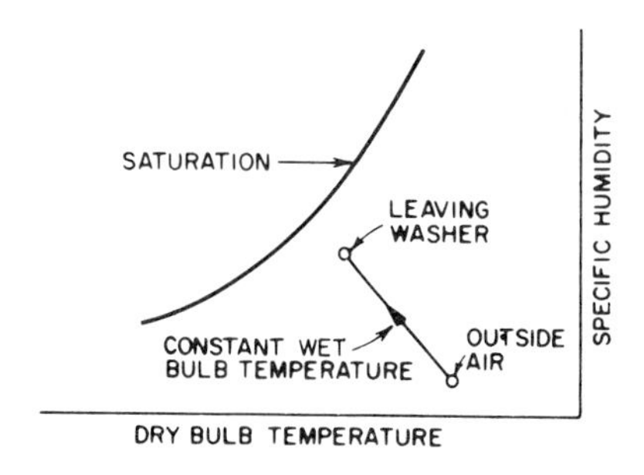

Figure 6-28 Psychrometric chart for evaporative cooling.

6.6.1 Two-Stage Evaporative Cooling Two-stage evaporative cooling may be used as an alternative to mechanical refrigeration, when outdoor conditions allow. This system provides lower dry bulb temperatures and relative humidities than can be obtained with ordinary evaporative cooling. Figure 6–29 shows the arrangement and control of a two-stage evaporative cooling system. Figure 6–30 is the psychrometric chart of the cycle. The dashed lines on the chart show a single-stage cycle as in Figure 6–28.

The cooling tower and precooling coil constitute the first stage. This sensible cooling reduces both the wet and dry bulb temperatures of the air, so that in the second stage a lower dry bulb temperature may be obtained. The room thermostat will be a two-position type, with two-stage control optional. The controllability of the system depends on the condition of the outside air.

6.6.2 Air Washer with Preheat

About the only control that can be applied to the ordinary air washer is to turn the spray water (or pump) on or off. If a minimum humidity

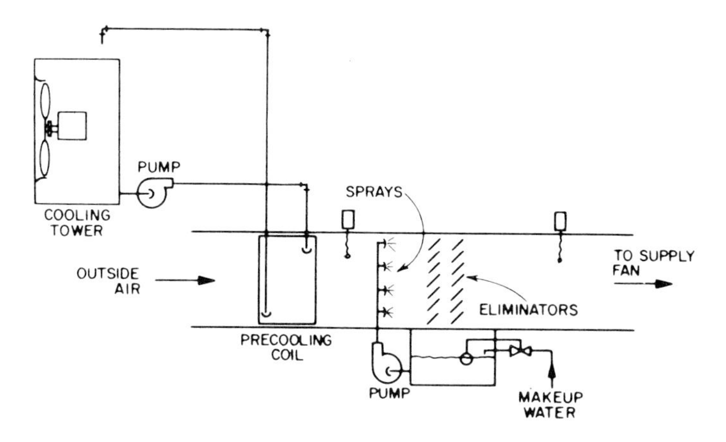

Figure 6-29 Two-stage evaporative cooling.

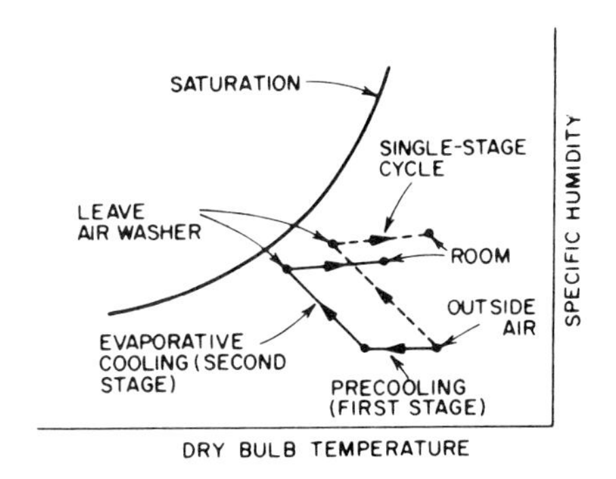

Figure 6-30 Psychrometric chart for Figure 6-29.

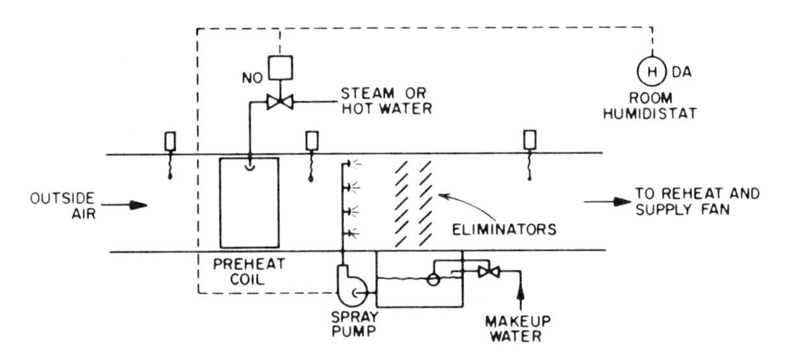

Figure 6-31 Minimum humidity; air washer with preheat.

is required, it is sometimes necessary to preheat the air to the desired wet bulb temperature. Such a control system is shown in Figure 6–31. The room humidistat senses low humidity and turns on the spray pump and then opens the preheat coil supply valve. As room humidity increases, the preheat valve is first closed, then if the increase continues the sprays are shut down. Under high outdoor humidity conditions cooling capacity is limited. Final room temperature control is provided by reheat coils. The psychrometric chart in Figure 6–32 shows this cycle.

6.6.3 Air Washer with Preheat and Refrigeration

Obviously, such a system will not control the upper limits of humidity. To accomplish this two choices are available, both requiring refrigeration.

1. Heat and/or cool the spray water by means of shell and tube heat exchangers (Figure 6–33). By allowing the humidistat to control the heat

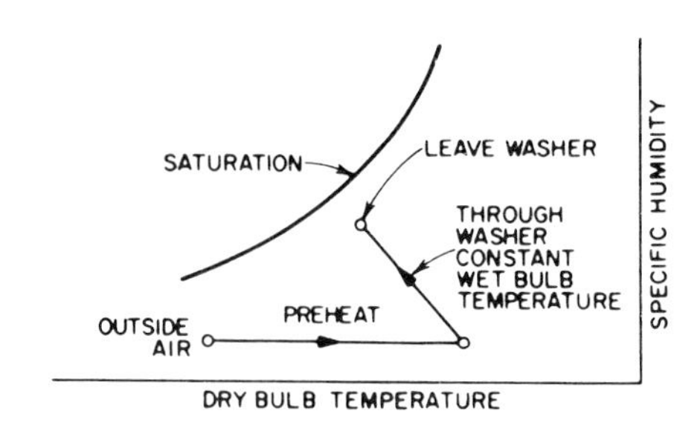

Figure 6-32 Psychrometric chart for Figure 6-31.

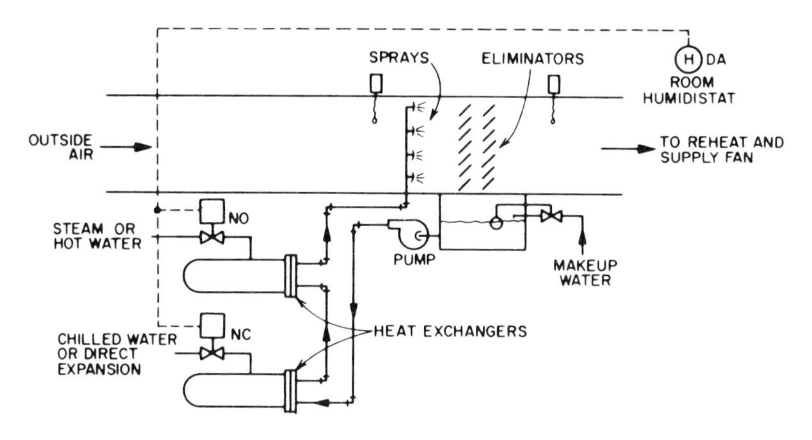

Figure 6-33 Humidity control by heating or cooling spray water.

exchanger supply valves it is possible to obtain very accurate control of the humidity as well as the air temperature leaving the washer. Figure 6–34 is the psychrometric chart diagram of this process. Reheat is required because the temperature leaving the washer is constant and is not a function of the building load.

2. Add a cooling coil in the air washer, creating a "sprayed-coil dehumidifier" (Figure 6–35). The term is not incorrect, since at high humidity conditions dehumidification does take place, but it also acts as a humidifier if necessary. Since the addition of the coil increases the saturation efficiency to between 95% and 98%, it is possible to do away with the humidistat and use a simpler, less expensive dry bulb thermostat (T1) in the air leaving the coil (the so-called "dew point" thermostat). This is set to maintain a fixed condition of dry bulb and relative humidity. Preheat will be required at low mixed-air humidities. When the mixed-air wet bulb is above the required wet bulb temperature of the air leaving the coil, then refrigeration must be used. Reheat is nec-

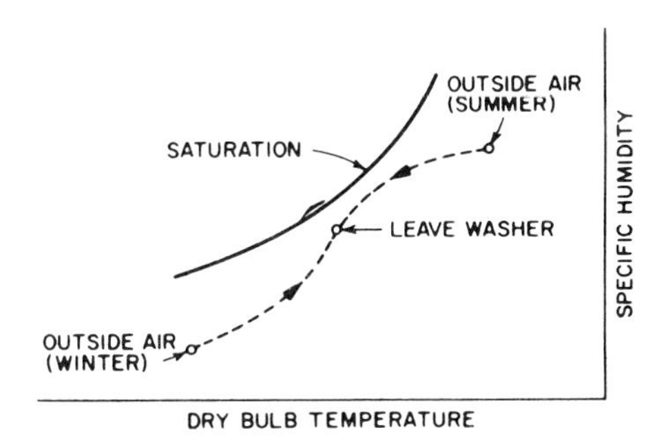

Figure 6-34 Psychrometric chart for Figure 6-33.

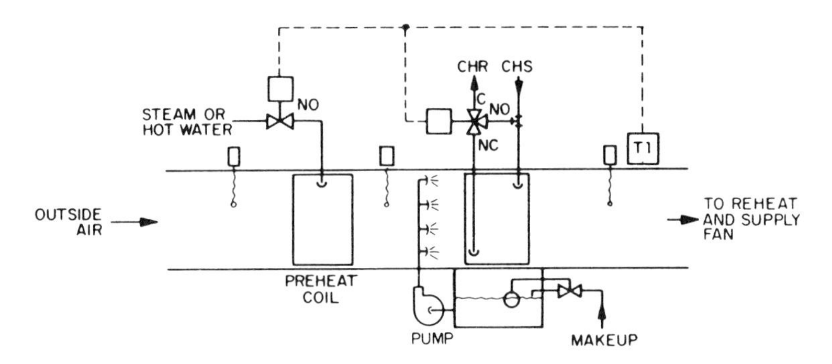

Figure 6-35 Humidity control, sprayed coil and preheat.

essary for final space temperature control. If all control settings are properly selected, it is possible to maintain very accurate control of space temperature and humidity. Figure 6–36 shows the cycle on a psychrometric chart, assuming cool, low-humidity outside air. The dew point thermostat senses a decrease in coil leaving-air temperature and shuts down the chilled water flow to the coil and then opens the preheat coil valve. Evaporative cooling is used. As the coil leaving-air temperature increases, the preheat coil valve is gradually closed. A further increase in the controlled temperature above the thermostat setting will cause the chilled water valve to open, using refrigeration for cooling. Figure 6–37 is a psychrometric chart showing what happens with high-temperature outside air. So long as the entering-air wet bulb is above the dew point thermostat setting, chilled water will be required and the coil leaving-air temperature will be maintained by varying the flow of chilled water through the coil. Direct expansion may be used instead of chilled water.

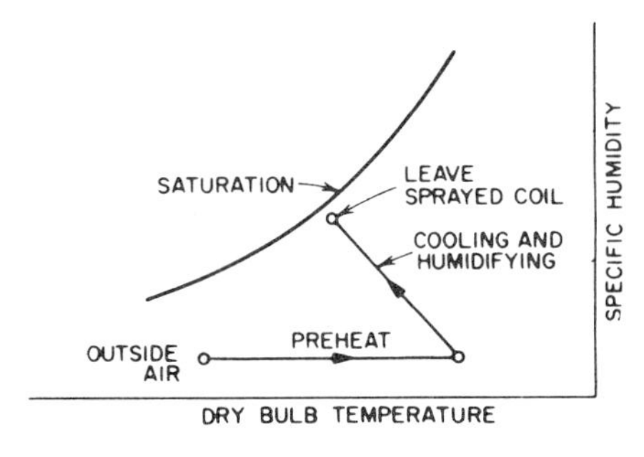

Figure 6-36 Psychrometric chart for Figure 6-35 (winter).

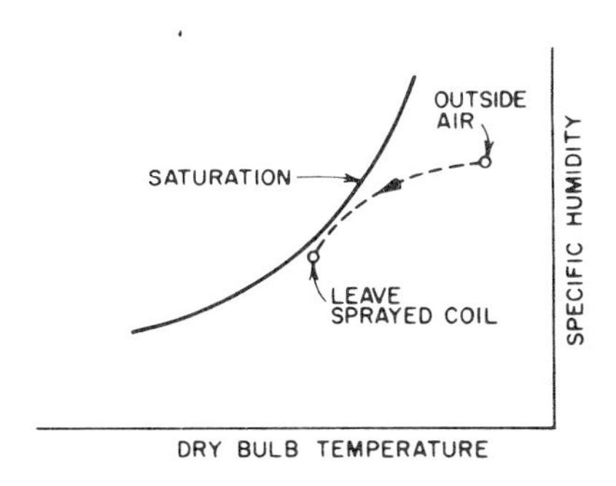

Figure 6-37 Psychrometric chart for Figure 6-35 (summer).

6.6.4 Air Washer with Mixed Air and Refrigeration

Mixed air can be used rather than 100% outside air if the control system is altered as shown in Figures 6–38 and 6–39. As long as the outside air wet bulb temperature is below the coil leaving-air set point of T1, the desired condition can be maintained by adjusting the mixing dampers to obtain a mixture which falls on the coil leaving-air wet bulb line, and utilizing the evaporative cooling effect of the sprayed coil. When the outside air damper is fully open and the coil leaving-air temperature increases above the set point of T1, then refrigeration must be used, either chilled water as shown, or direct expansion. Again, reheat is required for space temperature control.

All of these systems using open sprays require chemical treatment of the spray water to minimize solids deposition. The systems with finned coils pose a severe maintenance problem due to deposition. The only satisfactory solution is to use demineralized water in the spray system. Sprayed coils may not be allowed under some codes because of the potential for bacterial contamination.

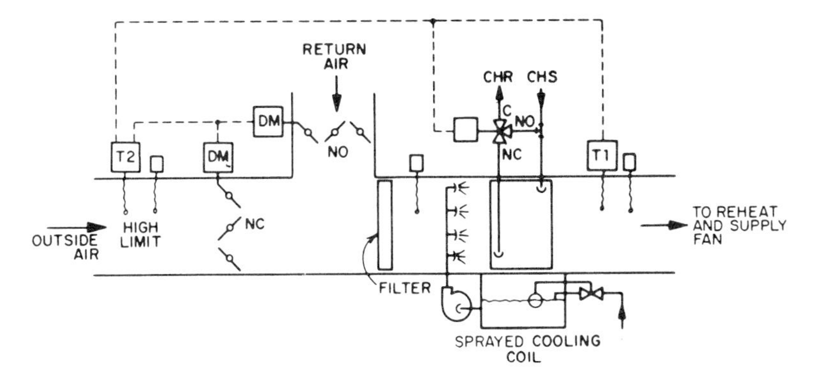

Figure 6-38 Humidity control; sprayed cooling coil with mixed air.

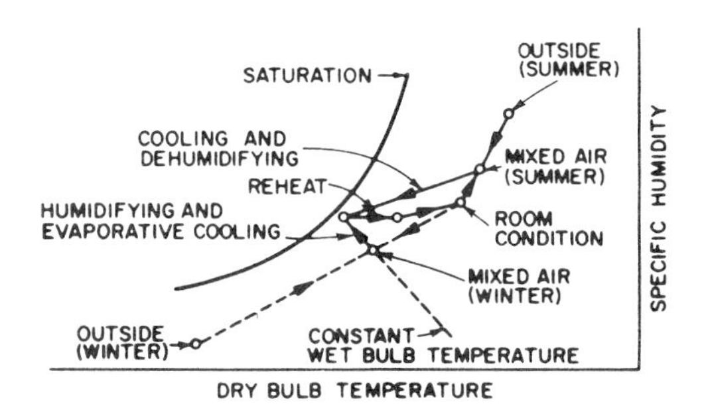

Figure 6-39 Psychrometric chart for Figure 6-38.

6.6.5 Steam Humidifiers

Steam humidifiers are often used because of their simplicity. A piping manifold with small orifices is provided in the air duct or plenum (Figure 6–40). The steam supply valve is controlled by a space or duct humidistat. Avoid the use of steam which has been treated with toxic chemicals. Almost any humidity ratio, up to saturation, can be obtained in the supply air stream. If a space humidistat is used a duct high-limit humidistat must be provided to avoid condensation in the duct.

6.6.6 Pan Humidifiers

Evaporative pan humidifiers vary in capability and controllability. The old-fashioned evaporator pan found in many residential warm-air heating systems is not very efficient, and lacks any kind of control. The addition

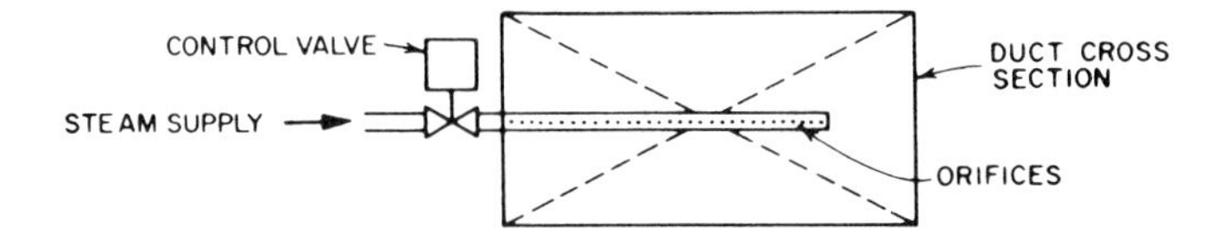

Figure 6-40 Steam grid humidifier.

of a heater, of immersion or radiant type, improves efficiency and allows a limited amount of control. Even with heaters, however, this type of humidifier will not usually provide space relative humidity of over 40%.

6.6.7 Atomizer Humidifiers

Slinger or atomizer humidifiers may be controlled on-off by a humidistat or manually. Again, space relative humidity will seldom exceed 40%. During "off" periods the lime deposits left from the water evaporation may be entrained in the air stream as dust. When using evaporative or atomizer humidifiers the water supply should be distilled or deionized. Mineral deposits from evaporation constitute a serious maintenance problem. "Rental" deionizers are very satisfactory for small installations. (See also paragraph 7.2.)

6.7 DEHUMIDIFIERS

As noted above, a sprayed cooling coil may serve as a dehumidifier if the entering air is sufficiently humid. However, to maintain very low humidities at normal air conditioning temperatures it is necessary to employ chemical dehumidifiers or low temperature refrigeration.

6.7.1 Chemical Dehumidifiers

Chemical dehumidifiers use a chemical adsorbent, usually with provision for continuous regeneration (drying) to avoid system shutdown. One form of dehumidifier (Figure 6–41) uses a wheel containing silica gel which revolves first through the conditioned air stream, absorbing moisture, and then through a regenerative air stream of heated outside air which dries the gel. In the process, a great deal of heat is transferred to the conditioned air and recooling is necessary. The dehumidification efficiency is largely a function of the regenerative air stream temperature. Figure 6–42 shows the

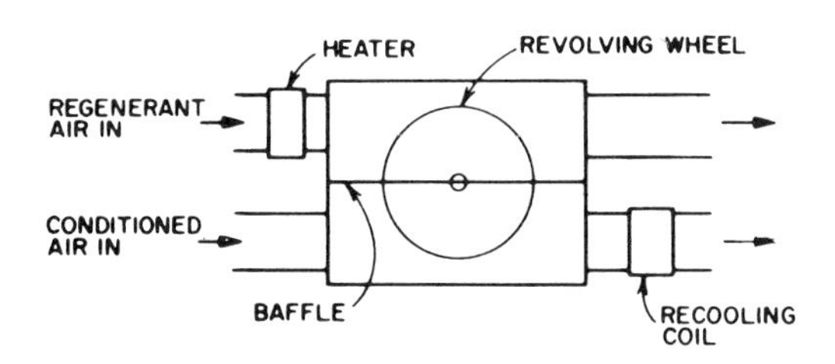

Figure 6-41 Dehumidifier; chemical adsorbent.

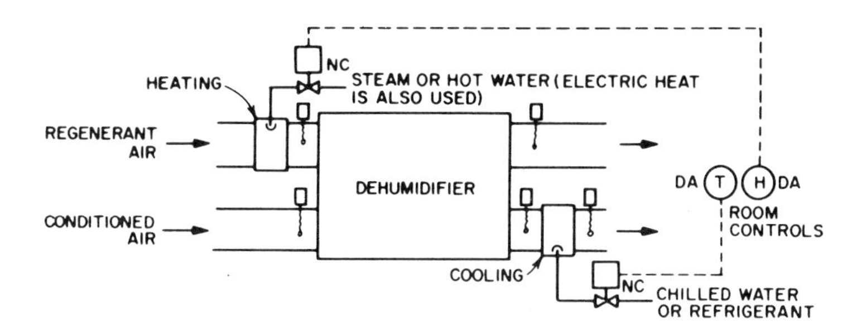

Figure 6-42 Control of chemical dehumidifier.

control system. The space humidistat controls the heating coil in the regenerative air stream. Final control of space temperature is accomplished by the room thermostat and the recooling coil. Very low humidities may be obtained in this way.

Another type of system uses a water-absorbing liquid chemical solution which is sprayed over a cooling coil in the conditioned air stream, thus absorbing moisture from the air. A portion of the solution is continuously pumped to a regenerator where it is sprayed over heating coils and gives up moisture to a scavenger air stream which carries the moisture outside. Control of the specific humidity of the leaving conditioned air is accomplished by controlling the temperature of the solution. Since this is a patented, proprietary system, the control system recommended by the manufacturer* should be used.

6.7.2 Dehumidifying by Refrigeration

Low temperature cooling coils may also be used to reduce humidity to low values. Since coil surface temperatures may be below freezing,

*Ross Engineering Div. of Midland-Ross Corp.

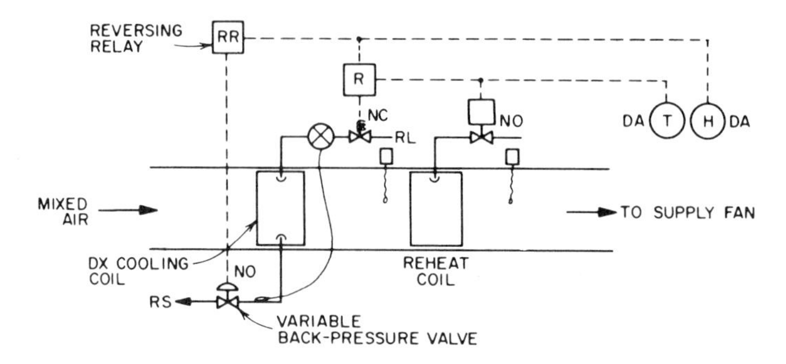

Figure 6-43 Dehumidifying with low-temperature cooling coil.

with ice formation resulting, special DX coils with wide fin spacing must be used, and provision must be made for defrosting by hot gas, electric heat or warm air. This approach tends to be inefficient at very low humidities and requires intermittent shutdown for defrosting, or parallel coils so that one may operate while the other is being defrosted. Reheat is necessary for control of space temperature. Since space humidity is largely a function of coil temperature, fairly good control may be achieved through humidistat control of a variable back-pressure valve (Figure 6–43). The selective relay allows the room thermostat to operate the cooling coil at minimum capacity when the humidistat is satisfied.

6.8 STATIC PRESSURE CONTROL

Static pressure controls are used to provide a positive or negative pressure in a space with respect to its surroundings. For example, a clean room will be positive to prevent infiltration of dust, while a chemical laboratory or plating shop will be negative to prevent exfiltration of fumes. Many nuclear processes require careful space segregation with fairly high pressure differentials. For most ordinary spaces it is impractical and unnecessary to design for pressure differentials greater than 0.1 in. water gage. Swinging doors are difficult to open and/or close even at this small pressure, and special sealing methods are necessary to maintain higher pressures. Air locks are often used.

6.8.1 Single Room

A pressure-controlled space may be served by its own air conditioning unit. Outside air may have minimum or economy cycle control by

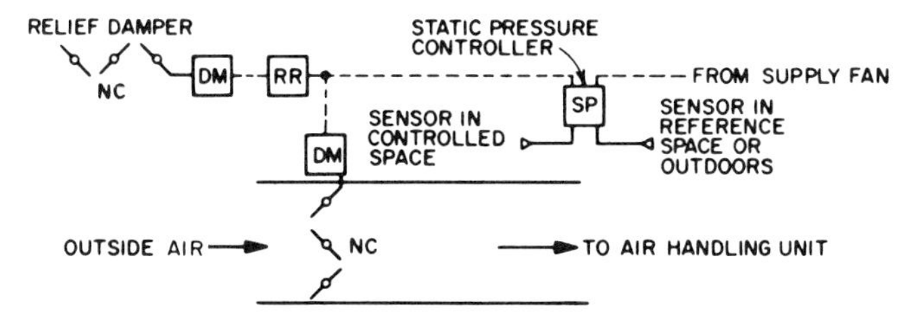

Figure 6-44 Outside air; static pressure control.

temperature, or may be simply 100%. Relief air is usually provided by power exhaust, although positive pressures in the space allow "gravity" relief. In either case, the static pressure controller controls the relief damper, and if 100% outside air is used, may also control the intake damper. Such a control system is shown in Figure 6–44. (See also Figure 6–7.) The reversing relay is needed so that both dampers can be normally closed.

6.8.2 Multiple Rooms

If several rooms are to be served by a single air conditioning unit then variable-volume dampers on supply and relief to each room will be operated by the static pressure controller for that room. Individual exhaust fans, or one large central exhaust may be used. (See discussions in Sections 9.4.2 and 9.4.3.)

6.9 ELECTRIC HEAT

Electric heaters may be controlled on the same basic cycles as other heating devices—two-position, timed two-position and proportional. Because of the use of electricity as the energy source, certain special considerations are necessary. Any electric heater must be provided with a high-limit control. Some codes require both automatic reset and manual reset high limits. Forced-air heaters should have airflow switches to prevent the heater from operating when the fan is off or when air flow is below the minimum rate required to prevent overheating of the electric element.

6.9.1 Two-Position Control

Two-position control of small-capacity heaters may be provided by a heavy-duty line voltage thermostat. More common, however, is the

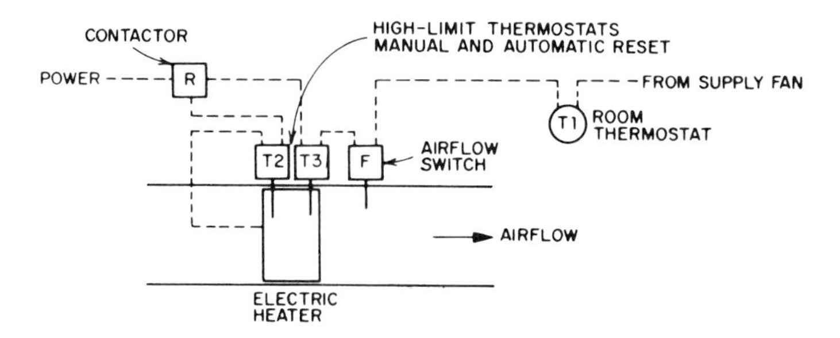

Figure 6-45 Electric heater; piloted control, two-position.

piloted system (Figure 6–45) with the thermostat operating a contactor. For large heaters it is common to use a multistage thermostat or sequencing switches with several contactors, each controlling current flow to a section of the heating coil.

6.9.2 Proportional Control

True proportional control may be obtained through the use of a saturable core reactor or variable autotransformer (Figure 6–46). A very small change in the DC control current can cause a large change in the load current. Since the controller must handle the full-load current it may become physically very large. Efficiencies are poor at part load.

6.9.3 Timed Two-Position Control

Timed two-position control can be achieved by means of a timer and the same contactors used for ordinary two-position control (Figure

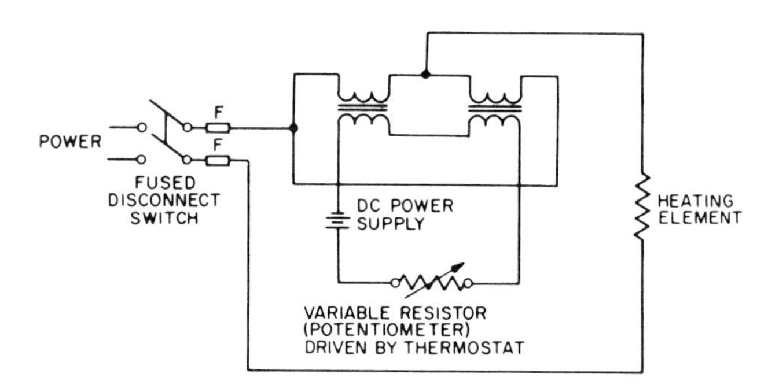

Figure 6-46 Electric heater; saturable core reactor control.

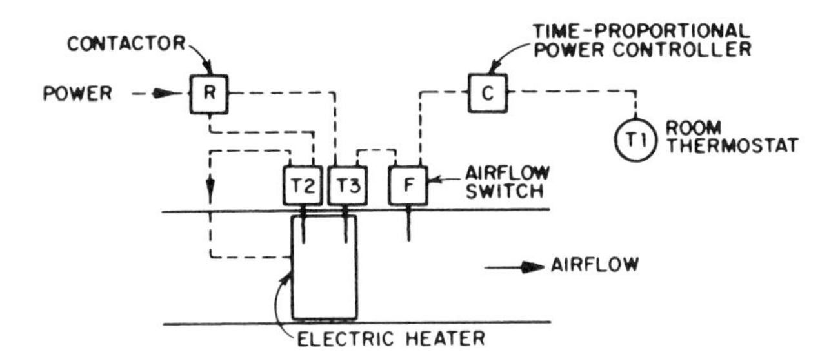

Figure 6-47 Electric heater, time-proportional control.

6–47.) The timer may be mechanical or electronic and provides a fixed time base (usually adjustable) of from one-half minute to five minutes. The percentage of on-time varies in accordance with the demand sensed by the room thermostat. The contactors will cycle on and off once in each time base period but the length of on-time will be greater if the room temperature is below the thermostat setting. However, rapid cycling, even with mercury switch contactors, may lead to maintenance problems. The preferred method is to use solid-state controllers; SCR's (silicon controlled rectifiers) or Triacs. These devices can provide extremely rapid cycling rates, so that control is, in effect, proportional (Figure 6–48). The solid-state controller has an electronic timer, which provides an extremely short time base. The load current is handled directly by the controller through semiconductor switching devices. The thermostat demand will vary the percentage of on-time. Power regulation may be accomplished by phase control or burst control. The SCR may be piloted by any type of proportioning thermostat, to vary the cycle time rate. The "Vernier control" system (Figure 6–49) combines sequence control with solid-state control to obtain near-proportional control. It is especially economical if the heater is large. The coil is divided into several

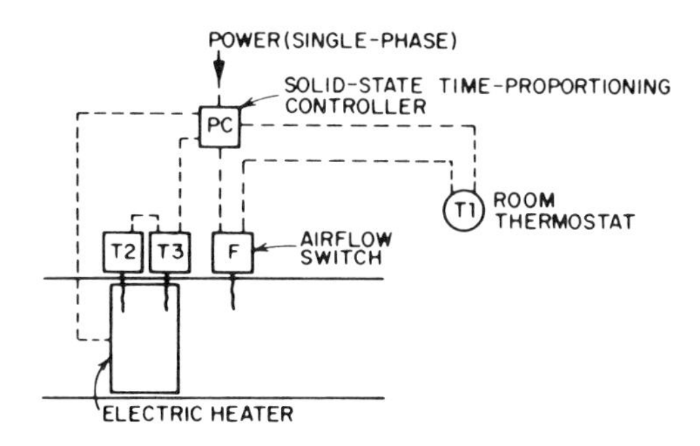

Figure 6-48 Electric heater; solid-state controller.

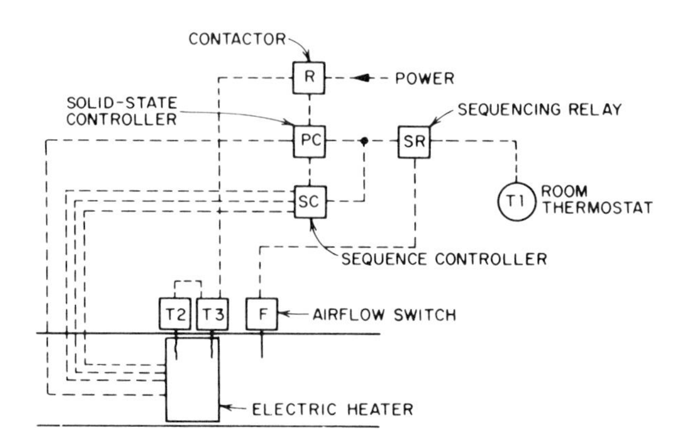

Figure 6-49 Electric heater; Vernier control system.

small sections with one controlled by the solid-state unit and the others by sequencing step controller. As the space temperature decreases below the proportional thermostat setting the solid-state unit modulates its section of the coil from off to 100% on. If heating demand continues to increase the first section of sequence controlled portion is turned on and the power to the solid-state section may be cut back. This continues until all sections are on, and, of course, the opposite sequence occurs on a decrease in heating demand. The modulated section should have 25% to 50% greater capacity than the sequenced sections to prevent short-cycling at changeover points.

6.10 GAS-FIRED HEATERS

Direct gas-fired heaters are, generally, packaged equipment complete with controls. Usually the complete package, including controls, is approved by The American Gas Association (AGA). Any change in the system, however minor, will void this approval. There are still a number of options among the approved packages.

6.10.1 Two-Position Control

Two-position controls, are most common (Figure 6–50). A variation of this is multistage two-position, with the gas burners sectionalized.

6.10.2 Proportional Control

Modulating gas valves are available. AGA approval for small systems is limited to self-contained types, with the sensing bulb mounted in the

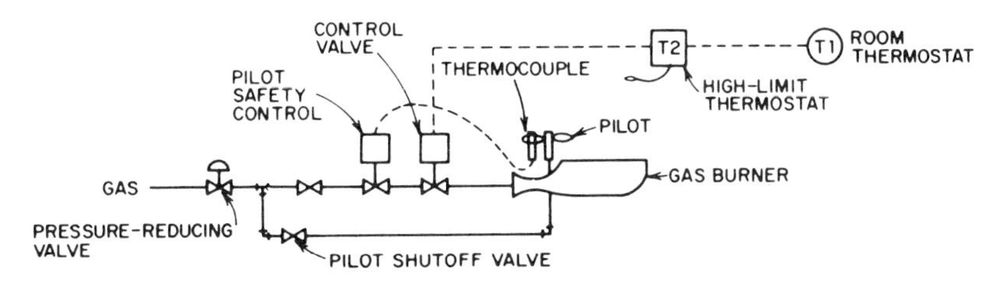

Figure 6-50 Gas heater; two position control.

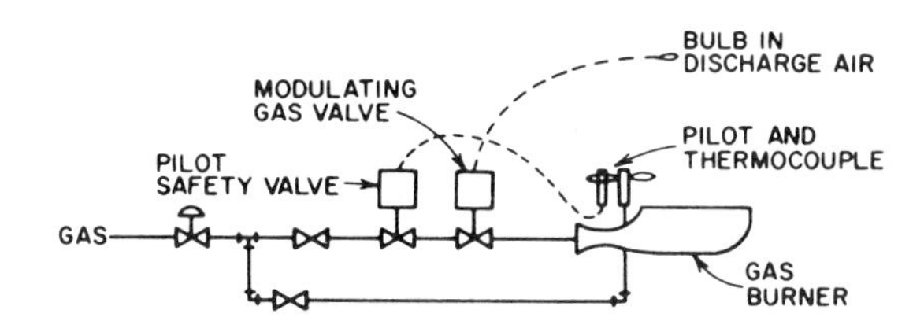

Figure 6-51 Gas heater; modulating control.

discharge air stream. Minimum capacity is about 30% of total capacity since the gas flame becomes unstable at lower rates (Figure 6–51). At this lower limit the valve becomes a two-position type.

6.10.3 Safety Controls

Any gas control system must include such safety controls as high-temperature limits and pilot flame proving devices. Forced-draft heating boiler controls may also include timed pre-purge and post-purge cycles to prevent gas accumulation and explosion. These specialized cycles are usually programmed by special-purpose electronic controls.

6.10.4 Forced Draft Burners

The above discussions deal with atmospheric-type gas burners. If forced draft or pressure burners are used the control systems will be furnished by the manufacturer to meet the requirements of one of the national insuring agencies such as Factory Mutual (FM). Many of the elements described above will be present in such systems. More efficient control of the combustion process is possible but discussions of these systems is beyond the scope of this book.

6.11 OIL-FIRED HEATERS

Oil burning heaters, like gas-fired units, generally come complete with a control package with UL or FM approval. Two-position or modulating controls are available, depending on the type and size of the equipment. This is a highly specialized area. Residential and small commercial control systems will be discussed in Chapter 7.

6.12 REFRIGERATION EQUIPMENT

Central refrigeration equipment may include compressors, condensers and chillers. Compressors may be reciprocating, positive displacement (screw-type) or centrifugal. Often, the compressor is part of a package water chiller as described below. Reciprocating compressors, up to about 50 tons capacity, may be used directly with DX coils. Accurate control of DX systems above 25 or 30 tons is difficult and it is unusual to find such systems above 50 or 60 tons. Larger reciprocating machines, positive displacement units and centrifugal compressors will be part of a package chiller system, complete with factory installed controls. Absorption machines also come as part of a package chiller except for small residential units. Condensers may be air-cooled, water-cooled or evaporative, and are sometimes part of the chiller package.

6.12.1 Reciprocating Compressors

Reciprocating compressors, except in very small sizes, have multistage capacity control. This is generally achieved by "loading" and "unloading" cylinders under control of a suction pressure or chilled water temperature controller, by raising the suction valve off its seat (Figure 6–52). Unloading devices may be mechanical or electrical and are always a part of the compressor package. The number of steps is determined by the manufacturer as a function of size, number of cylinders and machine design.

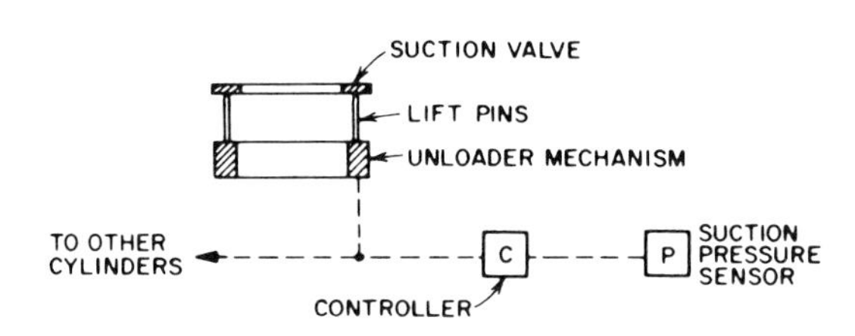

Figure 6-52 Refrigeration compressor unloader.

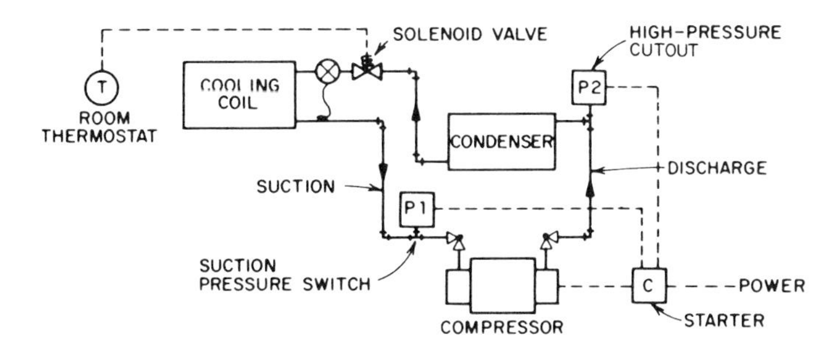

Figure 6-53 Refrigeration compressor; "pump-down" control.

Starting and stopping the compressor may be done directly by the room or chilled water thermostat, but is more often done on a "pump-down" cycle. On a rise in temperature of the controlled medium the thermostat opens a solenoid valve in the refrigerant liquid line to the chiller or cooling coil. Refrigerant flow raises the suction pressure. The low-pressure switch closes and starts the compressor. When the thermostat is satisfied and closes the solenoid valve, the lowering of the suction pressure as refrigerant is pumped out of the evaporator opens the low-pressure switch, stopping the compressor. Figure 6–53 shows a typical electric control system for a medium-sized reciprocating compressor with safety and operating controls.

6.12.2 Centrifugal, Positive-Displacement and Absorption Chillers

Since centrifugal and absorption machines always occur as part of a chiller package, the complete control system will be furnished by the manufacturer. These systems usually include fully modulating capacity controls with a low limit of 20% to 30% of maximum capacity, and provide elaborate safety and interlock controls to protect the equipment. A typical control system for a centrifugal package is shown in Figure 6–54. The chilled water and condensing water pumps are started manually (or the condensing water pump may be interlocked to start when the chilled water pump is started). When the chilled water thermostat calls for cooling the compressor will start, provided that flow switches and safety switches are closed. The thermostat will then modulate the inlet vane capacity controller, with the capacity limiting controller acting as a maximum capacity limiting device. When load falls below about 20% of machine capacity the thermostat will stop the compressor. A time-delay relay, not shown, is usually provided to prevent restarting the compressor at less than 30 min intervals. This is necessary to prevent damage to the motor.

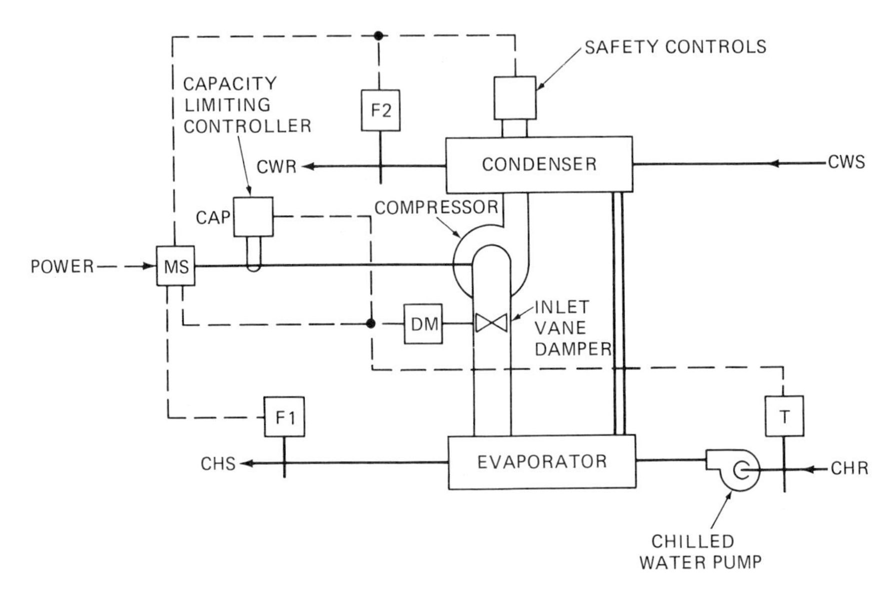

Figure 6-54 Centrifugal water chiller control.

6.12.3 Air-Cooled Condensers

Air-cooled condensers use ambient air forced across a condensing coil to cool and liquify the compressed refrigerant gas. The condenser fan is controlled to start automatically whenever the compressor runs. Condensing pressures at high ambient temperatures will be higher than with water-cooled or evaporative condensers, since these depend on wet bulb temperatures. But, at low ambient temperatures, this condensing pressure may fall so low as to cause operating problems in the refrigerant system. Any air-cooled system which must operate at low ambient temperatures must be provided with "head-pressure" control.

A simple head-pressure control system (Figure 6–55) uses modulating dampers to reduce airflow, finally stopping the fan when the damper is nearly closed. Alternatively, a variable-speed fan may be used.

A more elaborate and effective system uses the "flooding" principle, in which a throttling valve slows down the flow of liquid refrigerant, causing the condenser coil to be partially or wholly filled with liquid, thus reducing its capacity. There are several "flooding control" systems, mostly patented by manufacturers.

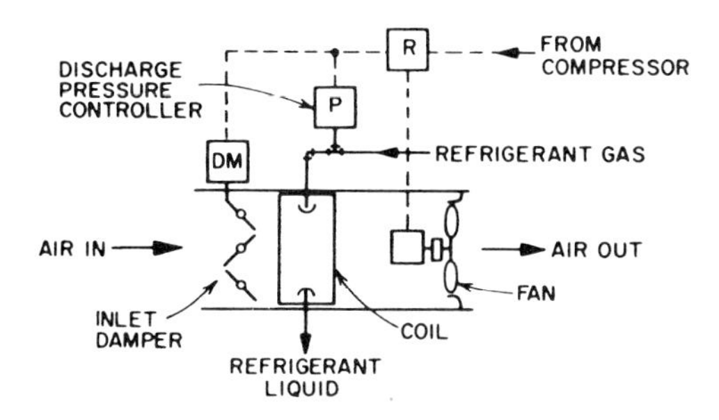

Figure 6-55 Air-cooled condenser.

6.12.4 Water-Cooled Condensers

Water-cooled condensers use water from a cooling tower or other source to cool and liquify the refrigerant gas. Condensing pressure can be closely controlled by modulating water flow or controlling supply water temperature by mixing condensing return water with supply water. If a cooling tower is used, water temperature may be controlled by cycling the tower fan or bypassing some of the tower flow (Figure 6–56). With the temperature of the condensing water supply (CWS) above the thermostat set point the flow valve V1 is open, bypass valve V2 is closed and the fan is running. As the CWS temperature decreases below the set point the fan is turned off. On a further decrease valve V1 closes and V2 opens. As the CWS temperature increases the reverse cycle takes place. A wide throttling range should be used since refrigeration system efficiency increases as the condensing temperature decreases. A typical control range is 70° to 85° F. Too low a condensing temperature can cause surge problems with the refrigerant compressor. Absorption chillers require close control of condensing water temperature, with "mixing" most often used (Figure 6–57). The CWS ther-

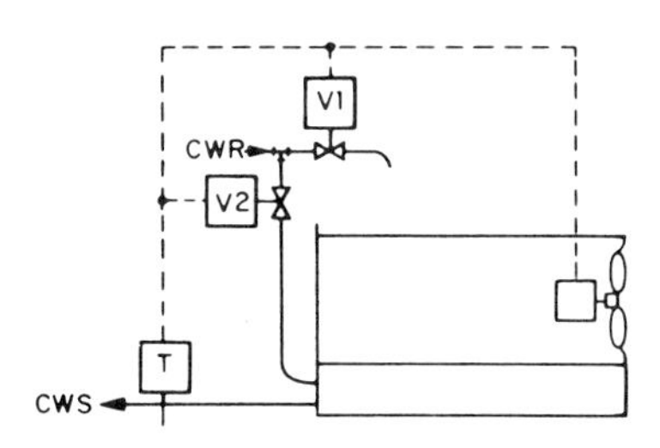

Figure 6-56 Cooling tower; bypass valve control.

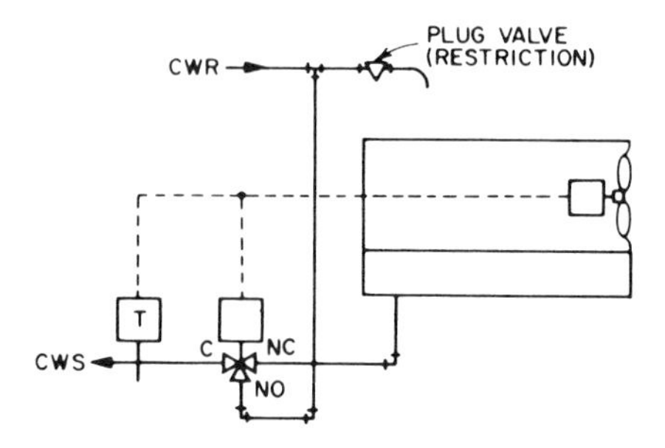

Figure 6-57 Cooling tower; three-way valve control.

mostat modulates the mixing valve to maintain a constant supply water temperature. When the valve is in full bypass condition, the cooling tower fan is stopped. In order for the bypass valve to function properly it is necessary to have several feet of gravity head above the valve to the tower. If a good static head is not available a diverting valve may be used, (Figure 6–58).

6.12.5 Evaporative Condensers

Evaporative condensers take advantage of wet bulb temperatures by spraying water over the refrigerant condensing coil. Air is forced or drawn across the coil and the resulting adiabatic saturation process provides efficient condensing. Control of condensing temperature is achieved by varying the airflow (Figure 6–59). The head-pressure controller modulates the outside air damper to provide constant head pressure. When the damper is nearly closed the fan is stopped. The condenser coil should be all "prime surface" (no fins) to minimize maintenance problems caused by mineral deposits.

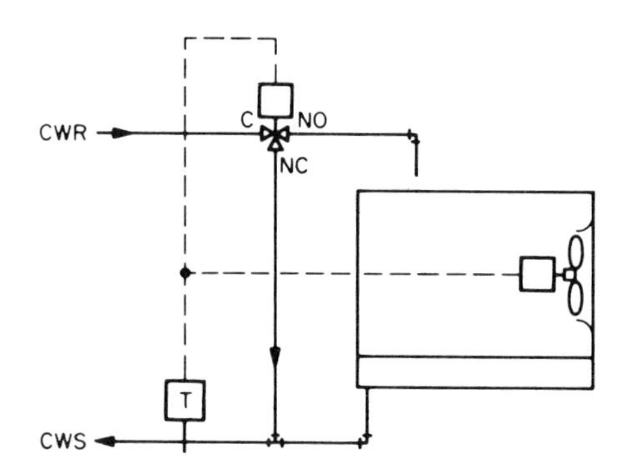

Figure 5-58 Cooling tower; three-way diverting valve.

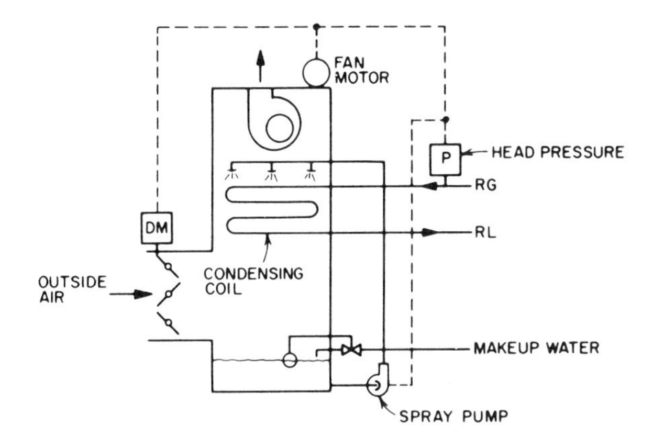

Figure 6-59 Evaporative condenser control.

6.12.6 Water Chillers

Chillers may not be part of a package. When installed as a separate piece of equipment two arrangements are possible. Flooded systems use a surge tank with low- or high-pressure float to control refrigerant feed. Direct-expansion systems use a thermostatic expansion valve. In either case refrigerant flow will be started or stopped by a solenoid valve in the refrigerant liquid line, controlled by a chilled water thermostat. Interlocks to prove chilled water flow and prevent freezeup are required. Figure 6–60 shows a direct-expansion chiller with water flow and low water temperature interlocks. The operating thermostat, T1, in the chilled water return will open the solenoid valve in the refrigerant liquid line, providing that the water is flowing and not near freezing.

6.12.7 Cooling Towers

Cooling towers which will operate only during the cooling season are usually provided with fan control only. A condensing water supply

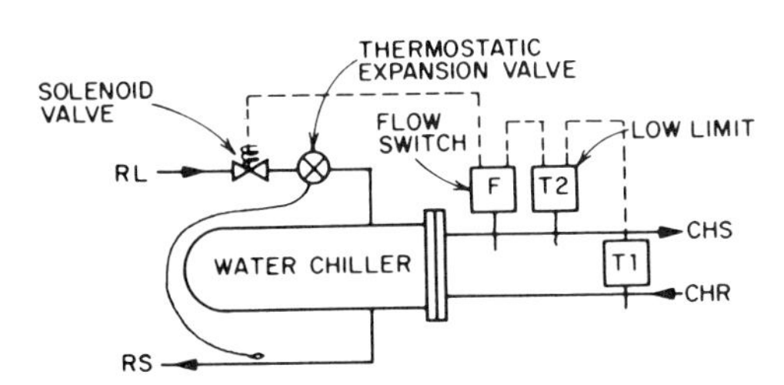

Figure 6-60 Water chiller control.

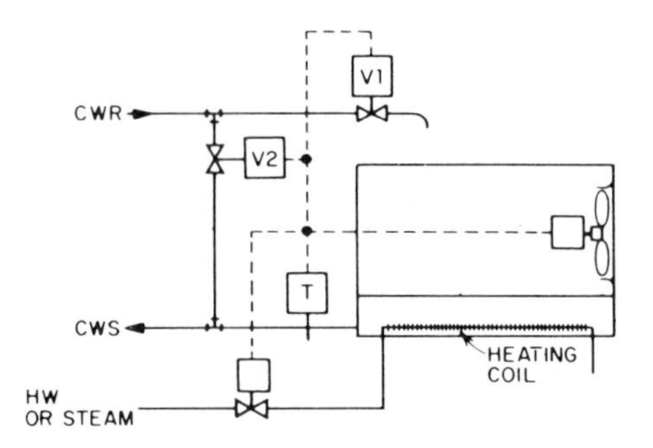

Figure 6-61 Cooling tower; winter operation.

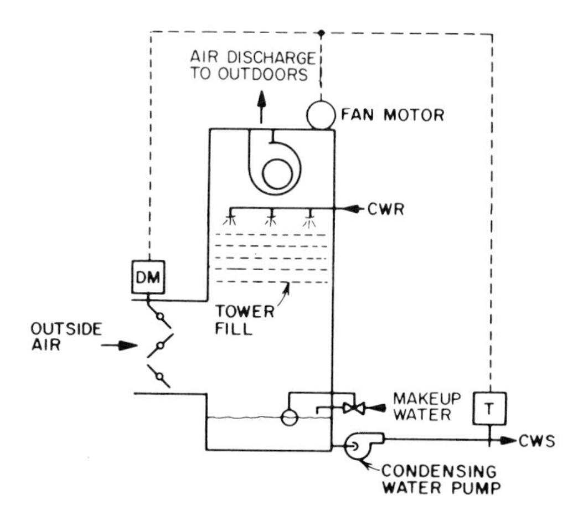

Figure 6-62 Cooling tower; indoor arrangement.

thermostat will start or stop the fan as the temperature rises or falls. On larger towers two-speed fans may be used, with two- or three-stage thermostatic control.

Towers which are used all year need more extensive control systems, including bypass valves, and heating to prevent freezeup. Figure 6–61 shows a year-round system with two-speed fan control, bypass and heating. On small or medium-sized systems (to about 200 tons) towers can sometimes be installed indoors so that modulating dampers to vary airflow will provide adequate control (Figure 6–62). When the damper is fully closed the fan will be stopped. An outdoor tower with an indoor sump may also be used. In any case a low water flow rate through the tower in freezing weather may cause ice build-up in the tower fill, with resulting damage to the tower. Modulating water flow should not be used.

6.13 FIRE AND SMOKE CONTROL

Motorized fire and smoke dampers are used for fire separation and for control and evacuation of smoke. The basic design of these devices is controlled by NFPA and local codes. The motor operators are installed to hold the dampers open, so that they "failsafe" to the closed position on loss of control air or power. Modern smoke control technology provides for opening and closing smoke dampers during a fire in such a way that the smoke generated will be evacuated from the building and not allowed to flow to areas within the building adjacent to the fire.

For example, in Figure 6–63 if smoke is detected in zone 2 the following sequence would occur.

Supply air damper 2-1 would close,
Supply air dampers 1-1 and 3-1 would remain open,
Exhaust dampers 1-2 and 3-2 would close,
Exhaust damper 2-2 would remain open.

The effect is to create a negative pressure in zone 2 and a positive pressure in zones 1 and 3, so that smoke will be removed from zone 2 to exhaust and replaced with clean air from zones 1 and 3. The controller is simply a logic device built up using relays, or, for a large number of zones, may be a programmable controller. A computer based supervisory control system can include smoke control among its many functions.

It must be emphasized that the typical HVAC system is not designed for

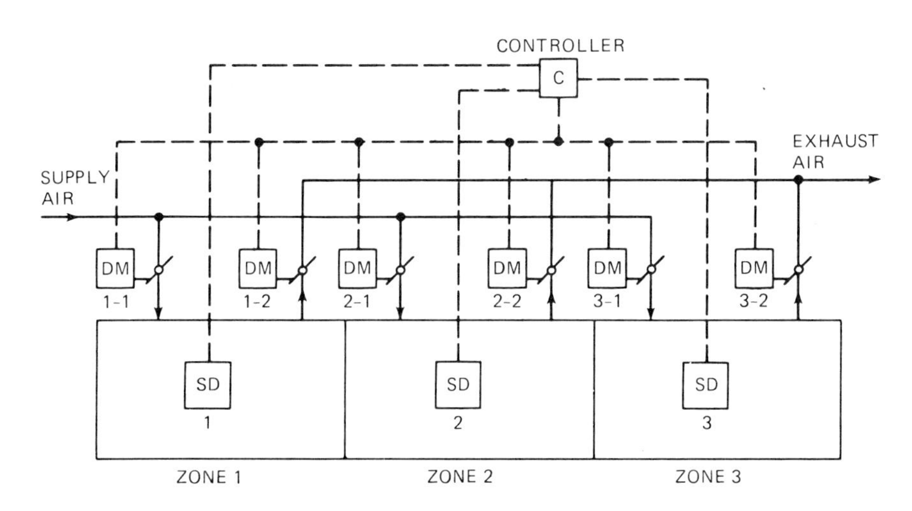

Figure 6-63 Smoke control system.

smoke control and must not be used for this purpose. Only when the HVAC system is an "engineered smoke control system" can true smoke control be accomplished. See bibliography reference 34.

6.14 ELECTRICAL INTERLOCKS

As has been indicated in several of the preceding examples, at some point it will be necessary for the temperature and pressure controls to interface with electric motor controls. Small motors may sometimes be operated directly by electric controllers. If the motors are large, or of a different voltage than the control circuits, or if pneumatic or electronic controls are used, then relays are required. This subject is discussed more fully in Chapter 8.

6.15 LOCATION OF SENSORS

The proper location for the sensor can best be determined by asking specifically what is to be controlled. If it is room temperature, then the location should be such that the sensor reads an average room temperature, with a minimum of exposure to supply air, drafts or radiant effects from windows or equipment. Near, or even in, return air openings is recommended.

In a large work area, you may want the best control at a particular work station. So, mount the sensor there. Sensors have been mounted on movable frames with long cables, and relocated as the critical work station changed.

The main thing to avoid is "side effects" which may prevent the sensor from seeing conditions correctly. For temperature sensors these may be radiation (cold or hot), drafts, lack of adequate air circulation or convective heat transfer through the mounting (as on an outside wall).

In systems with several rooms on a single zone, it is essential to select an "average" room for the sensor location. Conference rooms or rooms with large load variations should be on separate zones, but, if this is not possible, do not let them be the sensing point for the zone. Given a choice between large and small rooms, select the larger space (unless the smaller space is used by the manager's executive secretary!). All rooms on a single zone should have comparable outside exposures with a common orientation (north, south, east or west, but not a mixture of these).

6.16 SUMMARY

This chapter has discussed elementary control systems or, more correctly, subsystems. Any larger system is built up from combinations of these elements. In Chapters 7 and 9 it will be shown how this is done. It is hoped you will be able to recognize these smaller elements in the larger systems.

7 Complete Control Systems

7.1 INTRODUCTION

In the previous chapter we discussed elementary control systems. These are, in general, the pieces and parts which can be fitted together to produce specific applications solutions.

This chapter will discuss and detail many of these applications as they have been or could be solved. The discussions proceed from the simple single-zone concept to more elaborate arrangements. Some highly complex systems will be dealt with in Chapter 9.

7.2 SINGLE-ZONE SYSTEMS

Single-zone systems have one basic parameter: a single zone temperature controller. This controller may serve one or more air handling units and may be a simple room thermostat, a discharge thermostat or a complex averaging or reset controller. In current practice most large control systems use separate sensors and controllers. In all of the discussions in this chapter, a sensor and controller may be substituted wherever a thermostat or humidistat is shown.

7.2.1 Single Air Handling Unit

Consider first a simple and common type of system consisting of a single air handling unit (AHU) with a room thermostat as the basic controller. For year-round air conditioning this AHU will have heating and cooling coils and sufficient outside air for ventilation. If the coils use hot and chilled water then the simple system might look as shown in Figure 7–1. When the supply fan is started the minimum outside air damper opens and the temperature control circuit is energized. The room thermostat then opens and closes the hot and chilled water valves in sequence to provide heating or cooling as required. Notice that only one of these valves may be open at one time. The fire safety switch will stop the fan if a high (about 125° F) return air temperature is sensed. A smoke detector may be used instead. The diagram shows the heating coil preceding the cooling coil. The primary reason for this is to minimize coil-freezing possibilities. There is considerable disagreement as to which sequence (cold-hot or hot-cold) should be used. Generally the hot-cold sequence is favored except when maximum humidity control is required. Then the "reheat" position is necessary for the hot coil (See Figure 7–6.) The system just described will have a rather wide operating differential even with a very small control differential. For example, if the thermostat senses a rise in room temperature and opens the chilled water valve it will take some time for the water to flow into the coil, lower its surface temperature and cool the air passing over it. More time is required for the air to travel to the room, mix with room air and lower the temperature at the thermostat location. The thermostat then closes the valve, but by this time the coil is full of cold water, which causes a continuing decrease in room temperature. The same sequence occurs on heating except that the overrun is usually greater due to the greater initial temperature difference between the hot water and the air.

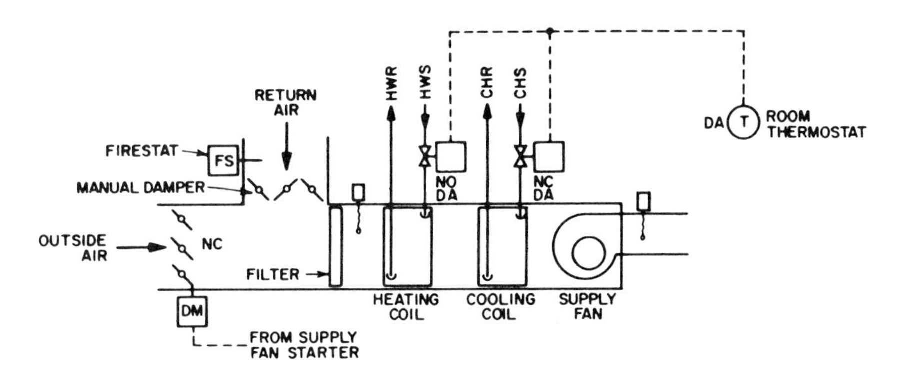

Figure 7-1 Single-zone AHU; minimum outside air.

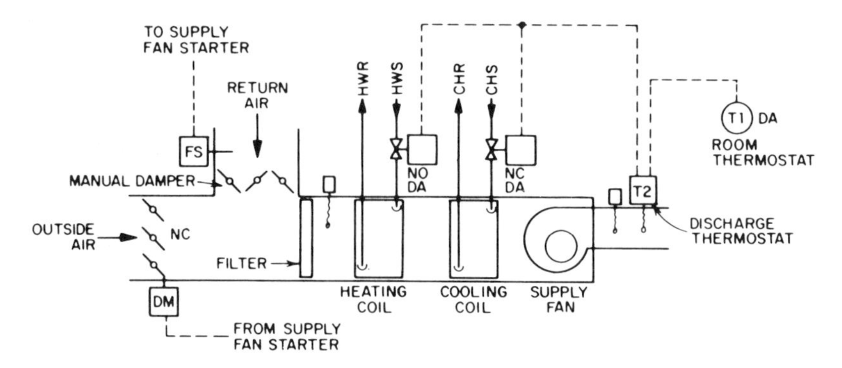

Figure 7-2 Single-zone AHU; minimum outside air, discharge thermostat.

Because this situation exists the system is frequently modified by adding a thermostat in the air stream leaving the AHU (Figure 7–2). The discharge thermostat may be used as a high-limit controller. If the discharge temperature exceeds the high-limit setting the discharge thermostat overrides the room thermostat and closes the hot water valve. This operation tends to limit the operating differential. A low-limit controller may also be used. The discharge thermostat may also be used as the basic "submaster" controller for the hot and chilled water valves, with set point being reset by the "master" room thermostat. This means that when the room thermostat senses the need for cooling, instead of controlling the valves directly, it causes a decrease in the set point of the discharge thermostat. And, when heating is required, the reverse is true. Since the discharge thermostat senses system response very quickly, system differential is decreased. The "reset schedule" may vary over a 30° to 40° range of the discharge temperature.

These "improvements" are not really needed for small zones, which respond quickly, but are highly recommended where a large area is served as a single zone.

7.2.2 Single-Zone Unit, Economy Cycle Outside Air

To this point in the chapter only fixed outside air has been considered. But, as noted in Chapter 6, a great many systems use the "economy cycle" control of outside air. If economy cycle control is added, with a separate relief exhauster, the system shown in Figure 7–3 is obtained. This is the system most commonly used today. Here, the fan is started manually or automatically with interlocks to start the relief fan and allow control operation when the fan is running. The outside, return and relief dampers modu-

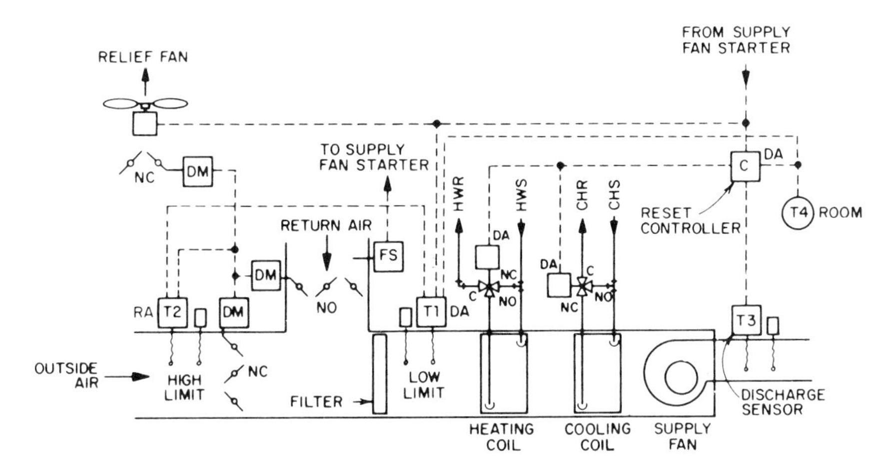

Figure 7-3 Single-zone AHU; economy cycle outside air, discharge controller.

late in response to the mixed-air controllers. The room thermostat can be used to reset the low-limit mixed air controller. This will provide greater energy conservation than can be obtained with a fixed low-limit set point. (See paragraph 6.2.2 and item 31 in the Bibliography.) The discharge thermostat controls the hot and chilled water valves in sequence. The room thermostat resets the discharge control point. Note that the diagram shows a single controller with two sensors. This is the way many manufacturers now provide this control, but it may also be two separate units in a master-submaster arrangement.

7.2.3 Single-Zone Unit, Static Pressure Control of Outside Air

By the simple addition of a static pressure controller to the above cycle it is possible to maintain a positive or negative pressure in the room (Figure 7–4). This controller operates the relief damper independently of the return and outside air dampers, thereby maintaining the set pressure. A throttling device in the control output prevents overshoot or hunting due to sudden changes such as opening a door.

It should be noted again here that room static pressures greater than ± 0.10 in. of water are difficult to maintain unless all opening to other spaces are weatherstripped. Such pressure differentials also cause difficulty in opening or closing doors and related problems. All pressurization requirements should be carefully studied, and the minimum suitable value should be used.

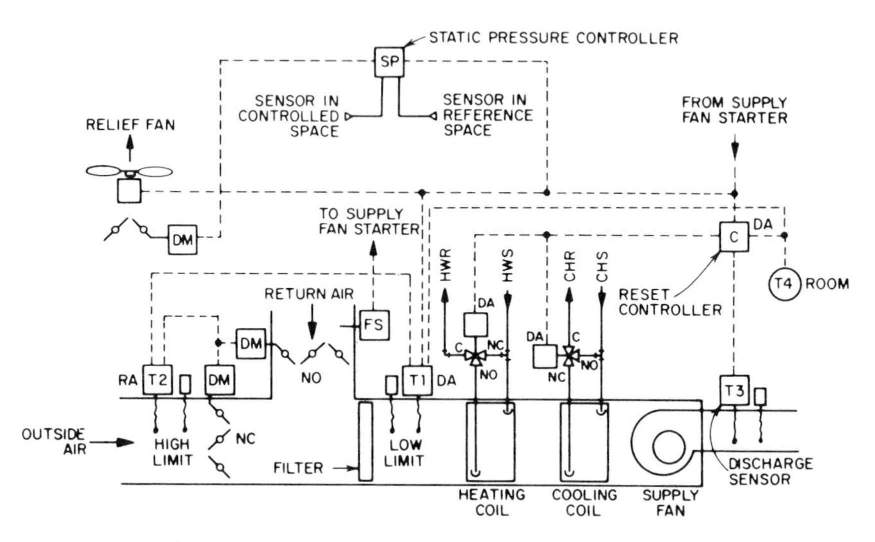

Figure 7-4 Single-zone AHU; economy cycle, discharge controller, static pressure control.

7.2.4 Return-Relief Fans

Large air handling systems, with lengthy return air ducts, should be provided with return-relief fans (Figure 7–5). Return air quantity is equal to supply air quantity less any fixed exhaust and exfiltration requirements. A regular economy cycle then modulates outside air, return air and relief dampers, and additional relief fans are not required.

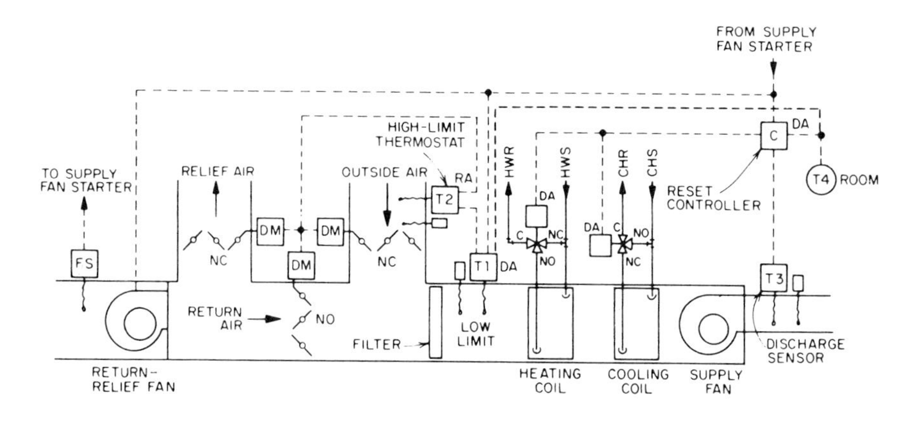

Figure 7-5 Single-zone AHU with return-relief fan.

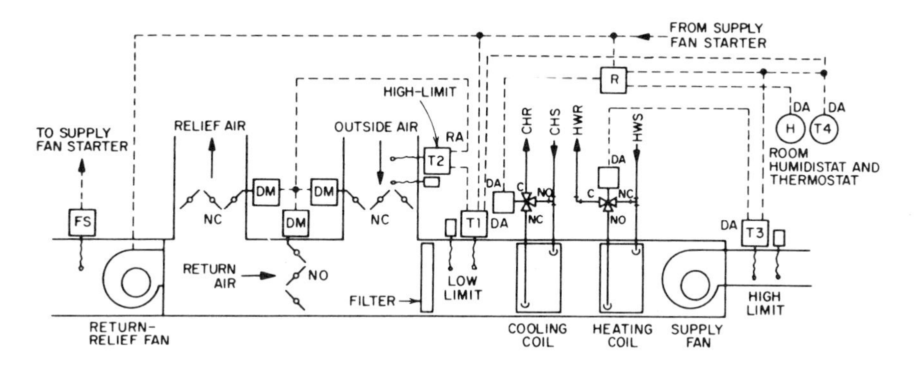

Figure 7-6 Single-zone AHU; maximum humidity control by reheat.

7.2.5 Single-Zone Humidity Control

As suggested earlier, control of maximum humidity can be provided by placing the cooling coil ahead of the heating coil in the airflow sequence. Then a space humidistat is added to the control sequence (Figure 7–6). Now, so long as room humidity is below the humidistat setting, the system operates as in Figure 7–5. If the humidity goes above the set point, the humidistat, through the selector relay, takes over control of the cooling coil, and calls for additional cooling, which provides dehumidification. If the room then becomes too cool, the thermostat will call for "reheating." The psychrometric chart in Figure 7–7 shows the cycle graphically.

Year-round humidity control within fixed high and low limits can be accomplished in several ways. One method used frequently in the past was the "sprayed-coil dehumidifier" described in Chapter 6. This is no longer

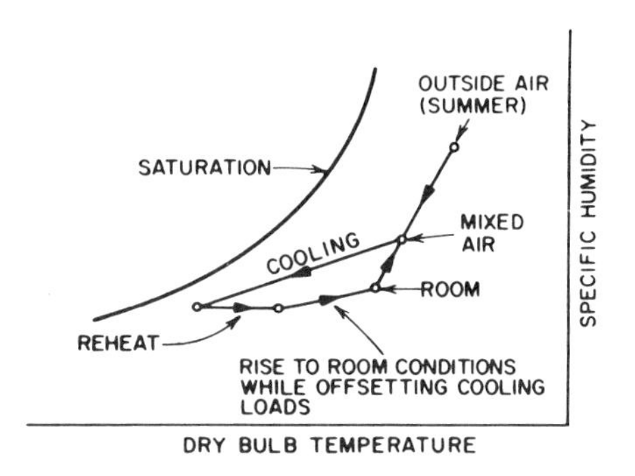

Figure 7-7 Psychrometric chart for Figure 7-6.

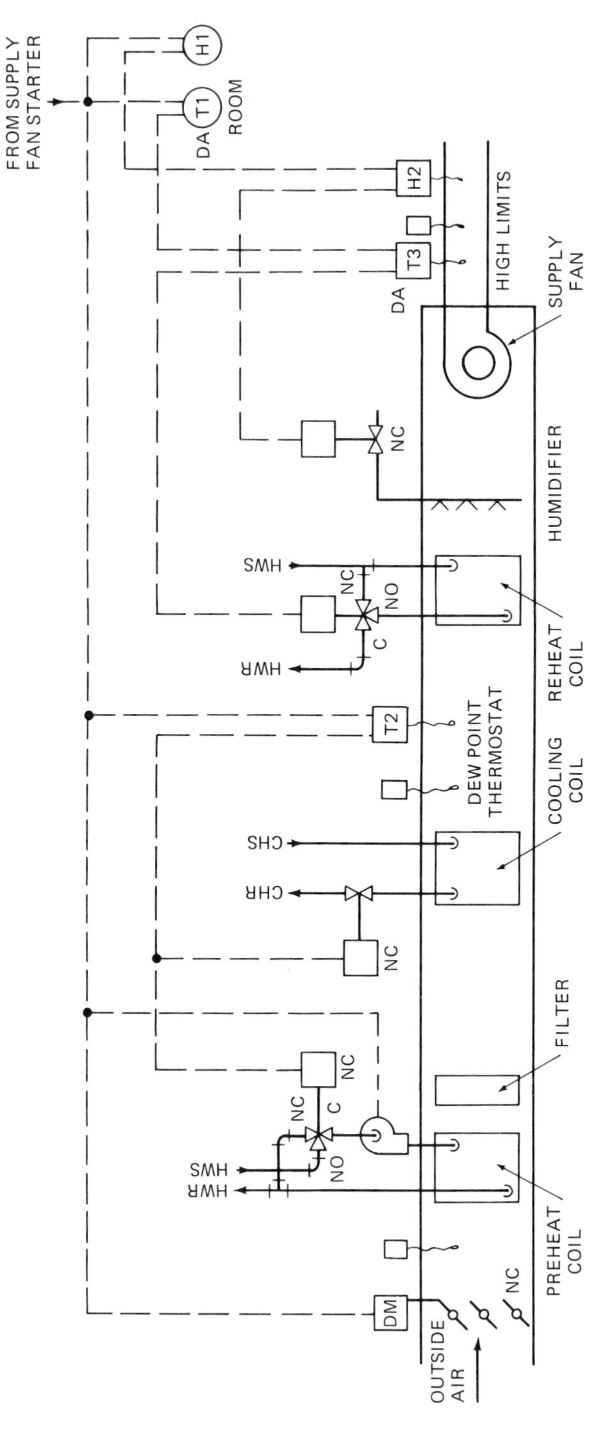

Figure 7-8 Single-zone AHU; humidity control, 100% outside air.

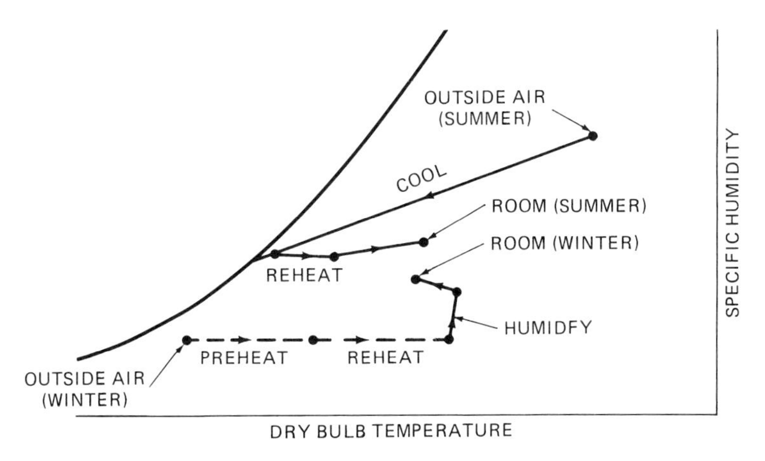

Figure 7-9 Psychometric chart for Figure 7-8.

used due to problems with solids deposition and bacterial contamination. Current practice is to use a humidifier to maintain the low limit, with high-limit control as described above or by means of a dew-point temperature sensor and controller.

Figure 7–8 shows a single zone system using 100% outside air, with controlled humidity. The humidifier is located in the AHU. It could be installed in the supply duct but must always be upstream of the duct temperature sensor since the humidifer adds some heat, as indicated in Figure 7–9, the psychrometric chart for the system. This system will easily maintain 50% to 60% RH maximum in summer and 35% to 40% minimum in winter. Lower maximum humidities may be obtained by decreasing the chilled water temperature or by using brine or direct-expansion cooling. Minimum humidities higher than about 45% are difficult and require the use of large steam grid humidifiers.

A similar control cycle can be applied if the system requires less than 100% outside air (Figure 7–10 and the psychrometric chart of Figure 7–11). The preheat coil is needed only if the minimum outside air is such a large percentage of outside air that the mixed-air temperature would be too cold. The low-limit mixed-air controller may be reset by the room humidistat to minimize the use of the humidifier. Otherwise, the outside air control is a standard economy cycle system.

A few words of advice regarding humidifiers in general would not be amiss here. Any evaporative system, such as a pan or sprayed coil, generates a severe maintenance problem due to deposition of solids from the evaporated water. It has been found, through experience, that a demineralizer on the makeup water will often pay for itself by saving the cost of cleaning up coils and pans. Steam humidifiers should not be used where the steam is treated.

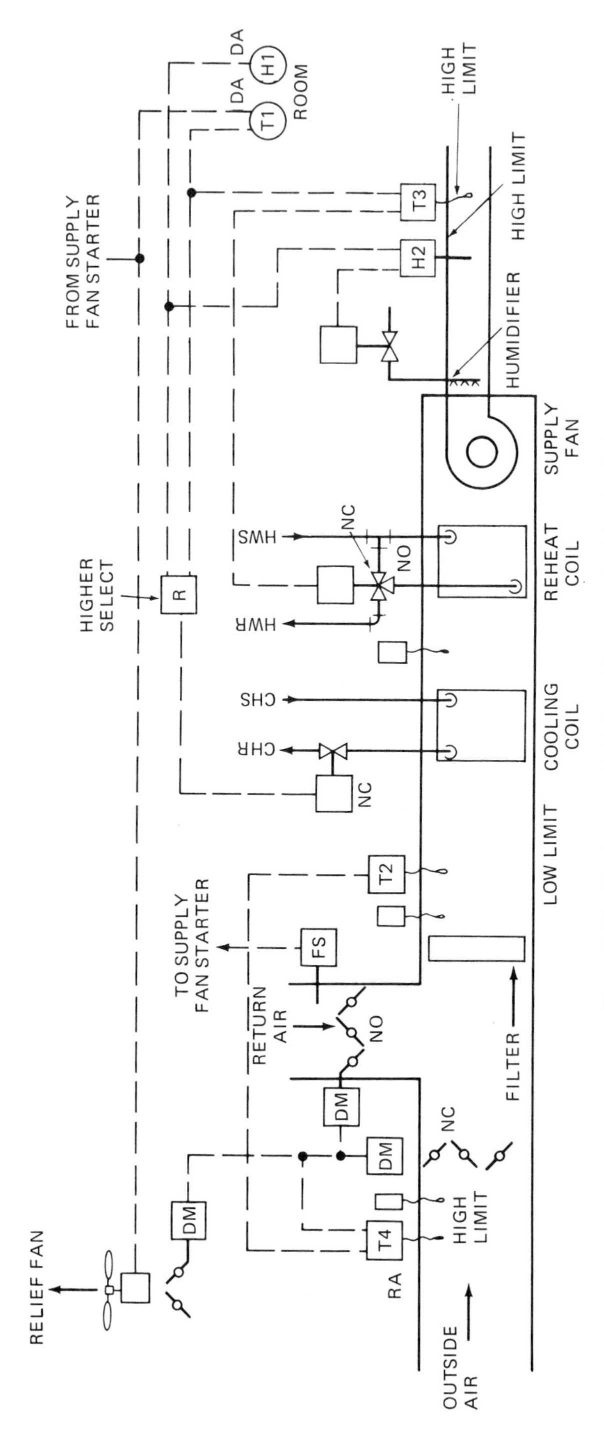

Figure 7-10 Single-zone AHU; humidity control with mixed air.

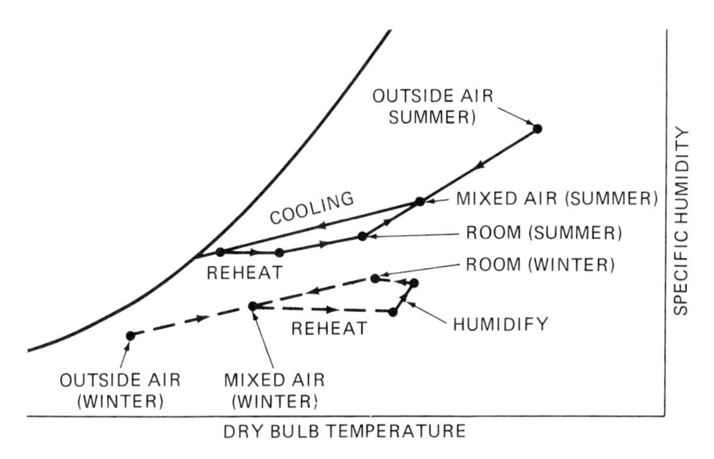

Figure 7-11 Psychometric chart for Figure 7-10.

Many steam additives are toxic, besides causing odors and unsightly deposits. Where a central steam or high-temperature water supply is provided and a steam humidifier is to be used, a small heat exchanger can be used to provide clean steam. Whenever a duct humidifier is used, a duct humidistat should be provided and used as a high limit to avoid condensation in the duct.

7.3 MULTIZONE AIR HANDLING SYSTEMS

In multizone systems there are two related and dependent control problems: control of air temperatures in hot and cold plenums, and control of temperatures in the zones. It is very easy to lose some control if the hot plenum temperature is too high, and the mixing dampers have a large leakage factor (over 10%). Modern, tight-closing dampers help to overcome the problem but it is still highly desirable to limit the hot plenum temperature to the lowest value required for control in the coldest zone. On the other hand, cold plenum temperature will not vary greatly even if uncontrolled, so that sometimes the control valve is omitted on chilled water coils. For optimizing of large systems a chilled water control valve is required.

An additional source of difficulty occurs at light-load conditions with modulating controls. Then, the temperature gradient across the face of the coil, from one end to the other, may be 5° or 10° or more. Obviously, this causes lack of adequate control, especially for zones which feed off the plenum near the ends of the coils. This can be solved by the use of circulating pumps with water coils, by the use of "distributing-tube" steam coils or by feeding both ends of long coils. With direct-expansion coils a two- or three-stage sequence, in which one or more rows *in the direction of airflow* constitute a stage, has been used successfully.

7.3.1 Multizone Control

Figure 7–12 shows a typical multizone control system. The return-relief fan and outside air controls have been described before. Usually the mixed-air high- and low-limit settings are such that no refrigeration is required in cold weather and no heating is required in hot weather. The bypassed mixed-air temperature is adequate for zone control. Hot and cold plenum temperatures are controlled by plenum thermostats with the hot plenum reset from outdoor temperature (cooling load is not a function of outdoor temperature). An estimate of the plenum temperatures required for heating or cooling at various outdoor temperatures is the basis for the reset schedules. Each zone thermostat controls the mixing dampers for its zone to maintain design zone conditions.

Figure 7–13 is the same as Figure 7–12 except for the hot plenum control. Here, in order to keep the hot plenum temperature at the lowest possible value it is necessary to use a relay or series of relays from the zone thermostats, so that control of the heating coil is given to the thermostat with the greatest demand for heating. This is the so-called *discriminator control.* Discriminator control can also be applied to dual-duct systems, to reheat systems, or to any system with more than one zone.

In theory, the use of discriminator control should result in the maximum conservation of energy. In practice there are problems. If one zone thermostat is inaccurate or has a set point significantly higher or lower than the other zone, this thermostat will drive the system to the exclusion of all others. The discriminator itself may provide a false output. Frequent checking and maintenance is needed to assure that the discriminator is performing properly.

7.3.2 Multizone Control, 100% Outside Air

Multizone units may need 100% outside air. In this case, preheat is required, usually to 50° or 55°, which provides a cold plenum temperature adequate for cooling interior zones. Figure 7–14 shows such a system.

7.3.3 Multizone with Humidity Control

Figure 7–15 shows a suitable arrangement of a multizone unit with humidity control. For maximum humidity control all of the air must be

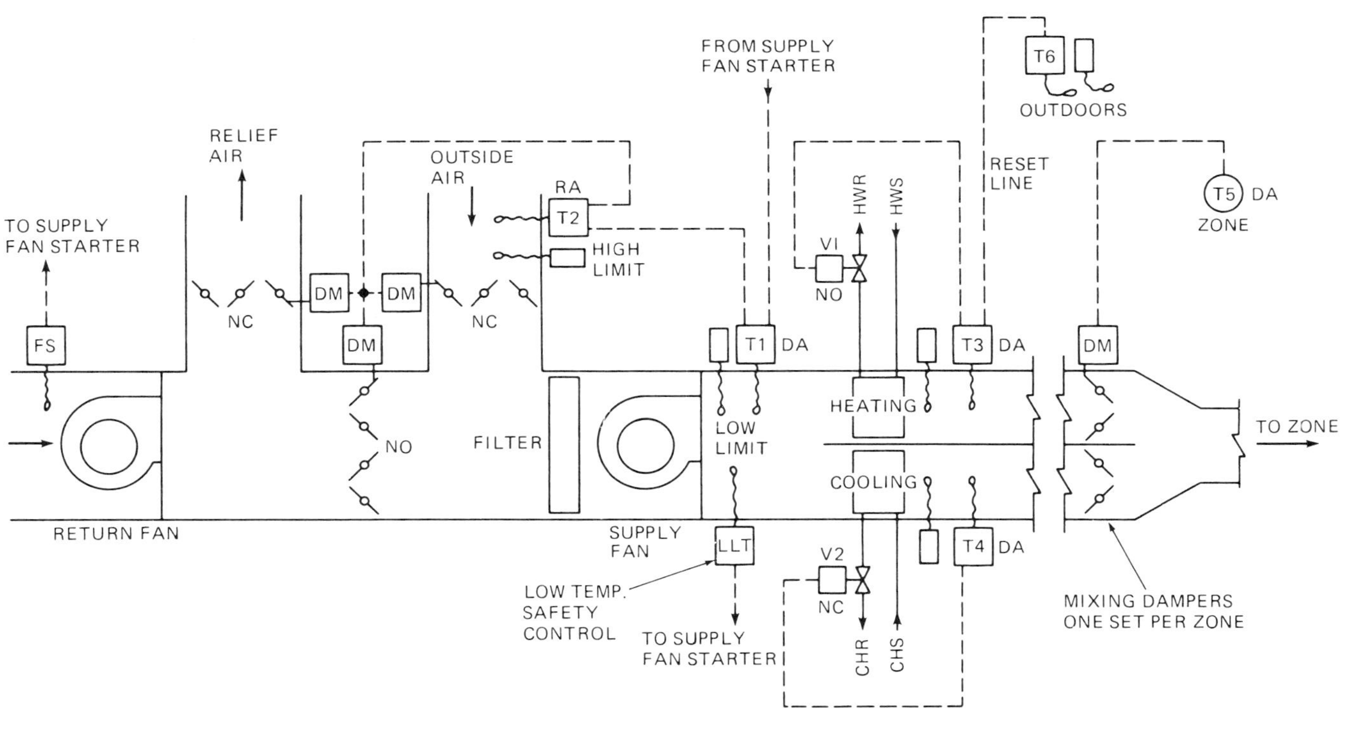

Figure 7-12 Multizone system; hot and cold plenum resel.

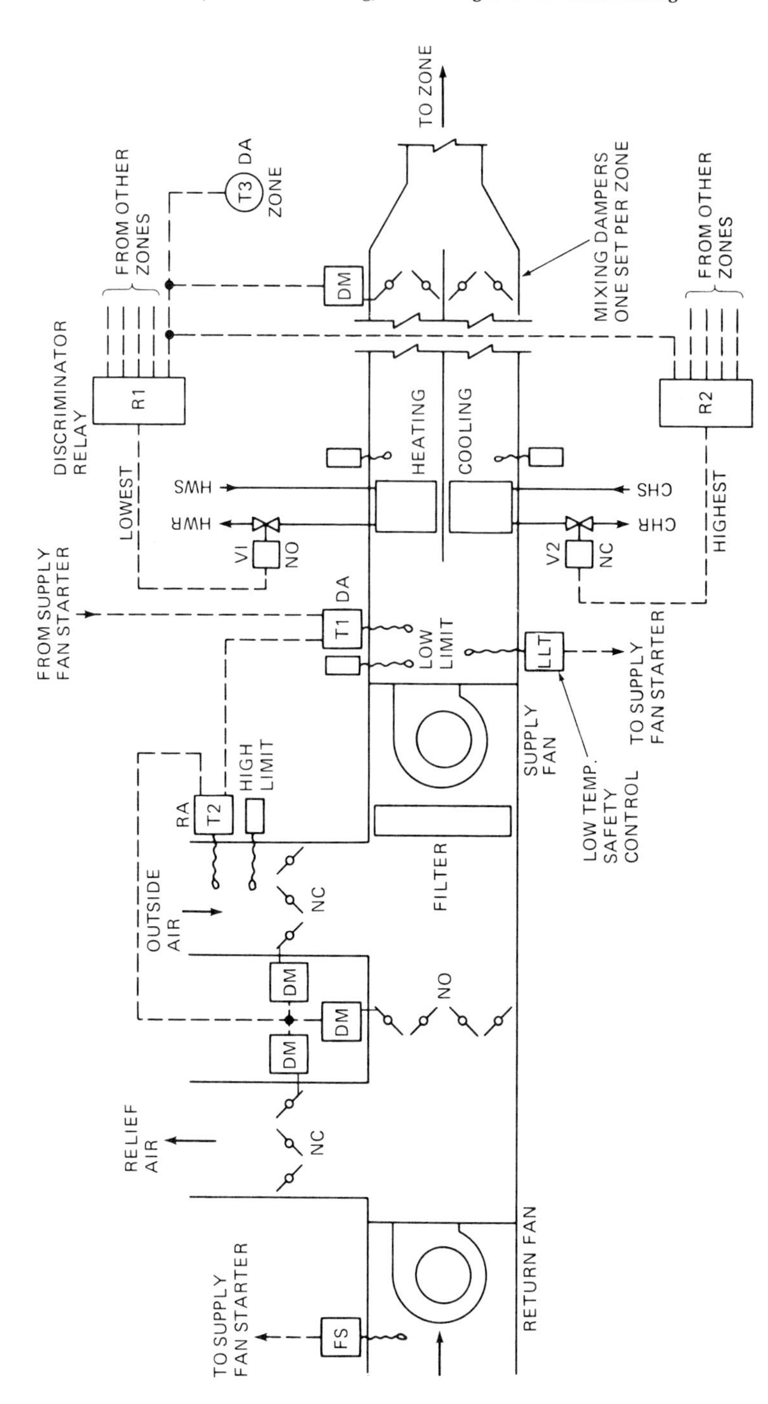

Figure 7-13 Multizone system; discriminator control of hot and cold plenum temperatures.

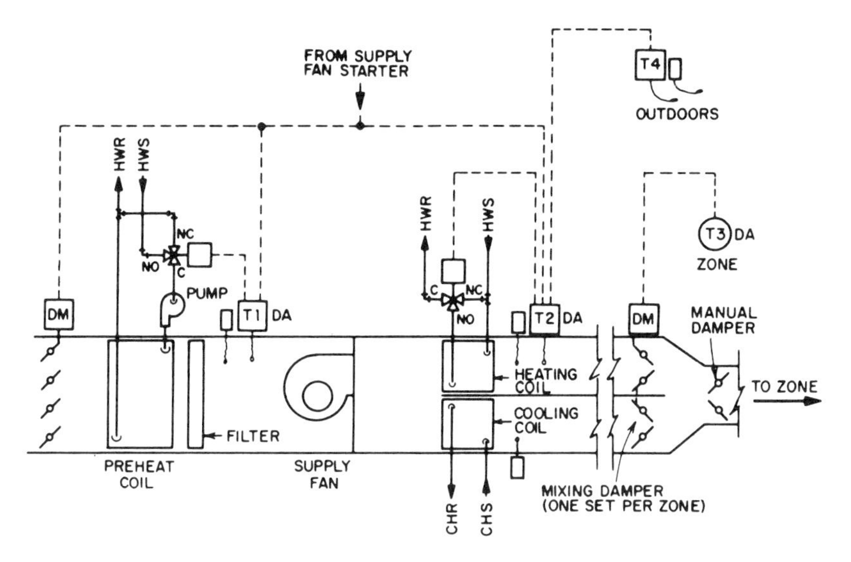

Figure 7-14 Multizone system; 100% outside air.

cooled to a dew point corresponding to the desired relative humidity. When humidity is to be added it must be supplied by humidifiers in the individual zone ducts. The psychrometric chart would be similar to Figure 7–11.

7.3.4 Three-Plenum Multizone System

Many energy codes now prohibit the use of "reheat" in air conditioning; this effectively disqualifies two-plenum multizone systems. The three-plenum multizone system eliminates the use of new energy for reheat by providing a bypass plenum in addition to the usual hot and cold plenums (Figure 7–16). The control sequence is as follows: At full heating load the hot damper is fully open and the bypass and cold dampers are closed. As zone temperature rises the hot damper modulates toward closed and the bypass damper modulates toward open. With the zone thermostat satisfied the bypass damper is fully open and both hot and cold dampers are closed. If the zone temperature rises above the set point the switching relay R2 causes the signal from the reversing relay to control the bypass damper so that it will close as the cold damper opens. Cold and hot plenum air streams are never mixed with one another.

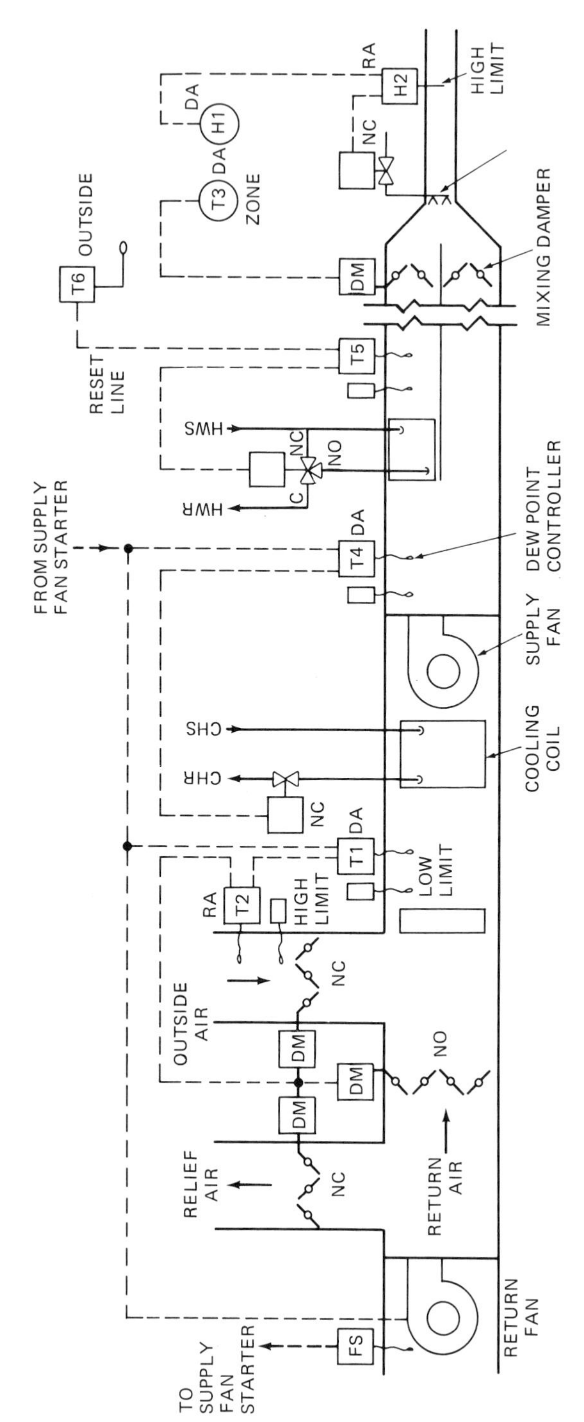

Figure 7-15 Multizone system; humidity control.

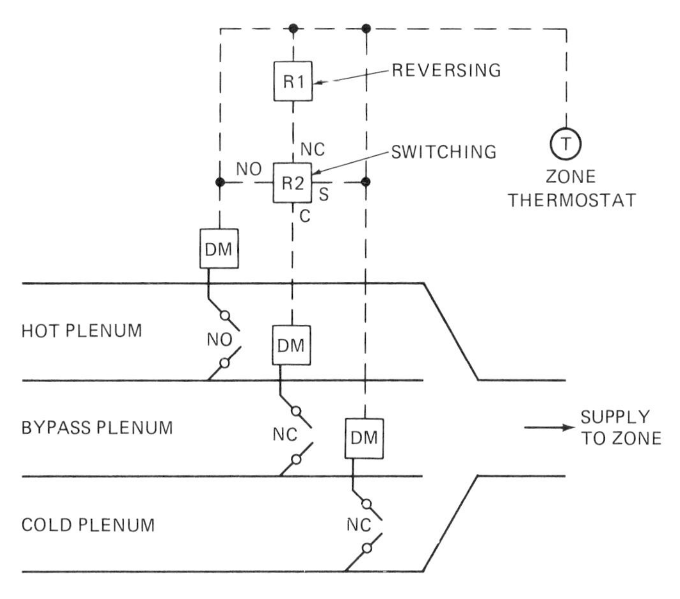

Figure 7-16 Three plenum multizone system.

7.4 DUAL-DUCT SYSTEMS

Dual-duct systems, whether high- or low-velocity, are simply multizone systems with the plenums extended and the zone mixing dampers located near the zone. Zones are generally smaller and response is faster. High-velocity, high pressure dual-duct systems also pose problems of large static pressure differentials between hot and cold ducts, which may affect the performance of the mixing dampers. This usually occurs during periods of maximum heating or cooling loads, when one side of the system is handling a large air quantity compared to the other side. This results in higher friction losses in the high-flow side and high static pressure differential at the mixing box.

High-pressure dual-duct systems are no longer considered good practice, due to the high fan work requirements. However, there are many systems in existence, installed in the 1950/60 period when the cost of electrical energy was very low.

7.4.1 Dual-Duct Static Pressure Control

Static pressure differentials may be minimized by static pressure control dampers in the main ducts, near the air handling unit. The sensor measures static pressure near the end of the duct and modulates the damper between full open and some minimum closure position to maintain the desired end-of-main pressure. This must, of course, be set high enough to properly serve the most remote mixing box. When both ducts have such controls the end-of-main differential should be small. It should be noted that many systems operate satisfactorily without static pressure controls, and the present-day mixing boxes work very efficiently even with large pressure differentials.

7.4.2 Mixing Units

Mixing boxes for high-pressure systems should be selected with a high enough pressure drop for good control, without being noisy. Between one and two inches water pressure drop seems to be optimum. Most mechanical constant-volume controllers do not give satisfactory operation below about 0.75 in. of water pressure drop. Many older systems used two-motor constant-volume controllers and these will operate satisfactorily at pressure drops as low as 0.25 in. of water. Most systems today use mechanical constant-volume controllers. These take various forms, but are all basically spring-loaded dampers which open on a decrease in upstream pressure and modulate toward a minimum closure position as pressure increases.

Figure 7–17 shows a one-motor, mechanical constant-volume mixing box control. The dampers are operated in parallel by the motor, so that as one closes the other opens. Pressure changes which occur during the cycle, as a result of pressure differential between the hot and cold ducts, are compensated for by the mechanical constant-volume controller. The room thermostat positions the damper motor to provide more hot or cold air as required.

Figure 7–18 shows a two-motor constant-volume controlled mixing box. The room thermostat modulates the hot damper motor toward open or closed position as required to maintain room temperature. The constant-volume controller senses the change in pressure drop through the mixing box, caused

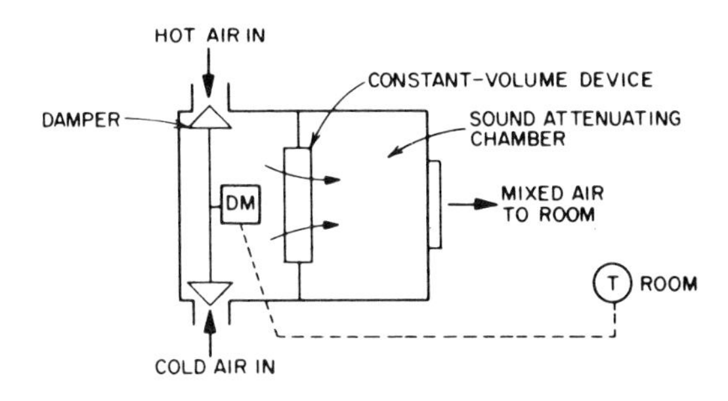

Figure 7-17 Mixing box; mechanical constant-volume.

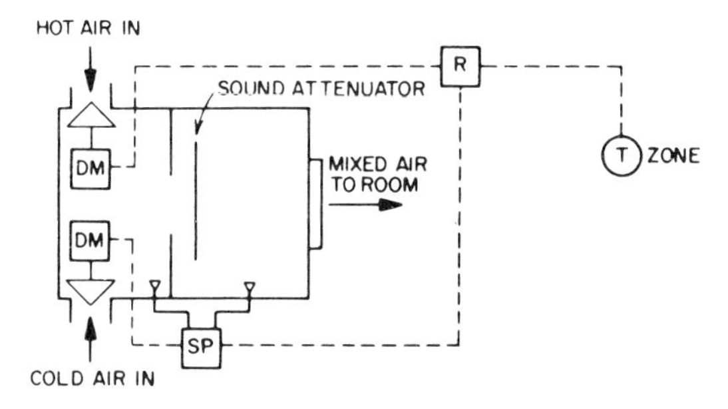

Figure 7-18 Mixing box; two-motor, constant-volume.

by the hot damper change, and positions the cold damper to compensate and thus maintain a constant flow through the box. The relay allows the volume controller to override the room thermostat and partially open the hot air valve when the cold air valve is full open but total volume is too low. This happens occasionally on high cooling demand.

This type of system is still used for zones larger than can be handled by a standard mechanical constant-volume mixing box.

Low-pressure mixing box systems are also used, though not widely, since they are usually limited in size, due to economics and space available for duct work. One-motor boxes without constant-volume control are normally used, since available pressure drops are not sufficient for constant-volume control.

7.4.3 Dual-Duct Air Handling Unit Controls

Controls for the air handling units in dual-duct systems are identical with those used for multizone units. Outdoor reset of hot and cold plenum temperatures was commonly used but is being replaced by discriminator control. (See paragraph 7.3.1 and Bibliography item 32.) Because of the large number of zones only a select few are used with the discriminator relay. (See Figure 7–19.) Sometimes the air flow rate in hot and cold ducts is measured and also used as an input to the discriminator relay, since these air flow rates reflect the building heating and cooling demand.

7.4.4 Two-Fan Dual-Duct System

A primary cause of energy waste in both multizone and dual-duct systems is the use of a common mixed-air plenum to supply both heating

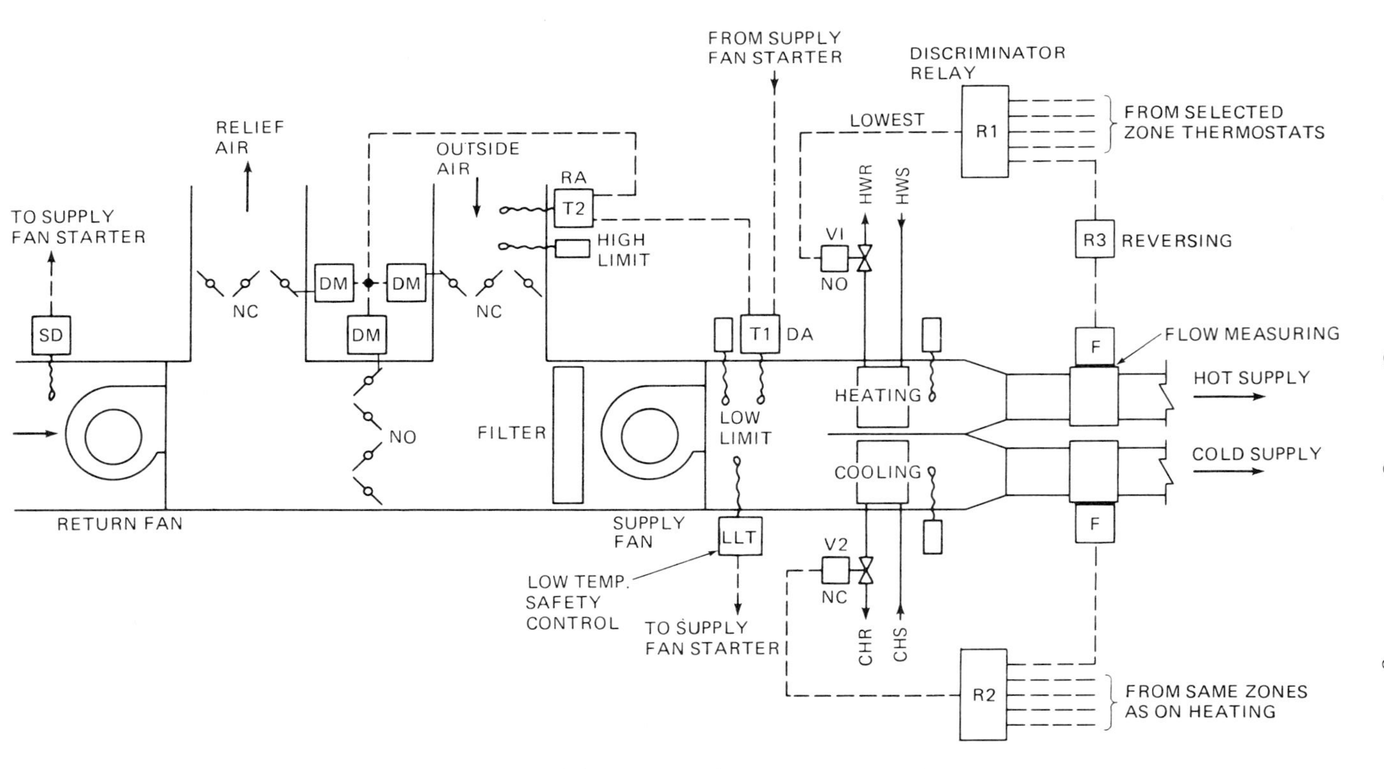

Figure 7-19 Dual-duct system with discriminator controls.

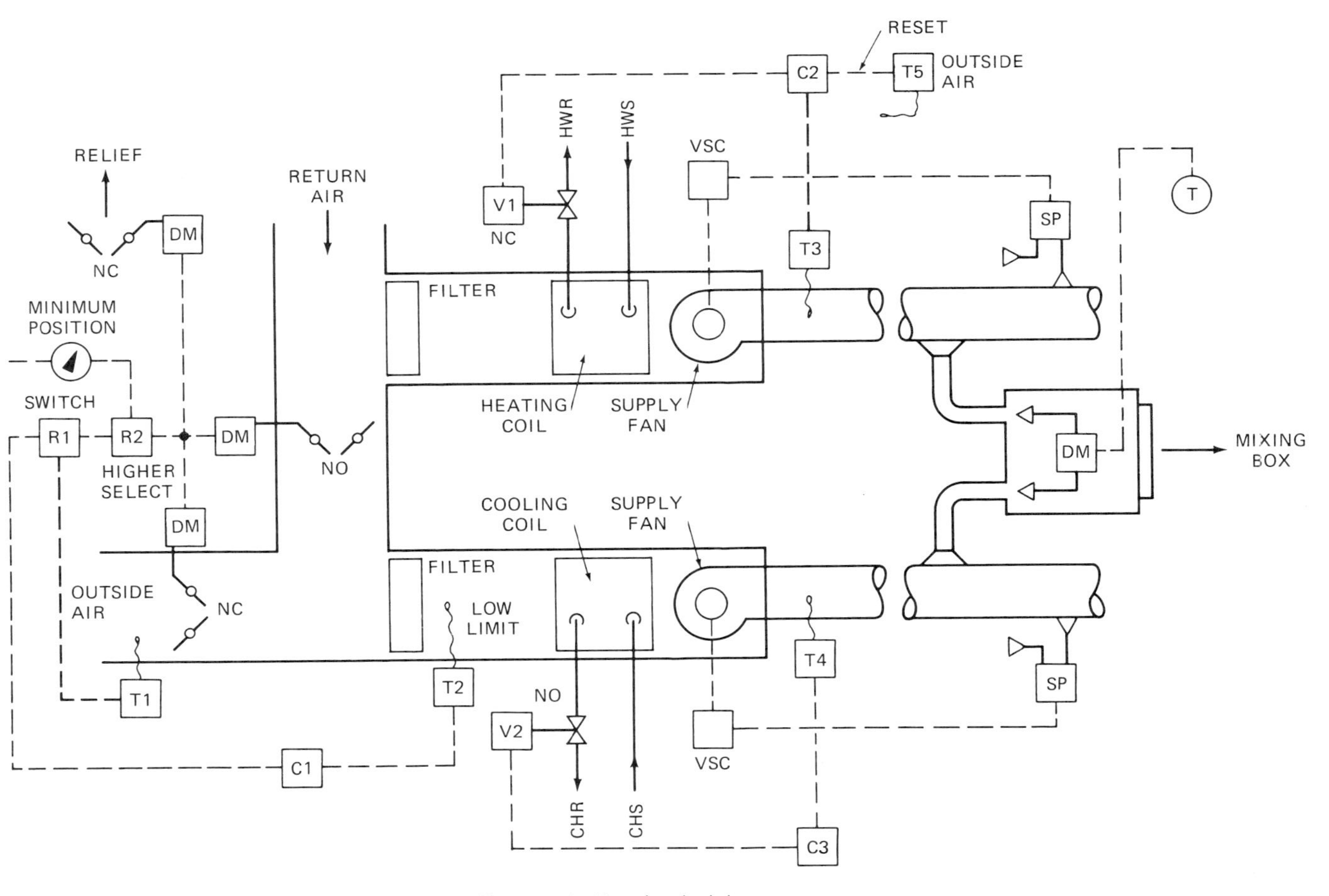

Figure 7-20 Two-fan dual duct system.

and cooling coils. This means that the refrigeration economy resulting from a 55° to 60°F mixed-air temperature is offset in part by the extra heating required in the heating coil. The two-fan dual-duct system shown in Figure 7–20 overcomes this problem. Most or all of the return air—usually warm—is handled by the hot-duct fan while most or all of the outside air—usually cooler—is handled by the cold-duct fan. The standard economy cycle control reduces outside air to a minimum in warm weather. The system energy consumption can be further decreased by providing variable volume control for both of the supply fans, using static pressure sensors in the supply ducts. (See paragraph 7.5.) Note that, since heating is never required in the interior zones, the boxes serving the interior zones need to be connected only to the cold duct.

7.5 VARIABLE-VOLUME SYSTEMS

Variable-volume (VAV) systems provide multizone control with only a single duct. The supply air is maintained at a constant temperature which varies with the season. The individual zone thermostat varies the air supply quantity to the zone to maintain the desired temperature condition. Minimum supply air quantity is usually not less than 40% of design airflow to provide sufficient ventilation. Many VAV systems do not include any heating function (except preheat when large amounts of outside air are required). Supplemental heating is used in all exterior zones. The zone thermostat controls the VAV damper down to its minimum setting and then starts to open the heating valve if heating is required.

7.5.1 Single-Duct Variable-Volume System

Figure 7–21 shows a typical single-duct VAV system. Each zone thermostat controls its zone damper to reduce the air supply to the zone as the cooling load decreases. Supply air temperature is controlled by a discharge temperature controller. In many existing systems this controller has a fixed set point. The preferred method, as shown, is to reset the controller set point as a function of cooling load. For this purpose a discriminator relay is provided, utilizing inputs from several selected zones. The same reset signal can be applied to the low-limit mixed air controller. (But see section 7.3.1.)

As the zone mixing dampers are modulated by the zone thermostats, the total air flow volume must vary. If no other provision is made the supply fan will adjust to the new air volume by increasing or decreasing its discharge pressure. This can cause excessive use of energy and objectionable noise at

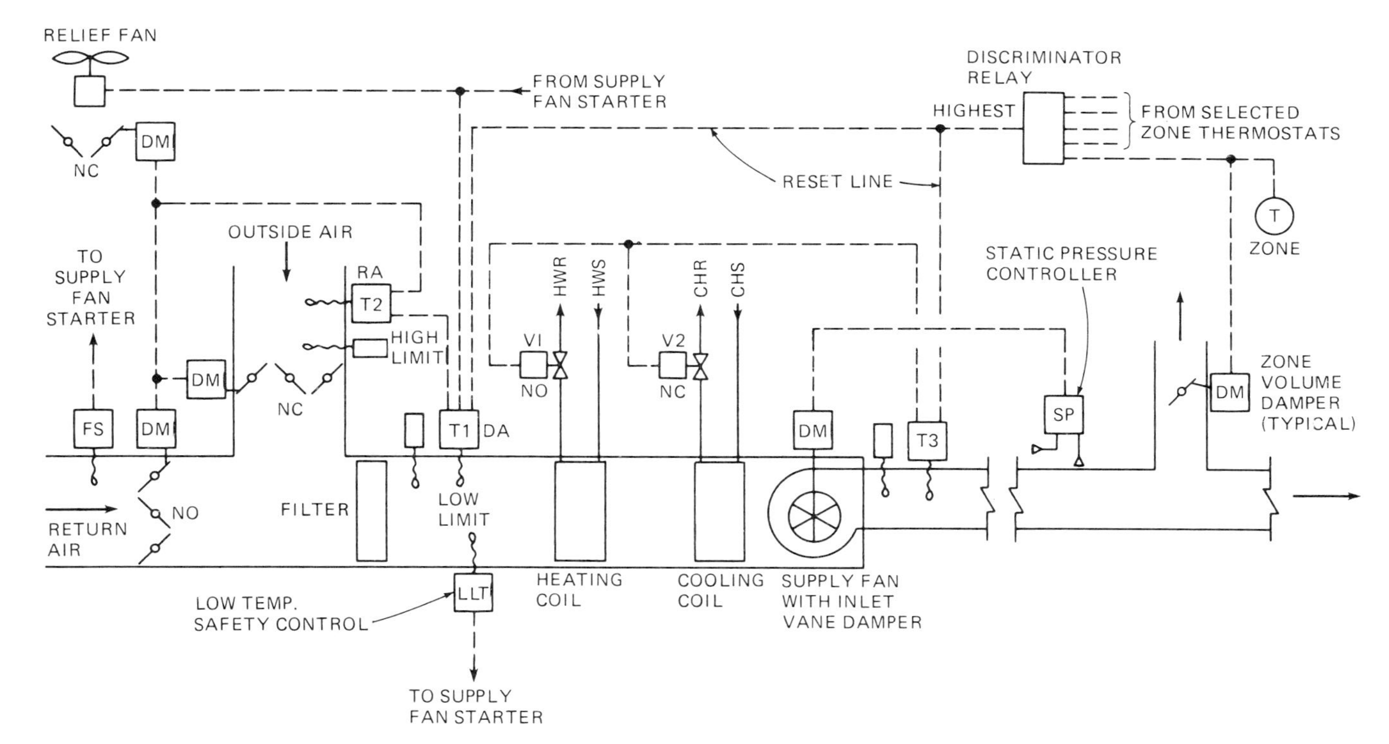

Figure 7-21 Variable-volume system with discriminator control.

low air flow when the zone dampers will have a greater pressure drop to deal with. Most VAV systems include some kind of volume control at the fan, usually inlet vane dampers. The dampers are controlled to maintain a constant static pressure at some point in the supply duct as shown in Figure 7–21.

Fan discharge dampers, or dampers in the duct, may also be used, but these are relatively inefficient in terms of both noise and energy use.

Fan motor speed control may also be used for air volume control and is most efficient in terms of energy use. The older method of speed control, involved the use of direct current (DC) motors with speed controlled by varying the current. Newer systems use standard AC motors with solid state SCR controllers and speed is controlled by varying the frequency. The motor speed varies directly with the frequency, but power input varies as the cube of the speed in accordance with the basic fan law. Thus, if a standard motor in a specific application runs at 1800 rpm at 60 Hz (cycles per second) and requires 10 KW of power, reducing the frequency to 40 Hz would reduce the speed to 1200 rpm and the power input to approximately 3 KW.

The motor speed controller would replace the inlet damper in Figure 7–21. No other change is required in the control system. The control signal would still come from the pressure controller.

Mechanical speed controllers may also be used. These include fluid drives, variable speed couplings and variable speed belt drives.

7.5.2 Dual-Duct VAV Systems

Dual-duct VAV systems can be used when both heating and cooling must be provided. This application is most often used for retrofit—to adapt an existing dual-duct system to variable air volume. Mixing boxes must have two damper motors so that hot and cold dampers may be controlled individually. The system is shown in Figure 7–22. The zone mixing box dampers are controlled in sequence as shown in Figure 7–23. When full cooling is required the cold duct damper is full open and the hot duct damper is closed. As the cooling load decreases the cold duct damper modulates towards its closed position but the hot duct damper remains closed until the cold damper reaches some minimum position—usually 25 to 40 percent of maximum. At that point the hot duct damper begins to open while the cold duct damper continues to close. As the heating load increases the cold damper will completely close and the hot damper will continue to open.

Figure 7–22 also includes discriminator controls for reset of hot and cold duct temperatures and reset of mixed air low-limit temperature.

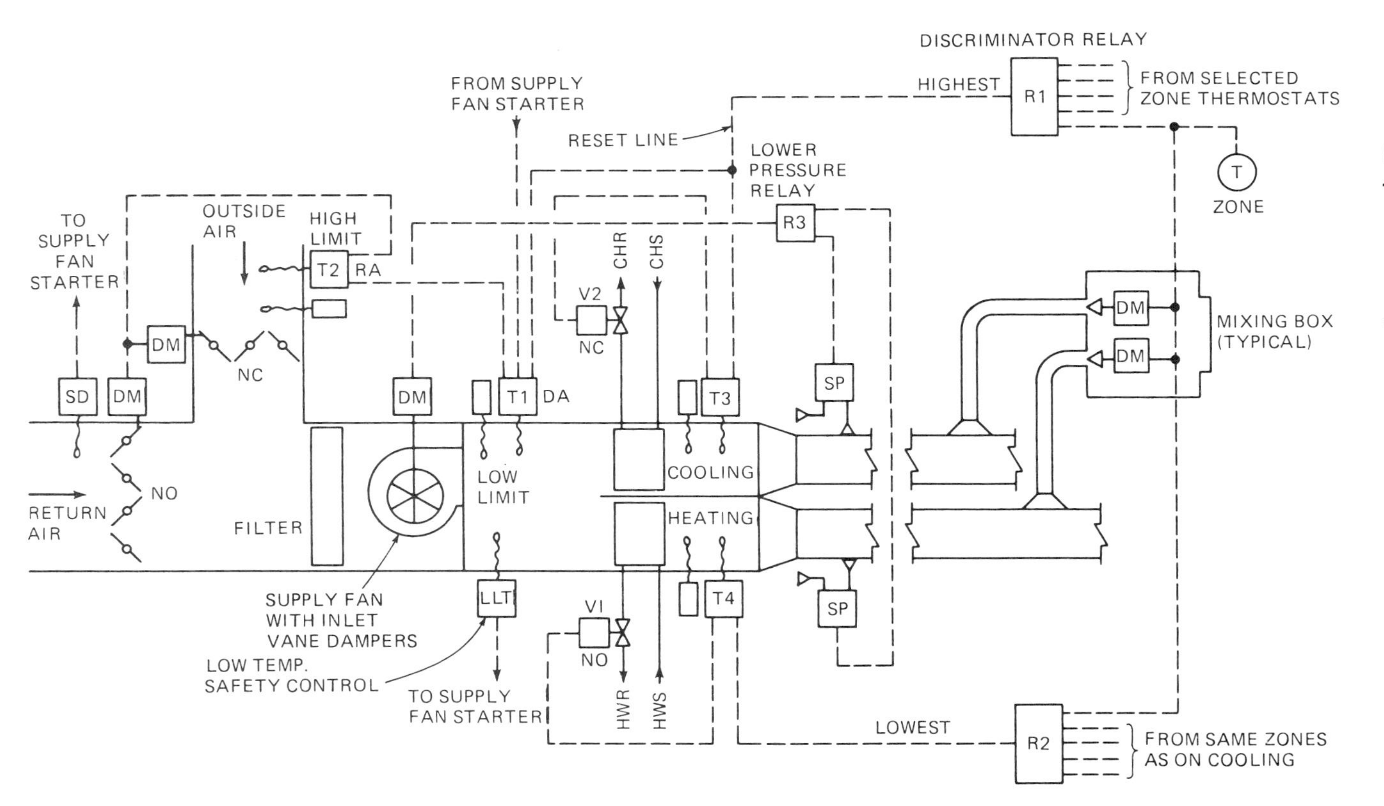

Figure 7-22 Dual-duct variable volume system with discriminator.

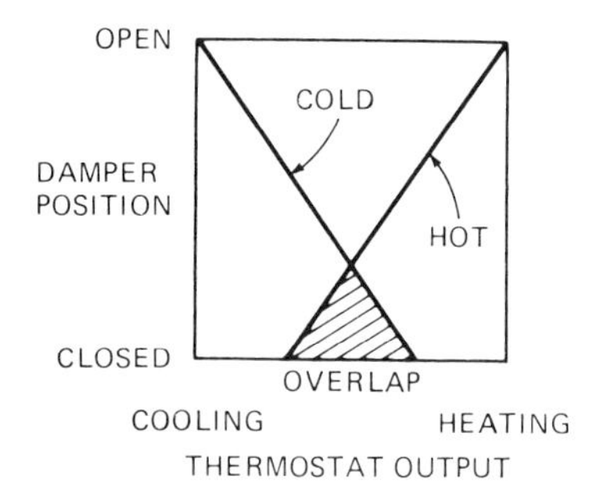

Figure 7-23 Mixing box damper operation; dual-duct VAV.

7.5.3 Return Air Volume Control

In a VAV system which includes a return air fan it is also necessary to control the return air fan volume. Usually there is some minimum outside air quantity which is required to make up exhaust and to pressurize the building to minimize infiltration. The return air flow rate should equal the supply air flow rate minus the minimum outside air flow rate. It is not sufficient to control the return air volume from the supply volume controller since the difference between supply and return volume would decrease as the absolute value of the supply volume decreased. Two methods are preferred for return fan volume control.

The first, shown in Figure 7–24, uses air flow measurement devices. The return fan volume is controlled to maintain an air flow rate which is less than supply air flow rate by a constant differential. Air flow rate measurement provides accurate control, but is somewhat expensive.

A simpler and less expensive method of controlling return air volume is by means of a static pressure controller in the mixed air plenum as shown in Figure 7–25*. If this plenum pressure is held essentially constant the minimum outside air will automatically be provided since pressure drop across the outside air damper will be constant. With this system the return air damper cannot be allowed to be completely closed. About 10% minimum return air must be used to maintain the control loop.

The difficulty with this concept is that of finding a stable static pressure sensing point in the turbulence of the mixing plenum. The best solution is to move the sensing point to the downstream side of the filter (dotted line in Figure 7–25). The pressure here will change somewhat over time due to filter

*A patent covering this concept has been applied for.

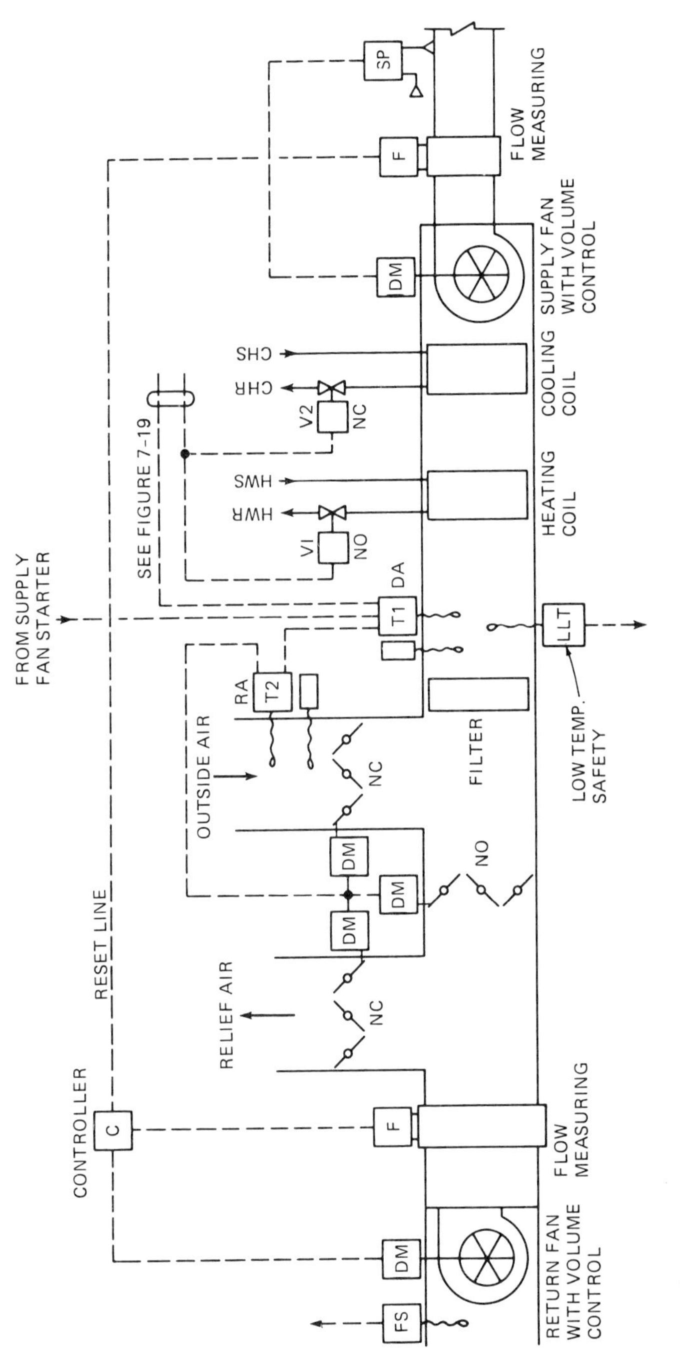

Figure 7-24 Return fan control in a VAV system; flow measurement.

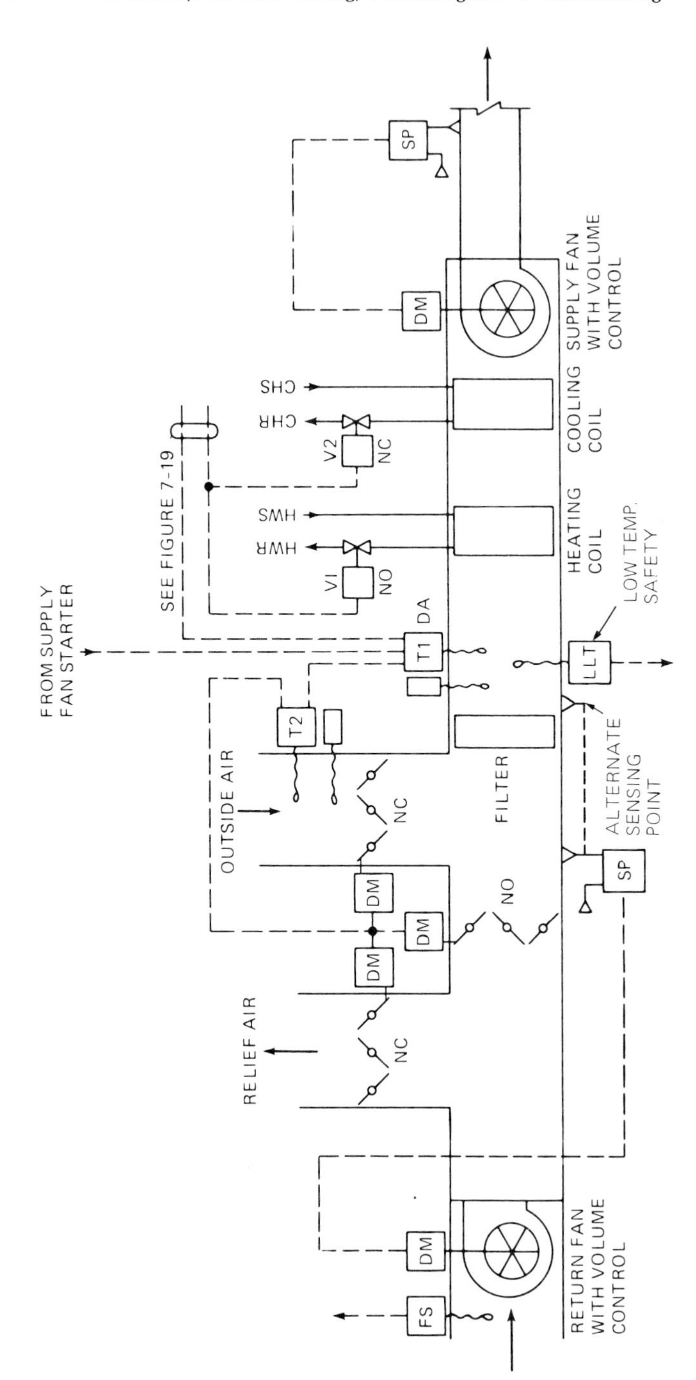

Figure 7-25 Return fan control in a VAV system; plenum pressure control.

loading but will be reasonably stable. It is preferable to use a return fan only when there is a high resistance (over 0.5 inches water) in the return air system.

7.6 REHEAT SYSTEMS

Reheat systems also provide multizone control with a single duct (Figure 7–26). The air supply temperature is essentially constant at a value suitable for cooling all year round. Reheat coils for each zone are controlled by the zone thermostat to satisfy zone temperature requirements. This provides more flexibility than variable volume since that system can now provide simultaneous heating and cooling. It also costs somewhat more to install and operate. The heating coil in the AHU may be omitted if minimum mixed-air temperature is about 50°. If discriminator controls are added, as in Figure 7–27, some heating and cooling energy can be saved. (But see paragraph 7.3.1.)

7.6.1 Reheat with Variable Volume

Some of this operating cost disadvantage may be overcome by combining reheat with variable volume (Figure 7–28). Air is supplied from the control unit at a constant temperature suitable for cooling, say 55°. This temperature should be reset by a discriminator as shown, though most older systems do not have this feature. When the zone air temperature is above the thermostat setting the volume damper is fully open and the reheat is off. As the zone temperature approaches the set point, the damper modulates toward some minimum position (usually not less than 40%). A further drop in zone temperature will cause the reheat coil to start heating. This system is more economical than simple reheat because less reheating is required, yet it provides simultaneous heating and cooling in adjoining zones. Radiation or some other means of heating may be used instead of reheat coils.

7.6.2 Reheat for Humidity Control

Zone reheat may also be used for controlling humidity when the dew point of the main supply air is held constant (Figure 7–29). This system is similar to the single-zone system discussed in section 7.2.5 and Figure 7–8 except that now there may be several zones, each with individual temperature control. Since the supply air dew point is the same for all zones,

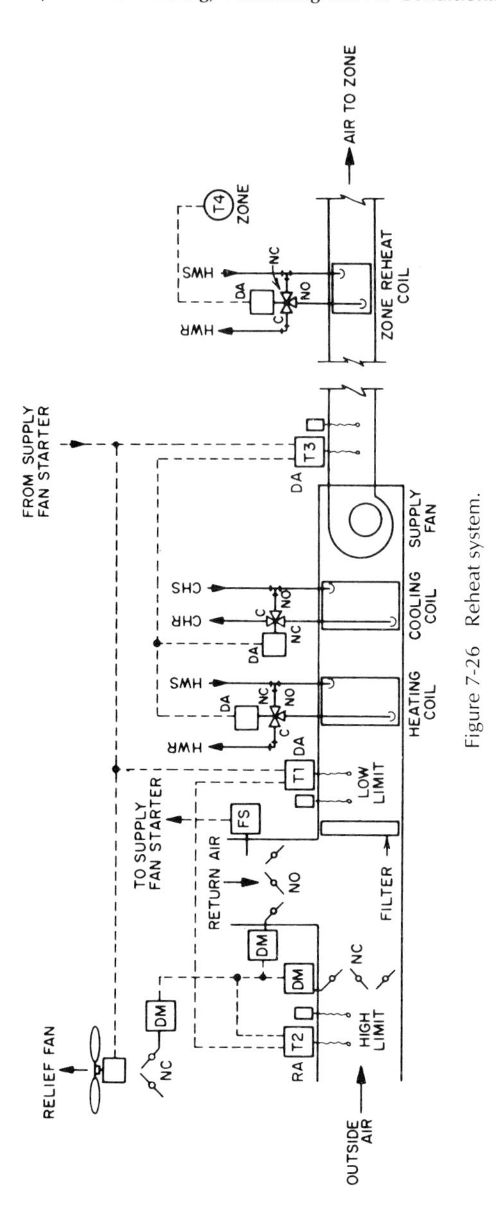

Figure 7-26 Reheat system.

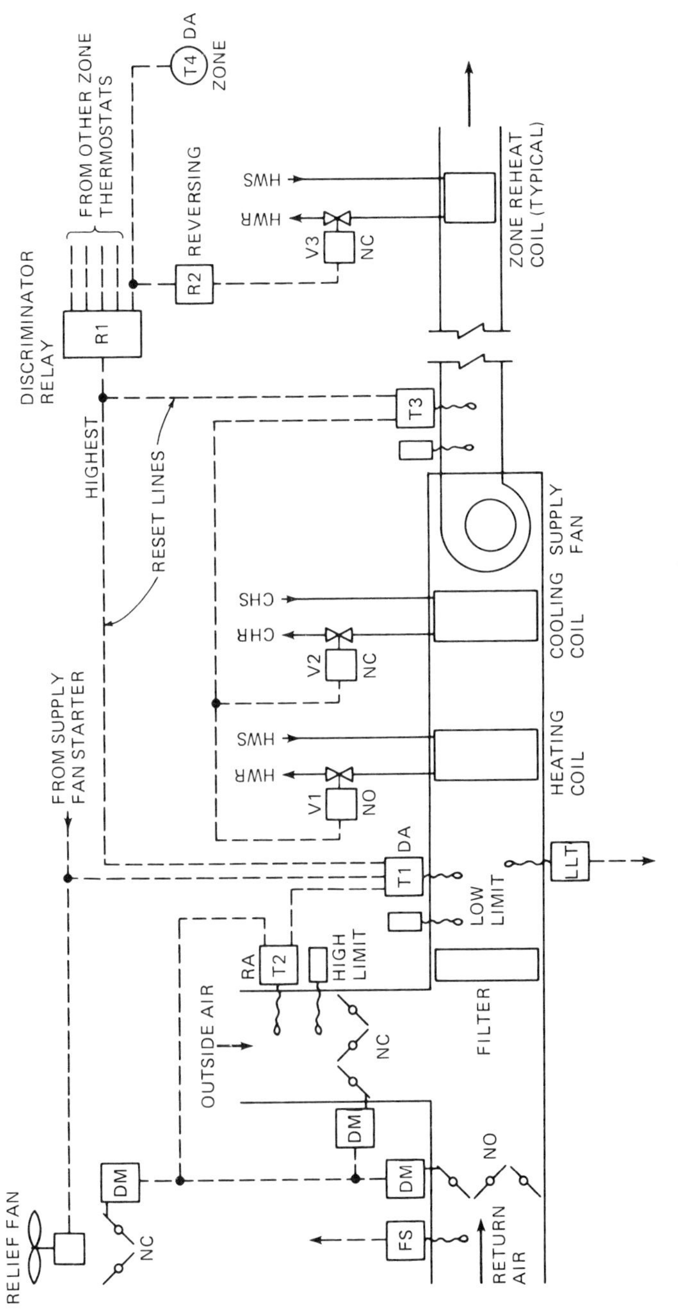

Figure 7-27 Reheat system with discriminator.

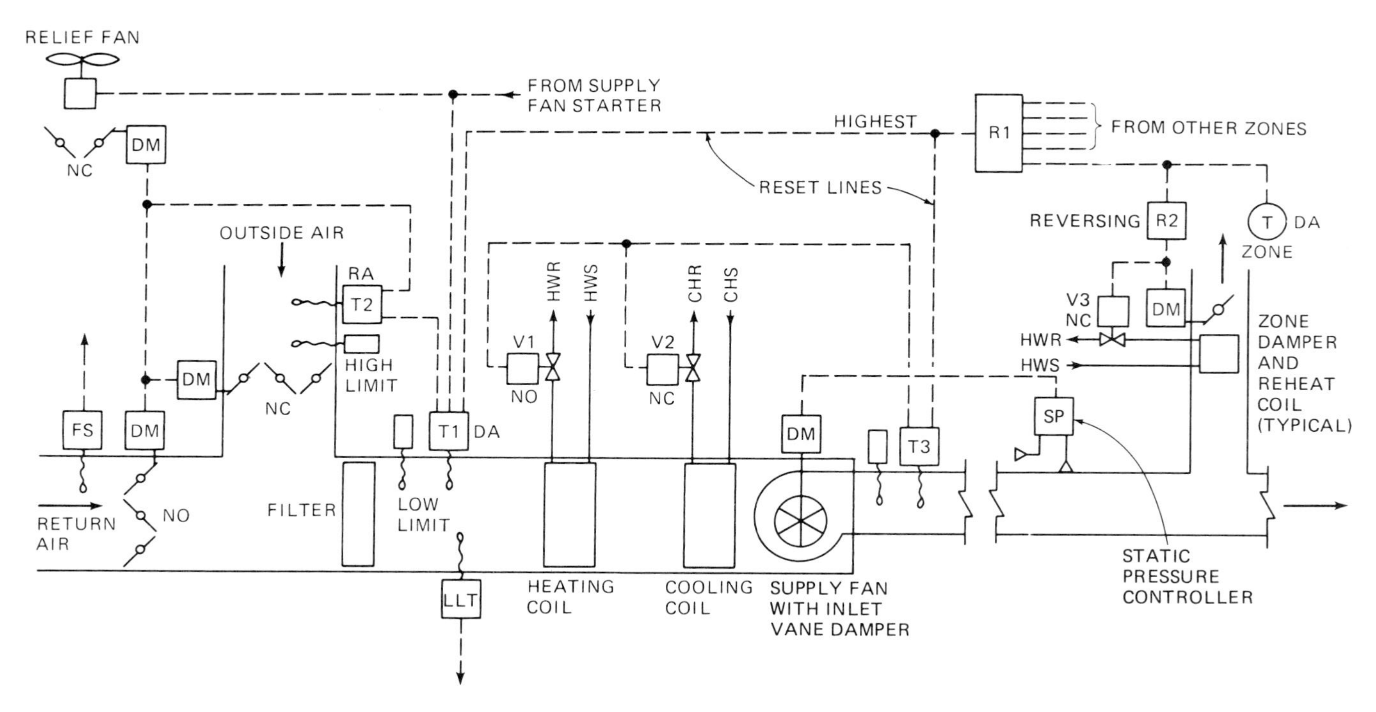

Figure 7-28 Variable volume system with reheat and discriminator.

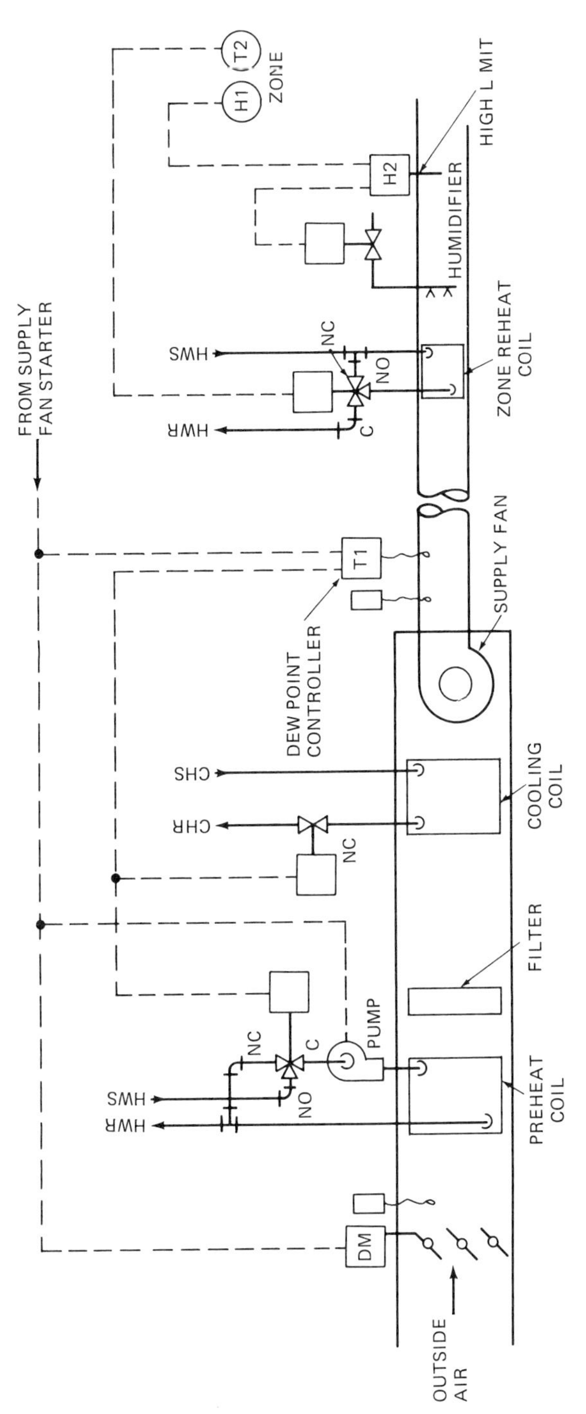

Figure 7-29 Humidity control; zone reheat.

the actual relative humidity in a zone is determined by its dry bulb temperature and internal latent load. It follows that these must be similar from zone to zone if relative humidities are to be uniform. Individual zone humidifiers or a single humidifier in the AHU may be used, depending on the accuracy of control required.

7.7 HEAT RECLAIM

In general terms, a heat reclaim system utilizes heat (or cold) that would otherwise be wasted. Heat reclaim systems thus may include heat pumps, but are by no means limited to them. A typical heat reclaim system may use the excess heat generated in the interior zones of a building to heat the exterior zones in winter, while at the same time using the cooling effect of the exterior zones to cool the interior. Or, where large quantities of exhaust air are required, with equivalent outside air makeup, the waste heat from the exhaust may be used to preheat (or precool) the makeup air.

7.7.1 Heat Reclaim—Heat Pump

Reclaim systems with heat pumps take advantage of the fact that almost all buildings have interior areas with little or no exposure to outdoor conditions. Since these spaces do contain lights and people and other heat-producing devices, such as office machines and computers, they require cooling at all times. Or, looking at it another way, they have excess heat which must be removed to maintain comfort conditions. Since the same building will also have exterior zones which require heating in winter it is possible to use the interior-zone excess heat as a heat source for a heat pump which supplies heat to exterior zones. Figure 7–30 shows an elementary schematic of such a system. Chilled water from the heat pump cools the interior zone, the rejected heat is carried by the refrigerant to the condenser

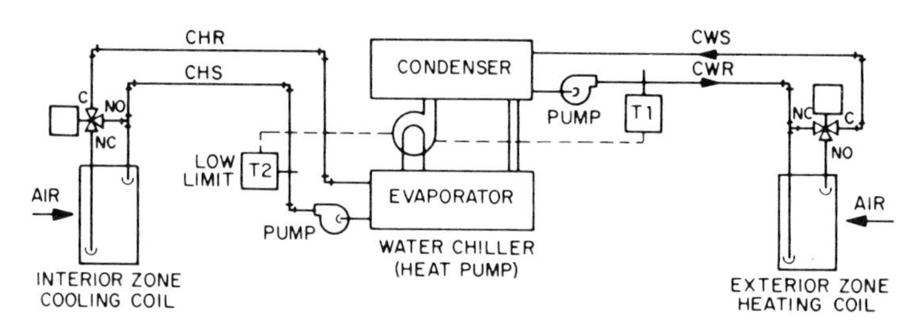

Figure 7-30 Heat reclaim; heat pump only.

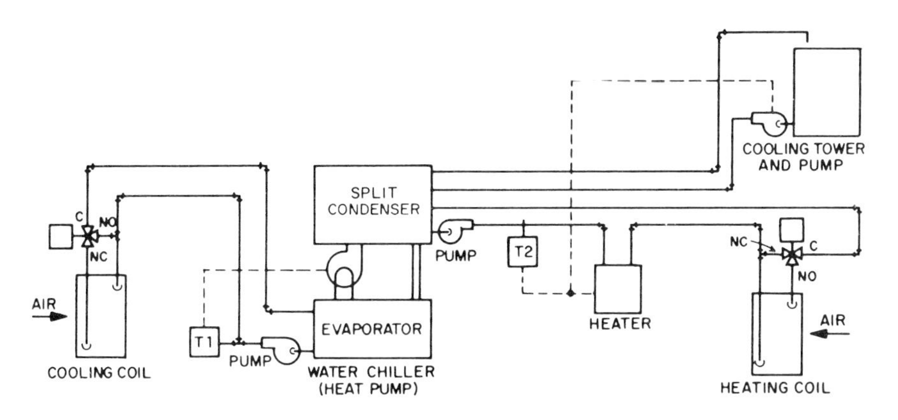

Figure 7-31 Heat reclaim; heat pump with auxiliary heat source and sink.

and the condenser water is used to heat the exterior zone. As can readily be seen, this elementary idea works only when there is a balance between interior cooling and exterior heating loads, so some additional provisions are necessary.

One common element is a "split" condenser (Figure 7–31). Now, that part of the interior heat which is not needed for exterior heating can be dissipated to a cooling tower or evaporative condenser. If a heat source, such as a boiler or electric heater, is added for those times when exterior losses exceed interior gains, a year-round system is obtained. It is desirable, but not essential, that separate air handlers be used for interior and exterior areas.

The interior-zone air handler, which could be any of the types previously discussed, provides cooling. The rejected heat, in the form of an increase in chilled water temperature, goes back to the heat pump, where chilled water supply temperature is the controlled variable. The rejected heat is conveyed by the refrigerant to the condenser water, which in turn is used as a heat source for the exterior-zone air handling unit heating coil. Water temperature leaving the condenser is controlled by a thermostat which operates as follows: When the water temperature is below the thermostat setting it calls on the auxiliary heat source for additional heating. As the water temperature increases the auxiliary source is shut down, and on a further increase, the other section of the split condenser is brought into play, dumping the excess heat to the cooling tower. In a small system this same chiller-heat pump could also supply cooling for the exterior zones, but in larger systems an additional chiller (not in a heat pump arrangement) will be provided.

This arrangement is used most often in high-rise office buildings or similar structures with interior areas which comprise a fairly large portion of the total floor area.

7.7.2 Heat Reclaim—Revolving Wheel

The revolving wheel heat reclaim system uses a large metal wheel which revolves slowly as exhaust air passes through one side and makeup air passes through the other. Heat is thus transferred by the turning wheel. The only control required is an interlock to start the wheel drive motor when the supply fan is operating. These devices are produced by several manufacturers and are very satisfactory where system geometry allows makeup and exhaust ducts to be contiguous.

7.7.3 Heat Reclaim—Runaround Coils

The "run around" heat reclaim system uses coils in exhaust and makeup air ducts, with piping connections and circulating pumps. While obviously more expensive than the revolving wheel, this system allows complete flexibility of system geometry and provides high efficiencies. Eight row coils will allow over 40% of the waste heat to be reclaimed in winter and about 25% of the waste "cooling" in summer. Where freezing air is encountered the system is usually filled with an ethylene glycol solution. Figure 7–32 shows such a system. The circulating pump is interlocked to run whenever the supply fan runs. The bypass valve and controller are used for low-limit control of the fluid temperature (35 °F) to prevent frost build-up on the exhaust coil.

7.7.4 Heat Reclaim—Runaround Coils in Humidity Control

A special application of runaround coils for heat reclaim can be found in an HVAC system for providing low humidity in a clean room or similar space. A typical application is in electron "chip" manufacturing where a typical space condition requirement is 75°F with 35 percent RH and a high air change rate is needed to maintain a stable and clean environment. The manufacturing process also requires large amounts of exhaust air. The high air change rate means that only about a 10 degree difference is required between supply air and space temperature; that is, supply air temperature would be 65°F. The leaving dew point temperature required to maintain the 35 percent RH is 45°F. Then 20°F of reheat would be required—approximately 65 percent of the total cooling load. The system shown in Figure 7–33 will save a great deal of that energy by means of a simple internal runaround reclaim system. Figures 7–34 and 7–35 show the processes on the psychrometric chart.

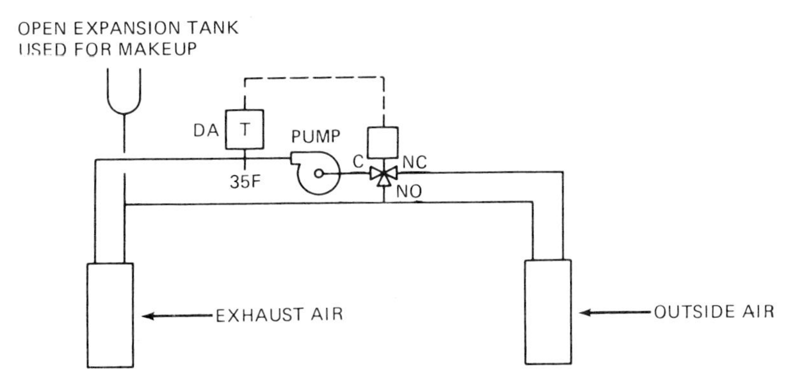

Figure 7-32 Heat reclaim; runaround coils.

For the example in Figure 7–34, a minimum of 40 percent outside air is assumed (for exhaust makeup), with summer design outside conditions of 95°F DB and 78°F WB. Then mixed air is at 83°F DB and 67°F WB. Cooling coil leaving air is assumed to be saturated at 45°F dew point. To accomplish this it may be necessary to provide low temperature brine rather than chilled water. It can be shown that a pair of ten row runaround coils can provide about 15°F of precooling and preheating for a saving of 30,000 BtuH per 1000 CFM over the conventional system without runaround.

At an intermediate condition of 60°F, 80 percent RH outdoors the same runaround system would provide somewhat less capacity, but the savings would still be significant (Figure 7–35). Below about 55°F outdoors the runaround system would cease to be effective and would be shut off.

In the control diagram (Figure 7–33) the runaround system is controlled by the discharge thermostat as a function of reheat requirements, with the discharge set point reset by the room thermostat. Additional conventional reheat is provided and this will be controlled in sequence so that maximum runaround capacity is always used.

An economy cycle is provided for winter operation. The high-limit changeover point is about 56°F to maximize mixed air temperature, since precooling furnishes the energy for reheating.

7.8 FAN-COIL UNITS

Fan-coil units are widely used in hotels, motels, apartments and offices. Essentially, they are small single-zone air handling units with a fan, a filter and a coil which may be used for both hot or chilled water or may be split, with one row used for heating and two or three used for cooling. The

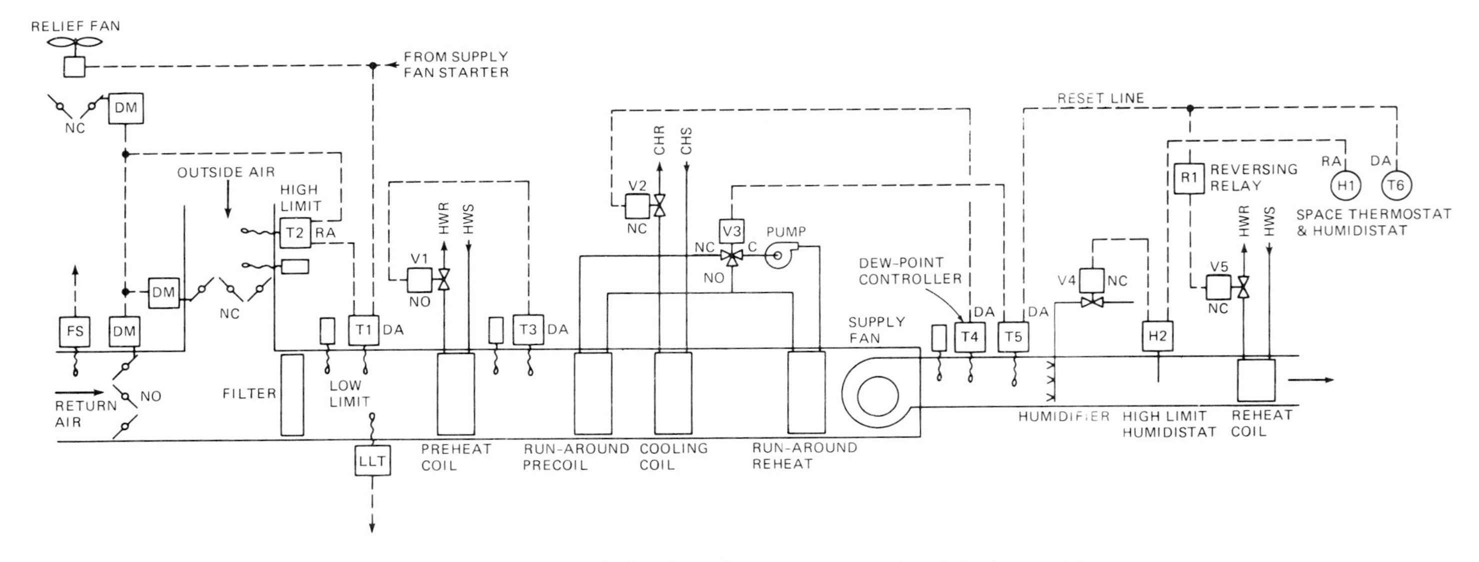

Figure 7-33 AC system for low humidity using runaround coils for heat reclaim.

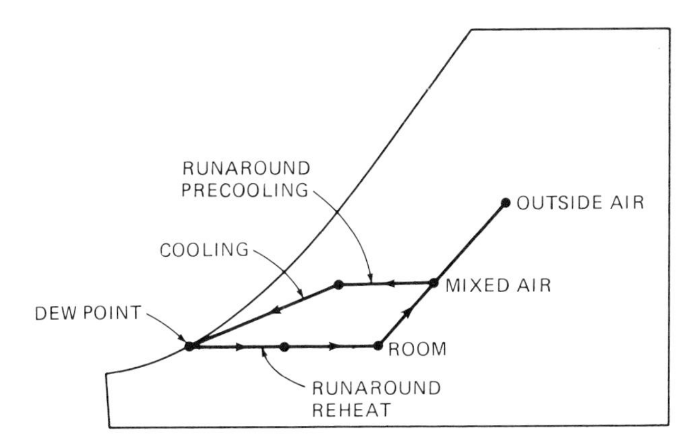

Figure 7-34 Psychrometric chart for Figure 7-33 (summer).

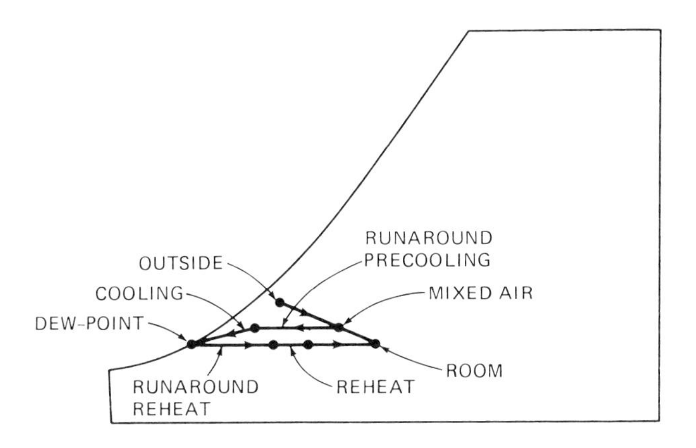

Figure 7-35 Psychrometric chart for Figure 7-33 (intermediate).

former arrangement is more common since it is less expensive and requires fewer controls.

The control system may be arranged for two-pipe, three-pipe or four-pipe supply, each of which has certain advantages and disadvantages. When a single coil is used for both hot and chilled water, hot water temperatures may be much lower than with standard heating coils. This is because of the extra surface required for cooling at the normally low temperature difference between chilled water and leaving air. Hot water temperatures of 110° to 140° are typical. Fan control is usually manual, typically using a two- or three-speed switch.

7.8.1 Two-Pipe System

Water is used, and hot or chilled water is supplied from a central plant in season. The room thermostat must be a summer-winter type—that is, it is direct-acting in winter, reverse-acting in summer when used with a normally open valve. Changeover from one to the other is accomplished by a general signal, such as changing main air pressure in a pneumatic system, or by sensing the change in supply water temperature. The thermostat may modulate the water flow valve or simply start and stop the fan while the water flow is uncontrolled. Figure 7–36 shows a common arrangement with manual multispeed control of the fan and thermostatic control of the water flow to the coil. With chilled water being supplied and the room temperature above the reverse-acting thermostat setting, the normally open valve is open, allowing the unit to provide cooling. As the room temperature falls the valve closes. On a further fall in room temperature, no heat is provided since hot water is not available.

With hot water being supplied the room thermostat is direct-acting and the reverse cycle takes place. Now, of course, no cooling is possible. It is apparent that this arrangement can cause loss of control in mild weather, when heating is required in morning and evening but cooling is required during the day.

A central plant for a two-pipe system includes a boiler or other heat source, a chiller, circulating pumps and changeover controls. Changeover may be manual or automatic, but in either case involves problems.

The principal difficulty arises on changeover from heating to cooling and vice versa. Warm water, even at 75° or 80° may cause chiller problems due to high suction pressure and overload. These same temperatures can cause thermal shock and flue gas condensation in the boiler. Also, flow rates in both chiller and boiler should be essentially constant.

The recommended system shown in Figure 7–37 has heat exchangers and

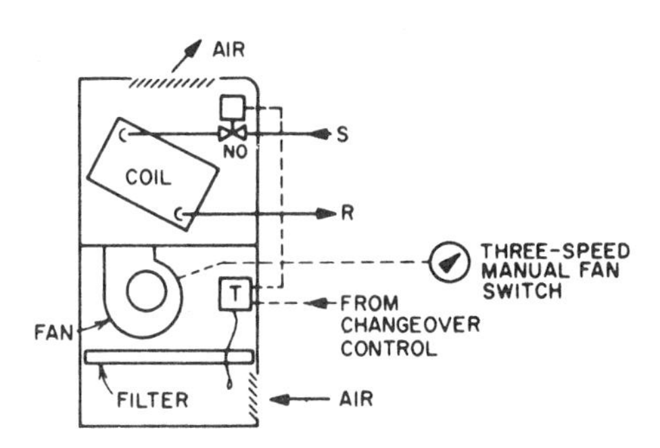

Figure 7-36 Fan-coil unit; two-pipe system, manual fan control.

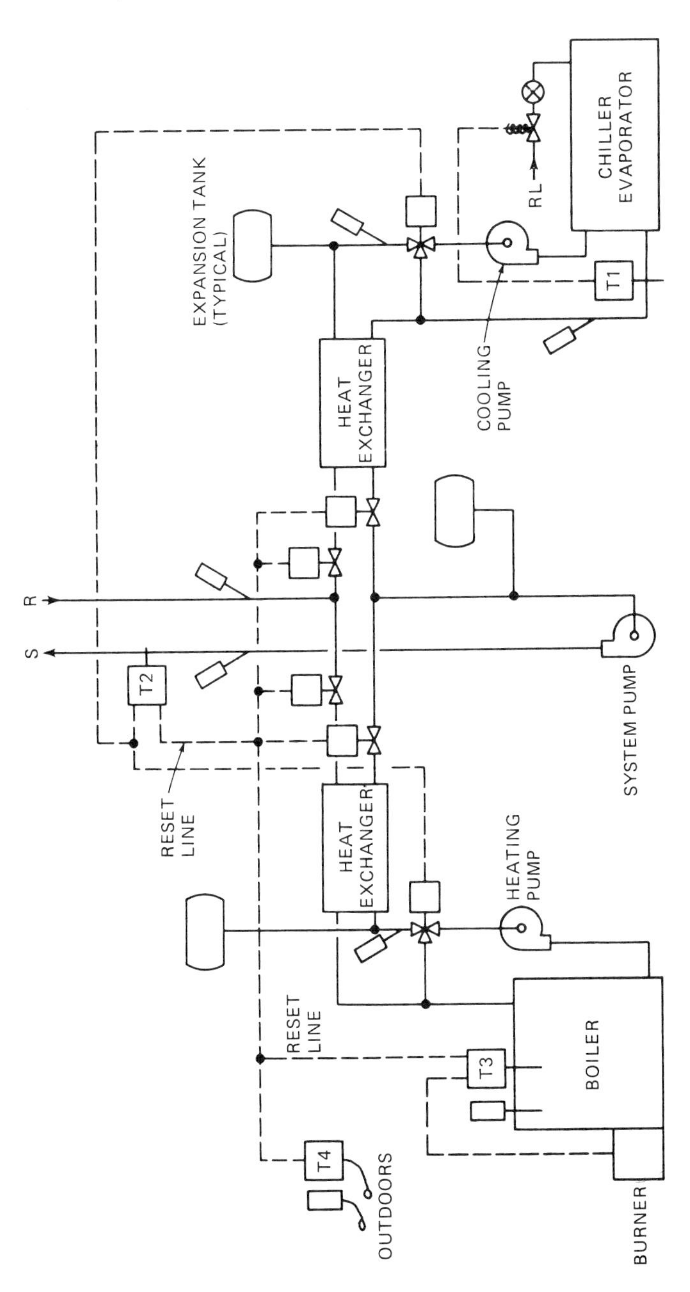

Figure 7-37 Two-pipe system; central equipment and control.

secondary pumps to isolate the chiller and boiler from the distribution piping. This allows the chiller and boiler to operate within safe limits while the distribution water temperature can be modulated through a wide range as outdoor temperature varies. This system is not very economical and is difficult to control properly in spring and fall when heating is needed in the morning and cooling is needed in the afternoon.

7.8.2 Three-Pipe System

Water is used, with separate hot and chilled water supply pipes and a common return pipe. Individual heating and cooling valves or a special three-pipe valve may be used. A standard direct-acting thermostat is required. Since it is possible, in season, to provide a choice of heating or cooling, this system is much more flexible than the two-pipe system. Mixing the return water, however, means that when both heating and cooling are being used the return water is at some intermediate temperature and a false load is placed on both chiller and boiler, with a resultant increase in operating cost.

Figure 7–38 shows a typical three-pipe fan-coil arrangement, using a special three-pipe valve. When the room temperature is below the thermostat setting the heating port of the three-pipe valve is open. As the room temperature increases the hot water port closes. There is then a "dead spot" over a small temperature range, during which both ports are closed. As the room temperature goes still higher the chilled water port opens. Chilled or hot water flows out to the common return main where it is mixed with chilled or hot water return from other units and conveyed back to the central plant.

A central plant for a three-pipe system includes a boiler or other heat source, a chiller, a circulating pump and temperature controls. Since the water flows through boiler and chiller are determined by the sums of the

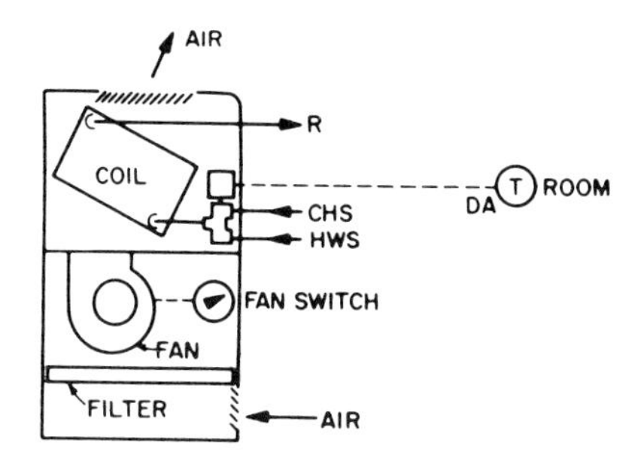

Figure 7-38 Fan-coil unit; three-pipe system, manual fan control.

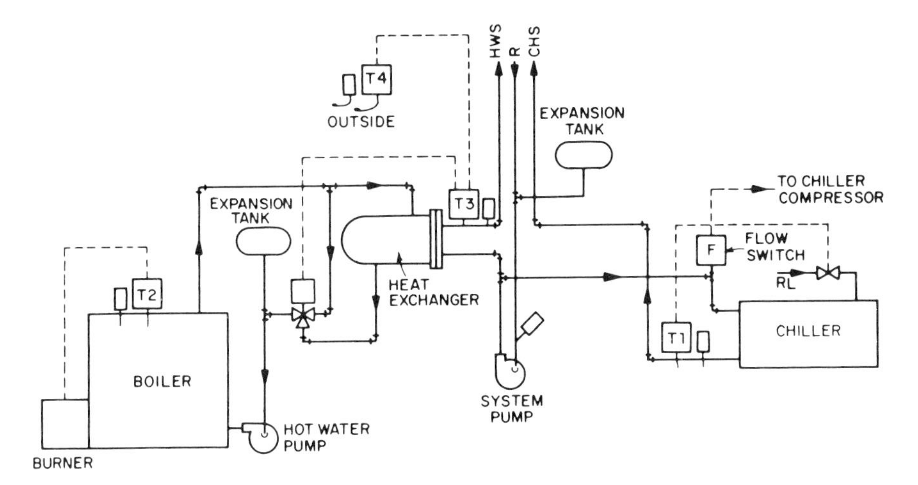

Figure 7-39 Three-pipe system; central equipment and control.

individual fan-coil usages, the system is self-balancing in this regard. Return water temperature is directly related to load, increasing when heating demand is increased and decreasing when cooling demand is increased. This seeming anomaly results from the fact that return hot water is hotter than return chilled water. Since the flow through the chiller will decrease as the return water temperature rises, there is little need for the elaborate temperature-flow controls discussed under two-pipe systems. However, a sudden increase in cooling load can cause problems. Also, low-temperature water going through the boiler can cause condensation of flue gases in the boiler, with resulting corrosion. For this reason, a heat exchanger is recommended for two- and three-pipe systems requiring hot water to be supplied at temperatures below 140° Figure 7–39 shows a typical three-pipe central plant control arrangement. The individual water supply thermostats control the heat exchanger and chiller capacities. A flow switch on the chiller will shut off the compressor if water flow falls below the minimum rate necessary to prevent freezing. To avoid this kind of shutdown it is preferable to add a heat-exchanger and secondary pump for the chiller, as shown in Figure 7–37.

7.8.3 Four-Pipe System

Four-pipe systems provide complete segregation of the heating and cooling media, thus avoiding the problems of two- and three-pipe installations. The normal coil configuration is a split arrangement with one row

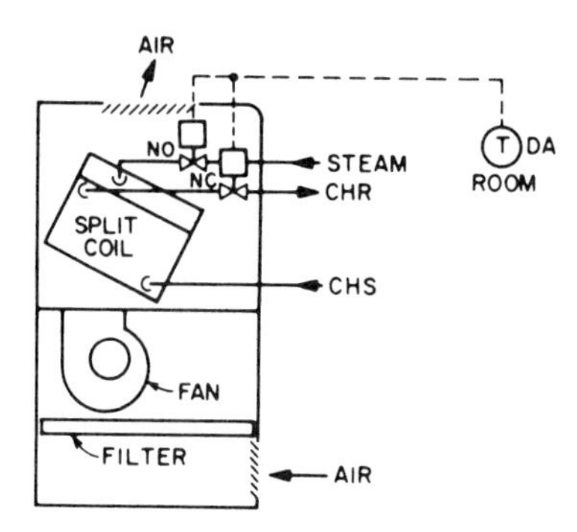

Figure 7-40 Fan-coil unit; four-pipe system, split-coil.

used for heating and two or three used for cooling. Sometimes two separate coils are used. With this set-up virtually any heating and cooling source can be used, including steam, hot or chilled water, direct-expansion cooling and electric resistance heating.

Figure 7–40 shows a typical split-coil control arrangement. Here, steam is used for heating and chilled water for cooling. With the room temperature below the setting of direct-acting thermostat, the normally open steam valve is open and the normally closed chilled water valve is closed. As the room temperature increases the steam valve closes. The controls should be adjusted to provide a "dead spot" where both valves are closed. As the room temperature rises still further, the chilled water valve opens. On a fall in temperature, the sequence is reversed.

It is also possible to use a single coil for either hot or chilled water, while keeping the piping systems segregated. This requires the use of two "three-pipe" valves.

Figure 7–41 shows such an arrangement. When the room temperature is below the setting of the thermostat the inlet and outlet valves are open to the hot water supply and return lines. As the temperature increases, the inlet valve heating port may modulate toward the closed position, but the outlet valve heating port will remain full open to minimize pressure drop. When the inlet valve is fully closed then the outlet valve will close, using two-position action. Again, there will be a period when both ports are closed and

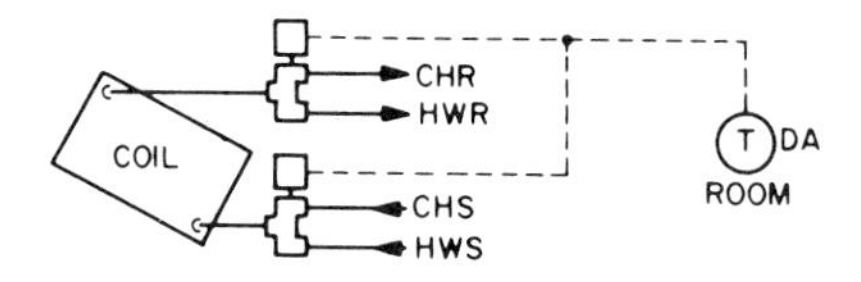

Figure 7-41 Fan-coil unit; four-pipe system, single-coil.

room temperature is at or near the thermostat setting. On an increase in room temperature above the setting, the inlet and outlet valve cooling ports will modulate toward open, or open using two-position sequence.

7.9 INDUCTION SYSTEMS

An induction unit depends for its operation upon a central source of temperature-controlled high-pressure air, which induces a secondary flow of room air across the unit coil. This coil is supplied with hot or chilled water through a two-, three- or four-pipe system. Thus it is necessary to control the primary air supply temperature, the temperature of the hot or chilled water and the flow of this water. Since condensate drains are not usually provided on induction unit coils, the chilled water supply temperature must be kept above the dew point of the entering air. Because a single coil is used for both heating and cooling, hot water supply temperatures can be comparatively low, in the order of 140° or less.

Therefore, most induction unit systems are arranged with water flowing first through the central air handling unit coils and then, in series, through the induction unit coils. Thus, chilled water may enter the central unit coil at 45° and leave at 55°, which is good for supplying the induction unit coil.

Figure 7–42 shows an induction unit system with four-pipe control. Heating and cooling coils in the primary air unit are supplied with hot and chilled water from a central boiler (or other heat source) and chiller. The primary supply air temperature is controlled at a value which varies with outdoor temperature according to some reset schedule. The chilled water leaving the cooling coil is returned to the chiller or passes into a secondary chilled water system which has its own pump and control. The supply water temperature in this secondary system is controlled at about 55° by a thermostat which operates a three-way valve to introduce water from the primary system coil or recirculate return water from the secondary system. The secondary hot water system operates in the same way.

The individual induction unit coil provides heating or cooling as required in response to the demand of a room thermostat as detailed in the discussion of the fan-coil unit. Notice that this heating or cooling may supplement or counteract the primary air action depending on the season and internal and solar loads.

Conventionally, induction units are commonly used on exterior walls and below windows; there they can furnish individual control for small zones with highly variable solar and other loads. They are not normally used for interior zones because of their small size.

A special type of induction unit is sometimes used in interior zones. This

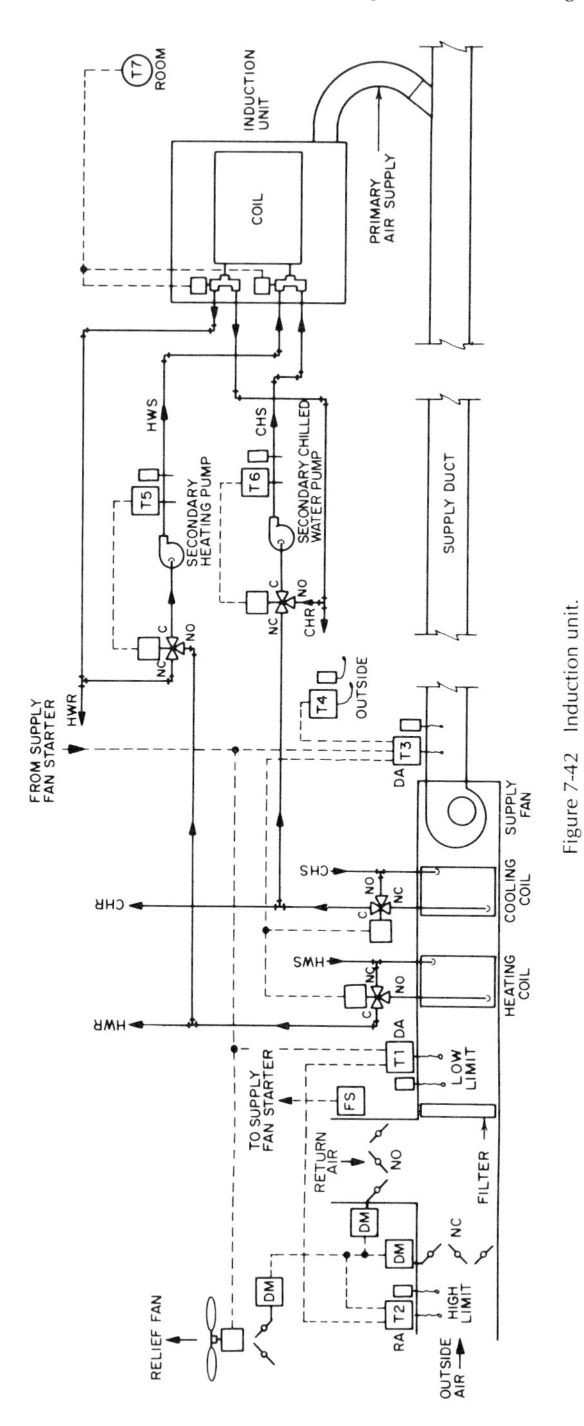

Figure 7-42 Induction unit.

unit is installed in a return air plenum above the ceiling and utilizes the heat of the return air and light fixtures to temper the primary cold air supply for interior zone control.

7.10 UNIT VENTILATORS

A unit ventilator is similar to a fan-coil unit but is usually larger and is arranged to provide up to 100% outside air through an integral damper system. This unit is designed primarily for heating, though some unit coils are sized for cooling also. Their primary use is in school classrooms or similarly heavily occupied spaces, where large amounts of ventilation air are required. Control systems for unit ventilators have been standardized by ASHRAE and are detailed in the HANDBOOK. The principal elements of the control system are the room thermostat, discharge air low-limit thermostat, damper motor and steam or water control valve. (Electric heat is sometimes used.) When cooling is also provided, the coil control will be similar to that described for fan-coil units with the addition of outside air control. This could be as shown in Figure 7–43.

On the heating cycle the thermostat is direct-acting. With room temperature below the thermostat setting the outside air damper is closed or open to a fixed minimum position, the return air damper is open and the normally open water valve is open. Hot water is being provided by the two-pipe system. As the room temperature increases the outside air damper modulates toward the open position. The discharge air low-limit thermostat will keep the valve open as required to maintain a minimum discharge air temperature of about 60°. When the room temperature reaches or exceeds the room thermostat setting the outside air damper will be fully open and the valve will be closed unless some heating is required to maintain the discharge air temperature low limit.

For cooling, the changeover control changes the room thermostat to re-

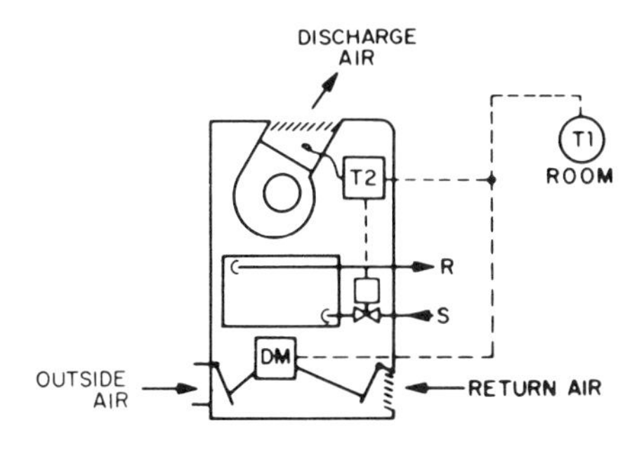

Figure 7-43 Unit ventilator.

verse-acting and chilled water is supplied by the central system. Now, with the room temperature above the room thermostat setting, the outside air damper is closed or open to a fixed minimum position, the return air damper is open and the water valve is open. As the room temperature decreases, the outside air damper modulates toward the open position and the valve modulates toward the closed position. The discharge air thermostat must be bypassed since it would try to open the valve at low discharge air temperatures, and this would of course nullify the effect of the room thermostat. Alternatively the discharge thermostat could be changed to reverse-acting, with the set point changed to about 75° as a high limit.

7.11 PACKAGED EQUIPMENT

Packaged equipment, by our definition, is HVAC equipment which is factory-assembled and ready for installation in the field with a minimum of labor and material. As such it includes a complete, ready-to-function, control system, usually electric but often with electronic components. Typical packaged equipment items are residential furnaces and air conditioners, rooftop units and direct gas-fired heating equipment.

Since most of these units include the controls, they will be discussed only briefly, to point out typical arrangements.

7.11.1 Self-Contained Fan-Coil Unit

The small, self-contained fan-coil unit is widely used in motels, hotels and small apartments. This is basically the same as a window-type air conditioner but with the configuration of a fan-coil unit. It contains a supply fan, direct-expansion cooling coil, refrigeration compressor, air-cooled condenser with through-the-wall ducts for the air, electric heating coil, filter and controls. A return air type of thermostat is often used, so that only power and outside air connections are needed to make the unit operable. Controls are simple, electric and conventional. The thermostat is direct-acting, two-position, calling for cooling as the room temperature increases above the set point and heating when it decreases below the set point. A fairly wide differential is provided to prevent short cycling of first heating and then cooling. A manual changeover summer-winter thermostat may be used. Fan control is manual, two or three-speed.

7.11.2 Residential Air Conditioning

A typical residential air conditioning system has a gas, stoker or oil-fired furnace within the house. This furnace has an add-on direct-expansion cooling oil installed in the discharge. Field-installed refrigerant pipes connect the coil to an air-cooled condensing unit outdoors. A two-position summer-winter thermostat with manual changeover is provided. Typically, this thermostat starts and stops the supply fan, as well as controlling the heating and cooling. A manual fan switch is sometimes furnished. On the heating cycle, the fan is started and stopped by a discharge plenum thermostat, to prevent blowing cold air. A high-limit thermostat acts to shut down the burner if the high-limit temperature setting is exceeded. Figure 7–44 shows a year-round residential air conditioning system with gas-fired furnace and air-cooled condensing unit. The thermostat has manual change-over from heating to cooling and also has an "on-automatic" fan switch. In the "cooling" position, when room temperature rises above the thermostat setting, the condensing unit and supply fan are started simultaneously, and the solenoid valve on the cooling coil is opened. The condensing unit is equipped with high- and low-pressure safety controls and thermal overload (not shown). Some units of this type and size do not use the pump down cycle. With the changeover switch in "heating" position, on a call for heat the gas burner will light (providing the pilot safety is satisfied). When the plenum temperature rises to its set point, the fan switch will start the supply fan. A high-limit thermostat (usually combined with the fan switch) will shut off the burner if necessary.

Figure 7–45 shows an oil-burning furnace. Cooling controls could be pro-

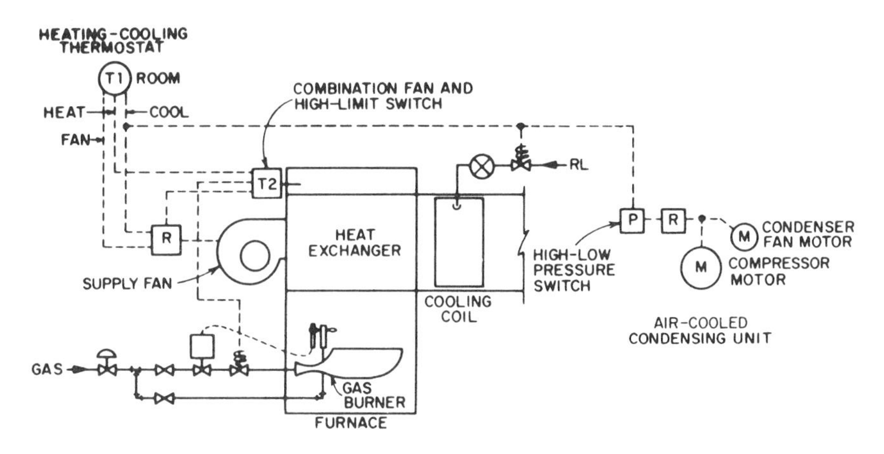

Figure 7-44 Residential gas-fired furnace with DX cooling.

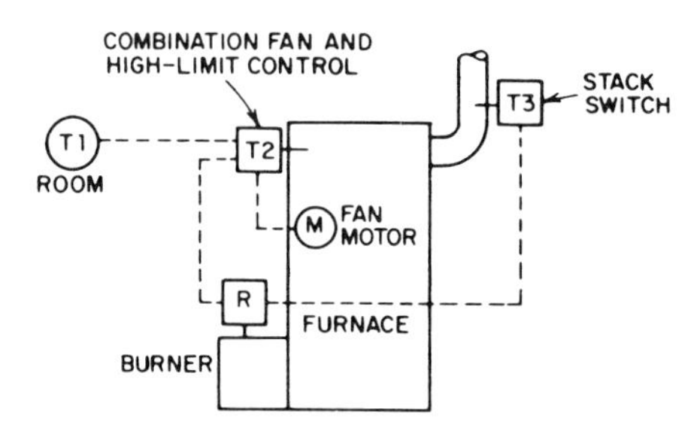

Figure 7-45 Oil-fired furnace.

vided as shown for the gas-fired furnace. On heating, a call for heat starts the oil burner motor (or, with a gravity-type burner, opens a valve) and energizes a time-delay relay in a "stack switch." This thermostat is mounted in the combustion vent stack and must be opened by the rise in temperature of the flue gas (indicating proper combustion) before the time-delay relay times out. Thus, if the oil fails to ignite the burner will be shut off. The switch must then be manually reset. Some models of this device provide for one recycle operation to purge unburned gases and try again for ignition before finally locking out.

Figure 7–46 shows a stoker-fired furnace. The stoker-fed fire must be ignited manually, and therefore must not be allowed to go out. Because of this, the stoker control system incorporates a relay which includes a timer. Regardless of heating demand, the timer operates the stoker for a few minutes each half-hour to ensure continuation of the fire. Again, cooling could be added as shown for the gas-fired furnace.

7.11.3 Residential Heat Pumps

Particularly in the south and in apartments, small (2 to 5 ton) residential heat pumps are very popular. A typical schematic and control

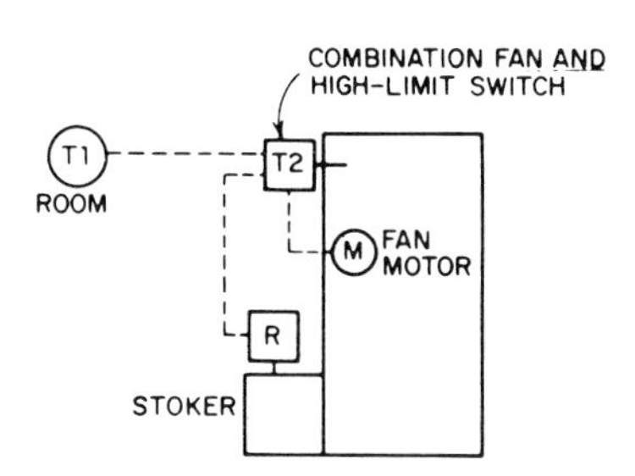

Figure 7-46 Stoker-fired furnace.

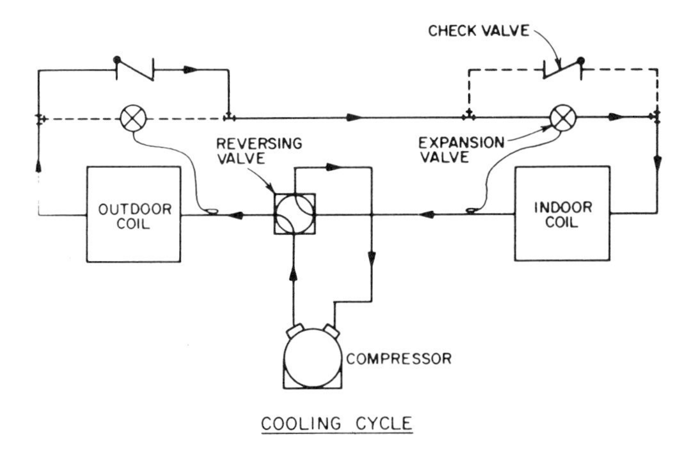

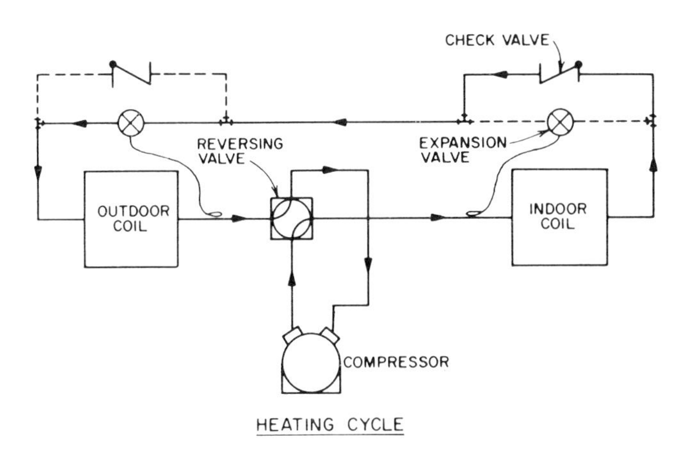

Figure 7-47 Residential heat pump.

system is shown in Figure 7–47. Controls are, of course, part of the package. Peculiar to this equipment are the reversing valve and defrost controls.

The reversing valve is basically a four-way solenoid. It must be very carefully designed to prevent leakage between the high- and low-pressure gas streams which flow through its two sides simultaneously. The sealing problem is further complicated by the large temperature differences between the two gas streams.

When the heat pump is operating on the reverse cycle to provide indoor heating, icing of the outdoor coil is common. Some kind of defrost cycle is therefore necessary to periodically remove the ice. Several methods are used, including:

1. A time clock provides a regular defrost cycle. This operates regardless of ice buildup.
2. A sensor notes the increase in air pressure drop through the coil due to icing and operates the defrost cycle until normal conditions are restored.
3. A refrigerant temperature or pressure sensor senses the decrease in suction pressure or temperature caused by icing.
4. A differential temperature controller senses the increasing difference between outside air and refrigerant temperatures caused by icing.

Most of these systems use air defrost, allowing the fan to run while the compressor is shut down. Units which may be required to operate in ambient temperatures below freezing must have electric heaters for defrost, or use occasional short periods of indoor cooling operation so that hot gas will defrost the outdoor coil.

7.12 OTHER PACKAGED EQUIPMENT

"Rooftop" units are packaged heating and cooling systems (or sometimes heating and ventilating only) which combine the simplicity of direct firing with the complexity of economy cycle outdoor air control, relief fans and even multizone supply. They are made in a wide variety of styles, fuels and arrangements, and with capacities up to 50 or 60 tons of cooling and several hundred thousand Btu's of heating. Controls are factory furnished and operate in many of the ways discussed in this book.

Boilers, and other direct-fired heating units, always include the controls as part of an overall package approved in its entirety by some national certification agency, such as the American Gas Association (AGA) or Factory Mutual (FM). Controls on the equipment cannot be changed in any way without voiding this approval.

Common features in control systems of large gas- and oil-fired boilers are: flame-failure safety shutdown; electric ignition with checking before startup; pre-purge, to remove any unburned gases from the combustion chamber before startup; post-purge, to do the same after shutdown; and low-fire startup to insure proper operation before bringing the burner to full capacity.

Controls may also include air-fuel ratio metering and measurement of O_2 or CO to improve combustion efficiency. Power boiler control diagrams typically employ ISA symbols rather than the HVAC symbols used in this book.

Boilers of any size are often interlocked to the water pump or a water flow switch, to ensure water flow before the burner is turned on.

Makeup air units are direct gas-fired units without vents. The products of combustion are mixed with the 100% outside ventilation air supplied and

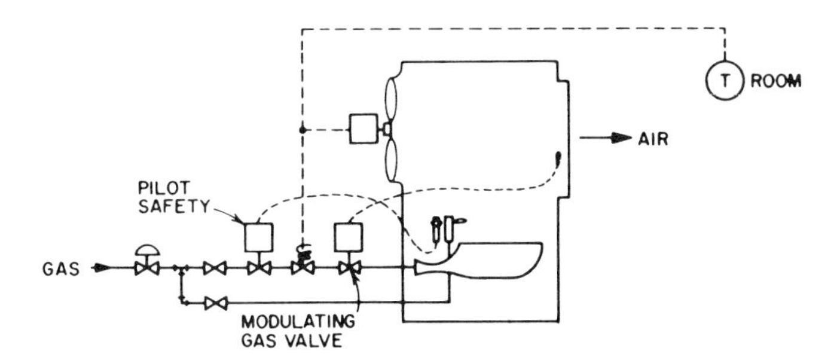

Figure 7-48 Makeup air unit; gas-fired.

are not considered hazardous because of the high dilution rate. Their use is restricted to industrial applications and commercial establishments with low personnel occupancy. Control is usually by means of a two-position room thermostat and gas valve, combined with a self-contained modulating valve with the sensor bulb in the discharge air. (See Figure 7–48.) It is often required that they be interlocked with companion exhaust systems, and in some areas they may not be used as the primary source of heating.

7.13 RADIANT HEATING AND COOLING

By definition, radiant heating and cooling depend for their effectiveness on the radiant transfer of heat energy. There are many varieties of radiant heating systems, ranging from low-temperature–high-mass concrete floors to high-temperature–low-mass outdoor systems with gas-fired ceramic or electric elements. A single type of control system for all radiant systems is, therefore, not possible. For discussion purposes several distinct groupings can be recognized.

7.13.1 Panel Heating

First, and most commonly used for general heating, are floor, wall and ceiling panels. Floor panels with hot water pipes or electric cables embedded in concrete have a large mass, and therefore, a large time lag. A conventional room thermostat will not provide adequate control. A room thermostat with outdoor reset, or even an outdoor thermostat alone, has proven more satisfactory. It is essential to limit the panel surface temperature to a maximum of 85°. This can most easily be done by limiting the water supply temperature. Figure 7–49 shows a floor heating panel controlled by

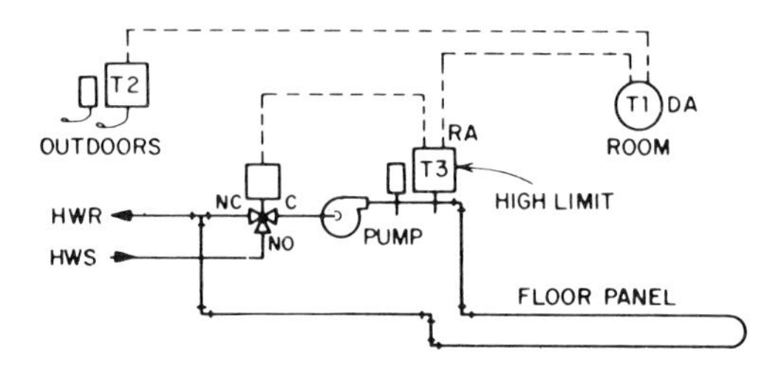

Figure 7-49 Floor panel heating; hot water.

a room thermostat with outdoor reset and a high-temperature–limit thermostat in the water supply.

Wall and ceiling panels generally have much less mass than floor panels and can therefore be controlled directly by a room thermostat. While surface temperatures up to 100° in walls and 120° in ceilings are acceptable, these still require supply water temperature control. The control system would look like Figure 7–49 without the outdoor reset (although reset could still be used to improve response).

Wall and ceiling panels can also be used for cooling. Water supply temperatures must be carefully controlled to avoid surface temperatures below the room dew point. Since this type of cooling is always supplementary to an air system, the water supply can be handled in the same ways discussed for induction units.

7.13.2 Direct Radiant Heaters

Direct-fired, high-intensity infrared heaters may consume gas or electricity, and are used for general heating. They are also used for spot heating, indoors and out. Except when they are used for general heating, control should be manual, because of the inability of a conventional thermostat to sense "comfort" when heat transfer is primarily radiant.

7.14 RADIATORS AND CONVECTORS

Convectors and convector radiators in various forms are widely used for basic heating and as supplementary heaters with air conditioning systems. This general classification includes fin-pipe and baseboard radiation, as well as the old-fashioned cast-iron column radiator and the modern fin-coil convector radiator. This equipment may use steam at low pressure or under vacuum, or hot water.

7.14.1 Low-Pressure Steam Systems

Low-pressure steam supply may be controlled by individual control valves on each radiator or by means of zone control valves which supply several radiators. In the latter case orifices are used at each radiator to ensure proper distribution of the steam, and two-position control is necessary. Steam pressure at the boiler is kept essentially constant by a pressure controller which operates the burner. Self-contained radiator valves, requiring no external power source, are widely used.

7.14.2 Hot Water Systems

Hot water may be controlled in the same manner as steam. Series flow through a group of radiators may be used if the design and selection allows for water temperature drop from the first radiator to the last.

7.14.3 Vacuum Systems

Vacuum systems are controlled by varying the steam pressure and temperature to suit the building heating requirements. In a typical system (Figure 7–50) the zone thermostat modulates the steam supply valve in response to demand. The vacuum pump is operated by a differential pres-

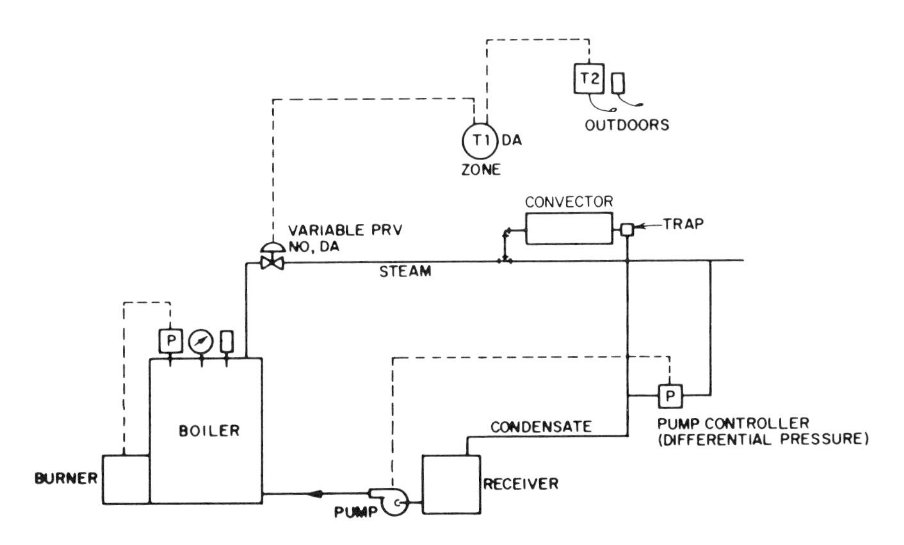

Figure 7-50 Vacuum heating system control.

sure-vacuum controller, to maintain a constant differential between supply and return mains. Outdoor reset of the zone thermostat may be provided to prevent overheating. A constant boiler steam pressure is maintained by a pressure controller. Vacuums as low as 25 in. or 26 in. of mercury are normal in mild weather. A zone may be the entire building, or individual zone supply valves and vacuum pumps may be used.

7.15 HEAT EXCHANGERS

Heat exchangers as heat sources have been mentioned several times. Heat exchangers are required when a large central plant supplies heat-temperature water or high-pressure steam, neither of which can be conveniently or safely used in the individual building HVAC equipment. As noted previously, heat exchangers may also be used to provide very-low-temperature heating water and thus protect the boiler from thermal shock and/or condensation of flue gases. Obviously, the control system varies with the function of the heat exchanger.

7.15.1 Heat Exchanger—Low-Pressure Steam

Figure 7–51 shows a simple heat exchanger control system, suitable for producing water in the 100° to 200° range using low-pressure steam or medium-temperature (up to 260°) boiler supply water as a heat source. The valve is controlled by the low-temperature supply water thermostat which may be reset by an outdoor thermostat if required. An interlock from water pump or water flow switch keeps the valve closed when water is not flowing. Modulating or two-position control may be used, with the latter usually adequate.

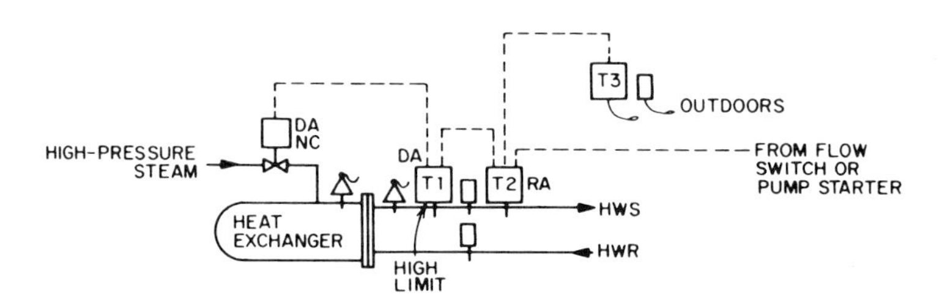

Figure 7-51 Hot water supply heat exchanger.

7.15.2 Heat Exchanger—High-Pressure Steam

When high-pressure steam is the heat source some additional considerations are necessary. The control valve must be made of materials suitable for the temperatures and pressures encountered. Relief valves should be provided on both the steam and water sides of the exchanger. And a high-limit water controller is required. (See Figure 7–52.) Two-position control is not satisfactory due to the high temperature differentials. Modulating control must be used, and even then there is a tendency to instability and hunting unless a wide throttling range is used. Some of the problems may be minimized by using a pressure-reducing valve in the steam supply line. This makes the flow rate easier to control but the large temperature differences remain. Care must be exercised in designing and applying these control systems.

7.15.3 Heat Exchanger—High-Temperature Water

High-temperature water (HTW) poses even greater problems to the designer of heat exchanger controls. A typical HTW central plant produces water at 400°F. In order to keep the water liquid at this temperature, the system must be pressurized to 350 psi or higher. These temperatures and pressures require careful selection of the control valves and piping materials.

An additional consideration is that, in order to save pumping and piping costs, the high-temperature water must have every possible Btu extracted from it at the point of use. That is, it must be cooled as much as possible, 250°F being a typical return water temperature. Heat exchangers are therefore selected for small final temperature differentials between the high-temperature water and the fluid being heated. "Cascading" is common, with

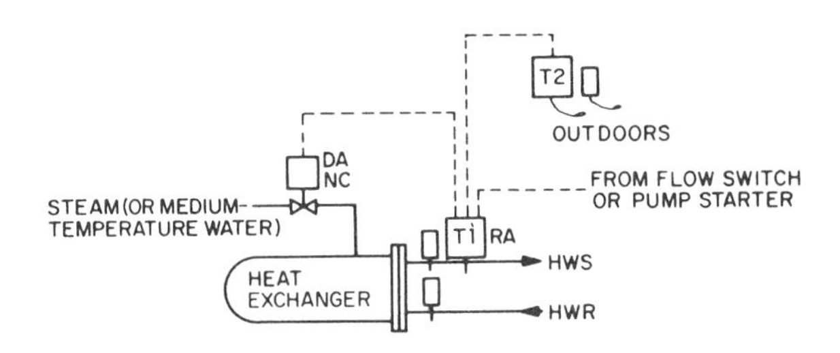

Figure 7-52 Heat exchanger; high-pressure steam.

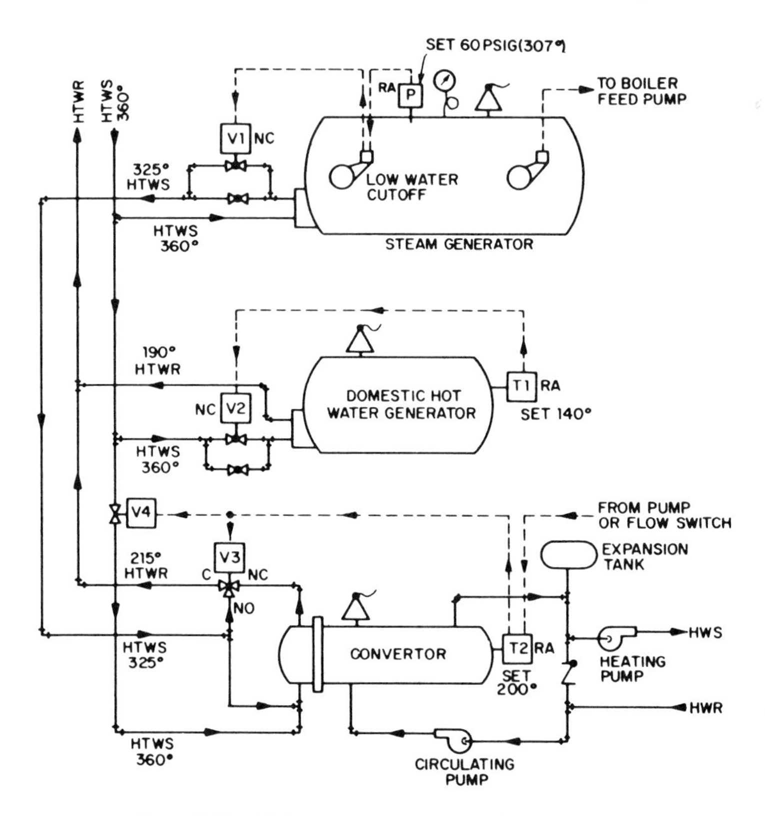

Figure 7-53 High-temperature water end-use controls.

low-temperature heat exchangers using the HTW last in a sequence which may start with low-pressure steam generators and proceed to building heating water exchangers.

Relief valves are required and control valves are commonly applied on the leaving side of the exchanger to minimize operating temperatures except for domestic hot water generators, where safety requires the control valve to be on the upstream side. Modulating control is used.

Figure 7–53 shows a typical set of HTW exchangers for a large building. The HTW temperatures shown are those that will exist at design loads and water flows. In operation they will, of course, vary somewhat. Notice the "cascade" effect from the steam generator to the building heating converter. So long as the steam generator is at full capacity, enough water will flow through valve V1 to provide full capacity to the convertor. If the convertor is at part load, some of the water will be bypassed through three-way valve V3, and the HTW return temperature will be greater than 215°. If the convertor load is high and the steam generator load is low, then T2 will sense the falling building hot water supply temperature and open valve V4, providing additional HTW. Valve V4 is selected with a high pressure drop at design flow to match the pressure drop through the steam generator and valve V1.

The domestic hot water generator is not in the cascade, since its loads are so different from those of the other devices.

The manual bypass valves may be used in case of difficulty with the automatic controls.

Control of expansion and maintenance of system pressure are critical. A drop in system pressure results in a lower water temperature, creating a cycle which causes shutdown of the plant. For a full discussion of this phenomenon and methods of control refer to the ASHRAE HANDBOOK, SYSTEMS volume. One method of controlling expansion, pressure and make up is shown in Figure 7–54. The expansion tank (or two or more tanks in parallel) is sized to suit the system, with a wide differential between high-level drain control and low-level makeup control. The steam boiler provides a steam cushion to maintain the required pressure. The makeup pump or pumps will have a small flow rate at a high head. The makeup water meter is useful for noting unusual makeup requirements which would indicate system leaks or malfunctions. A backflow preventer (not shown) is needed on the makeup water.

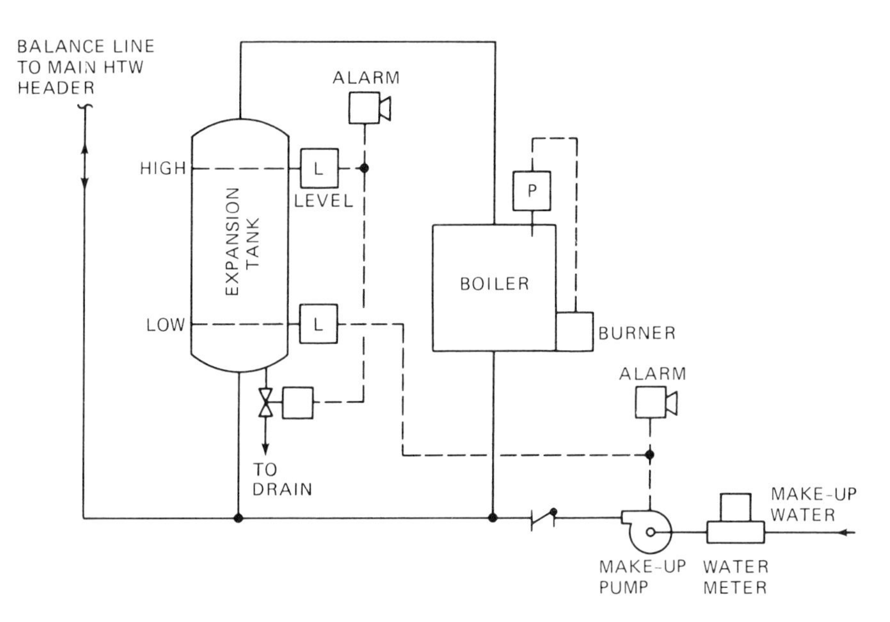

Figure 7-54 High temperature water expansion-pressurization system.

7.16 SOLAR HEATING AND COOLING SYSTEMS

Solar energy systems come in a variety of arrangements and usages but all have three essential elements in common—collector, storage, and distribution.

7.16.1 Elementary Solar Heating System

The elementary solar heating system is shown in Figure 7–55. A water-to-water system is shown, but air-to-air and water-air systems are also used. The differential thermostat allows the collection pump to operate only when the collector temperature is at least 4 to 5° warmer than the storage. In modern water-to-water systems the collector loop will use an ethylene glycol solution to prevent night-time freezing. If this is not done, an automatic collector drain and fill system must be used, which leads to air venting problems.

The space thermostat operates the distribution pump as required to maintain space temperature, while the low limit thermostat operates the auxiliary heating system when usage exceeds storage.

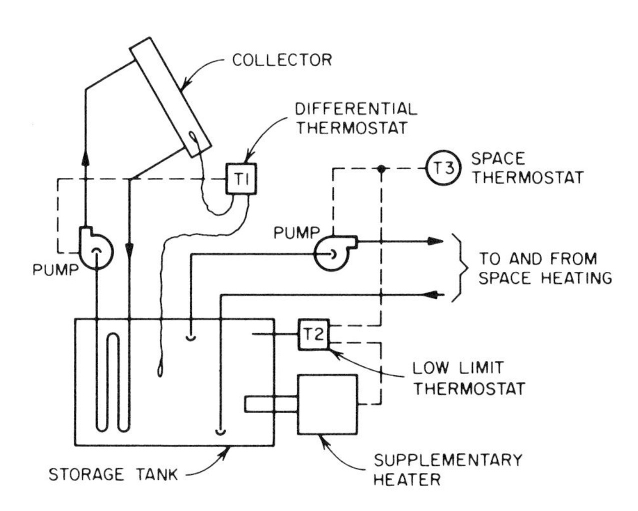

Figure 7-55 Elementary solar heating system.

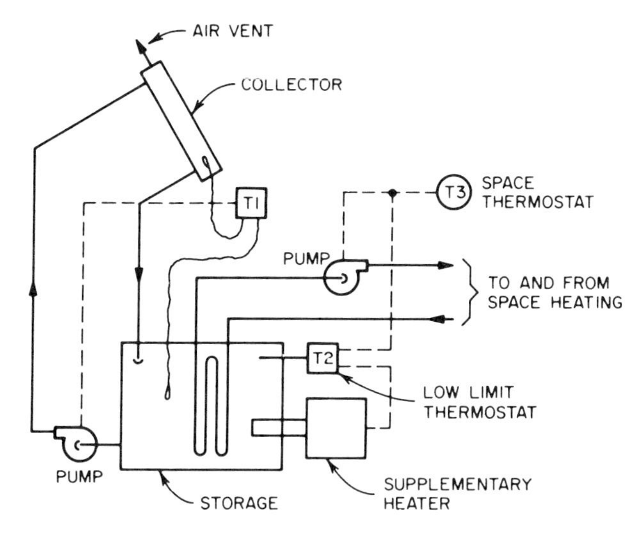

Figure 7-56 Self-draining collector.

7.16.2 Self-Draining Collector

The self-draining collector shown in Figure 7–56 is frequently used for heating swimming pools, where the pool itself is the storage tank. The collector is a covered tray across which the water flows in an open stream, flowing back to storage by gravity. This system can also be used for domestic water heating, using a heat exchanger as shown and eliminating the need for an ethylene glycol loop with its possible cross-contamination. The differential thermostat is the only control needed.

7.16.3 Absorption Refrigeration with Solar Heating

Figure 7–57 shows a combination heating and cooling system using absorption refrigeration. To operate efficiently the absorption refrigeration system needs water at 200°F or higher, therefore a "concentrating" collector is recommended. The heating-cooling type space thermostat operates the circulating pumps and positions the three-way valves for proper operation. Notice that the three-way valves allow the stored hot water to be used directly for heating the space, or through the refrigeration system to produce chilled water for cooling the space. The absorption systems in-

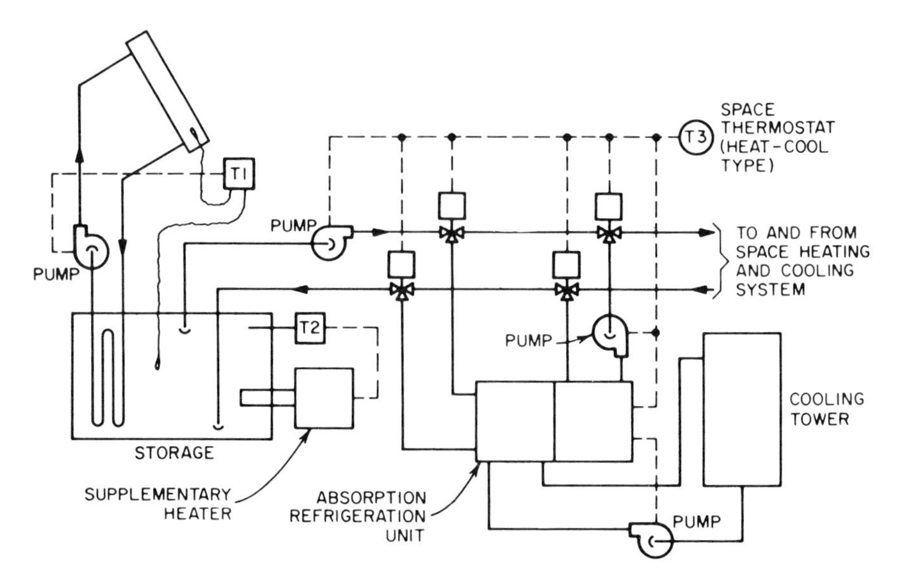

Figure 7-57 Absorption refrigeration with solar heating (two-pipe system).

cludes factory installed controls for control of chilled water temperature, condensing water interlock, and other operating functions. A two-pipe system is shown, but a four-pipe arrangement is also possible.

7.16.4 Solar Assisted Heat Pump

A very efficient way of utilizing solar energy for heating is the "solar assisted heat pump," shown in Figure 7–58. The elements of the system are the collector, heat exchanger, storage tank, package water chiller (the "heat pump") with water-cooled condenser, cooling tower for use on the cooling cycle, air handler heat exchanger coil, and the necessary pumps and valves. The table summarizes the status of the various elements on each control cycle.

Whenever the stored water is not warm enough to heat the building directly, this water is diverted to the chiller section of the heat pump where it is cooled and returned to storage. The higher temperature condensing water conveys the heat (plus the heat of compression) to the air handler. Use of the heat pump provides more storage by increasing the temperature range available and improves the efficiency of the collector by lowering the collector temperature. The system is effective with storage temperatures as low as 40°F. During the cooling season the system operates in a conventional man-

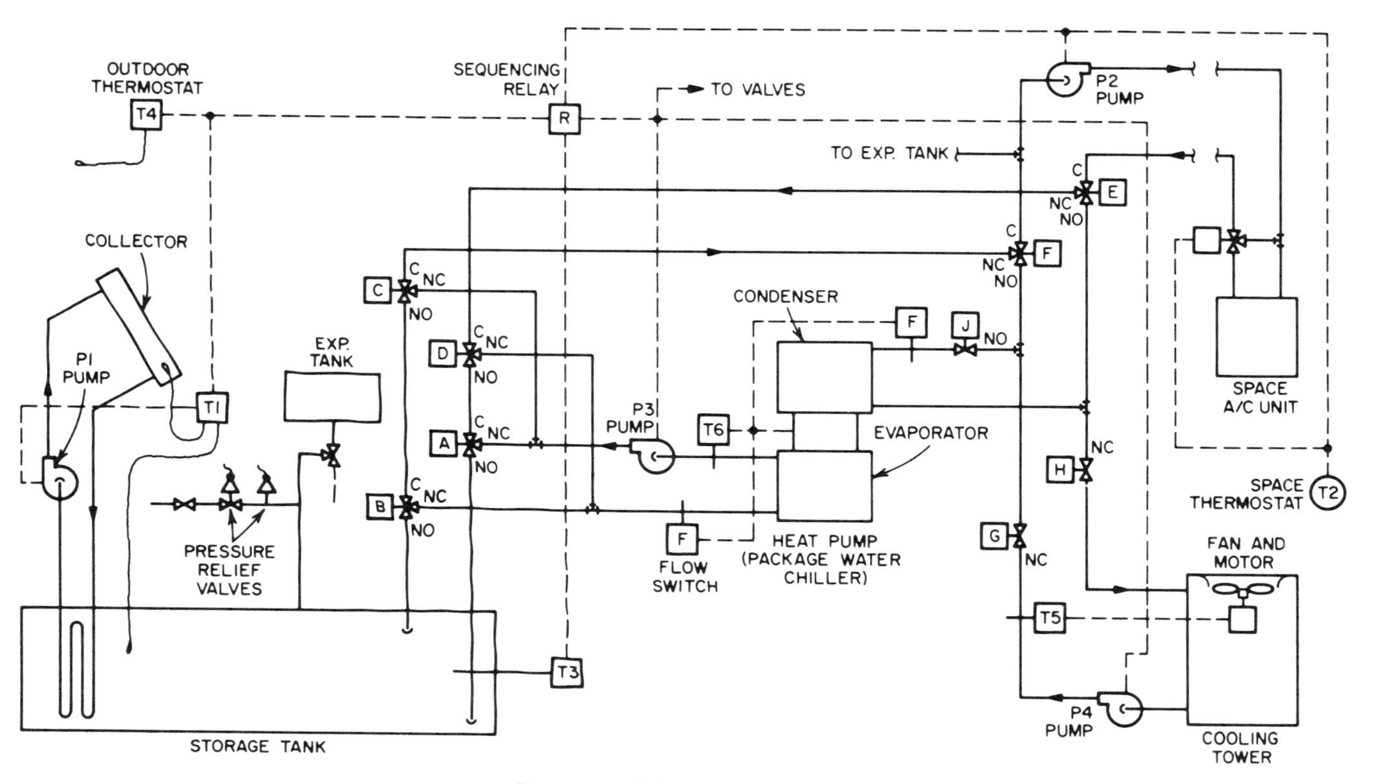

Figure 7-58 Solar assisted heat pump.

STATUS TABLE FOR FIGURE 7-58

	VALVE POSITIONS					PUMPS				
Cycle	A–B	C–D	E–F	G–H	J	P2	P3	P4	Tower Fan	Heat Pump
Solar heating	N	N	E	N	N	Note 3	OFF	OFF	OFF	OFF
Heat pump heating	E	N	N	N	N	Note 3	ON	OFF	OFF	Note 5
Cooling	N	E	E	E	N	Note 3	OFF	ON	Note 4	Note 5
Tower	N	N	N	E	E	Note 3	OFF	ON	Note 4	OFF

Notes
1. N = Normal, deenergized.
2. E = Energized.
3. Pump P2 controlled by T2.
4. Tower fan controlled by T5.
5. Heat pump controlled by T6 as a low limit controller.

ner, using the cooling tower for heat rejection. In the arrangement shown, the cooling tower water can also be used directly for cooling. Some variations on this allow storage of cooled water by running the cooling tower at night.

7.17 SUMMARY

In the preceding discussions the energy source has generally been ignored. Obviously, many of the functions, such as direct control of motors and solenoid valves require electrical energy. The next chapter considers the methods of describing electric control circuits and interfacing those circuits with other types of energy.

8 Electric Control Systems

8.1 INTRODUCTION

Sooner or later (usually sooner) the HVAC system designer comes to the necessity for starting and stopping electric motors for fans, compressors, boilers, pumps and accessories. And, all too often, he feels that these things are the responsibility of the electrical engineer. So, having graciously informed said electrical engineer of the location and size of these motors, the system designer goes away satisfied that he has done his part.

Unfortunately, this is not enough, as becomes evident when the equipment is installed without the necessary interlocks to the temperature control system.

Obviously, the electrical engineer must size and select the wire, conduit, starters, disconnects and switchgear necessary for supplying power and control to the motor. And this information must appear on the electrical drawings for the edification of the electrical contractor who is to install it. But these people have no way of knowing what the control system needs unless we, as system engineers, tell them explicitly and in language they can understand.

So we must learn how to express electrical equipment and wiring diagramatically using standard electrical symbols and interfacing between the electrical and temperature control systems with proper relays and transducers.

8.2 ELECTRIC CONTROL DIAGRAMS

Consider, then, some of the conventions and symbols used to convey electrical information. Figure 8–1(A) is a point-to-point or graphic schematic of a motor starter with momentary contact pushbuttons for manual start-stop control. The same schematic is shown in "ladder" form in Figure 8–1(B). Either of these forms is acceptable, but the ladder schematic is more frequently used and is easier to follow.

The symbol (1M) designates the solenoid coil in the starter, which, when energized, actuates the power and auxiliary contacts. The related contacts are identified by the same number, especially when "spread out" in the ladder schematic. The symbol —| |— indicates a "normally open" contact. That is, when the coil is not energized, the contact is open. When the coil is energized, it closes. The symbol —|/|— indicates the opposite or "normally closed" contact. The symbol —∿— indicates the heater portion of the thermal overload relay in the starter. Next to each in the graphic schematic is a nor-

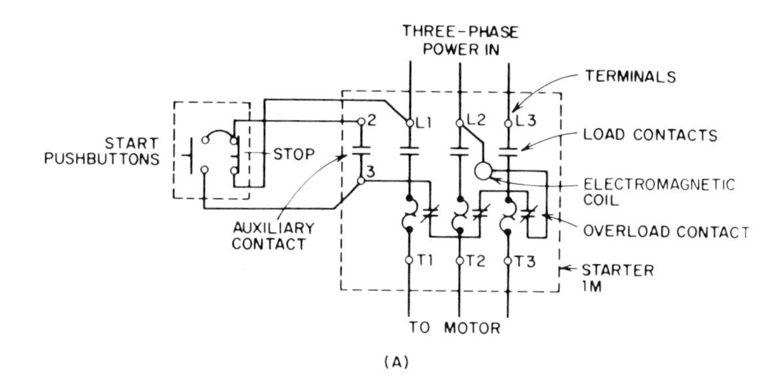

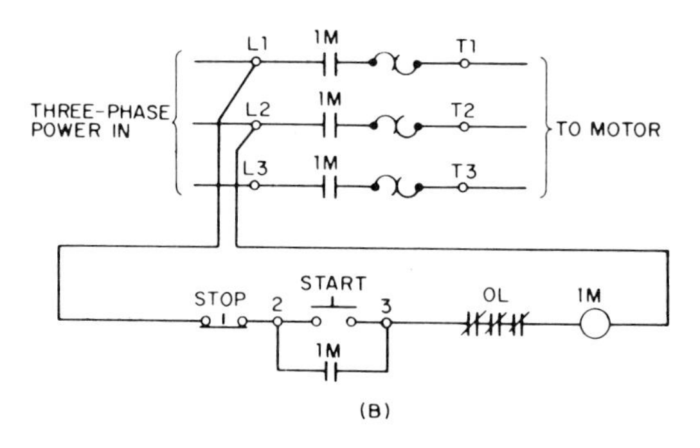

Figure 8-1 Motor starter with pushbuttons. (A) Pictorial diagram. (B) Ladder diagram.

mally closed contact. When excessive current is drawn by the motor, for any reason, this causes the heater to heat up and open the contact. After the trouble is found and corrected, the contact may be reclosed (reset) manually. Various types of overload relays are available, depending on the size and voltage of the motor.

Notice that the auxiliary contact in the starter is wired in parallel with the "start" pushbutton. In this relation it serves to maintain the circuit after the start button is released, and is therefore called a "maintaining" or "holding" contact. When the motor stops for any reason the holding circuit opens and the motor will not restart until the "start" button is pushed.

In this and the other diagrams in this chapter the overload contacts have been shown wired on the "hot" side of the coil in accordance with recommended practice. Most starters, however, are factory wired with the overload contacts on the "ground" side of the coil. This is satisfactory if there is no ground, as may happen with some three-phase circuits. When there is a ground side to the control circuit, then the wiring should be changed to be in accordance with the diagram.

Figure 8–1 showed a typical manual system, with control power taken directly from the motor power source. Since the motor power source is frequently high-voltage (230 V or 480 V for pumps and fans, sometimes much higher for large centrifugal compressors), it is often considered desirable to provide a low-voltage source for control power. This may be 120 V or even 24 V or 48 V and may come from a separate source or from a control transformer in the starter. The advantage of the control transformer is that opening the disconnect switch interrupts all power, whereas with a separate control power source it is necessary to open two switches to interrupt all power. With complex, interlocking control systems, a separate power source is sometimes necessary.

Figure 8–2 shows a typical starter with a control transformer for supplying low-voltage control power. This is almost identical to Figure 8–1(B), the only difference being the addition of the transformer. This is necessary for proper selection of relay and starter coils.

The motor may also be controlled by a hand-off-automatic switch (Figure 8–3). Thus it may be started and stopped manually as needed, but is normally operated by a pilot-device contact in the "auto" circuit. This arrangement is used where motors are interlocked with other motors, or with temperature, pressure or flow controls.

"Pilot" or indicating lights are often used to indicate the on or off condition of a device. For "on" indication it is desirable to use a flow or pressure switch in the circuit to give a positive assurance of operation, since the mere fact of power being delivered to the starter does not necessarily prove that the motor is running, or, if running, is effectively moving a fluid or otherwise performing as desired.

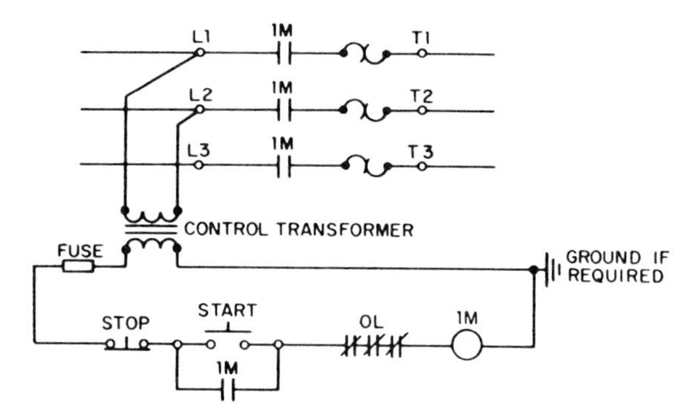

Figure 8-2 Motor starter with low-voltage control circuit.

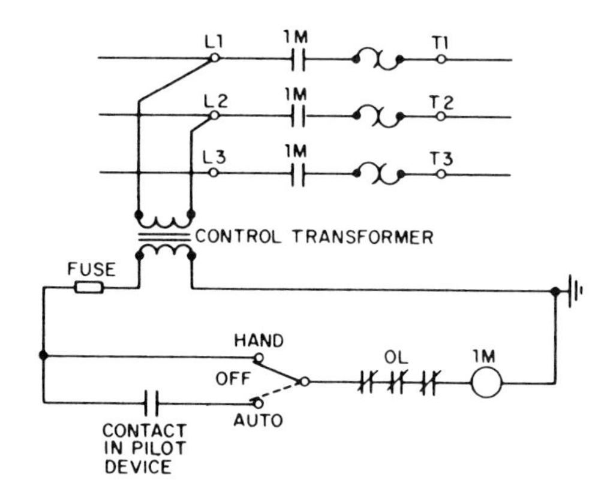

Figure 8-3 Motor starter with hand-off-auto switch.

Figure 8–4 shows the same control as Figure 8–3 but with the addition of a "running" pilot light. This figure also shows the fused disconnect switch and the motor.

8.3 ELECTRICAL CONTROL OF A CHILLER

To illustrate a relatively simple and basic interlock system consider the diagram in Figure 8–5. Here are shown a water chiller, a compressor, a circulating chilled water pump, a condensing water pump, a cooling tower fan and the necessary relays and safety controls. The power for the control circuit is obtained from a control transformer in the compressor starter, since this is the critical piece of apparatus. To place the system in operation the chilled water pump is started manually by means of a "start"

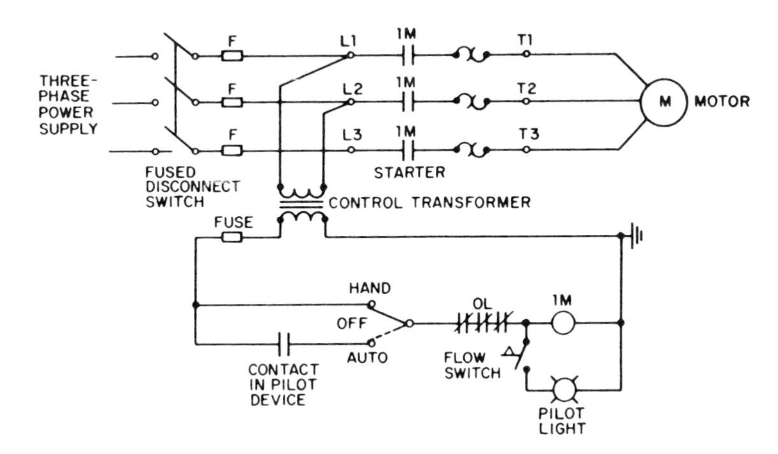

Figure 8-4 Motor starter with pilot light.

pushbutton. As the water flows in the system, its temperature leaving the chiller is measured by a two-position thermostat. When the water temperature is above the thermostat setting, the thermostat contact closes, opening the solenoid valve in the refrigerant liquid line to the evaporator. The resulting rise in the suction pressure will close the low-pressure switch, starting the condensing water pump. If water is flowing in both chilled and condensing water circuits (flow switches closed) and all safety controls are closed, then the compressor motor will start.

The cooling tower fan is started and stopped by a thermostat in the condensing water supply, so that the condensing water will not get too hot or too cold.

When the chilled water thermostat is satisfied, its contacts will open, closing the solenoid valve, but the condensing pump and compressor will continue to run until the system "pumps down," that is, until the decreasing suction pressure opens the low-pressure switch. This "pump down" cycle is accomplished by the pump down relay (1CR).

Numerous safety controls are provided to protect the equipment against operating under adverse and potentially damaging conditions. The oil pressure switch contains a heater which is energized when the compressor starts. It is necessary for increasing oil pressure to open a pressure switch in the heater circuit before the heater opens a thermally delayed contact in the compressor starter circuit.

The compressor may also be stopped by too high a condensing pressure, too low a suction pressure, high refrigerant temperature, low chilled water temperature or inadequate flow of chilled or condensing water; or, of course, by the thermal overloads in the motor starter.

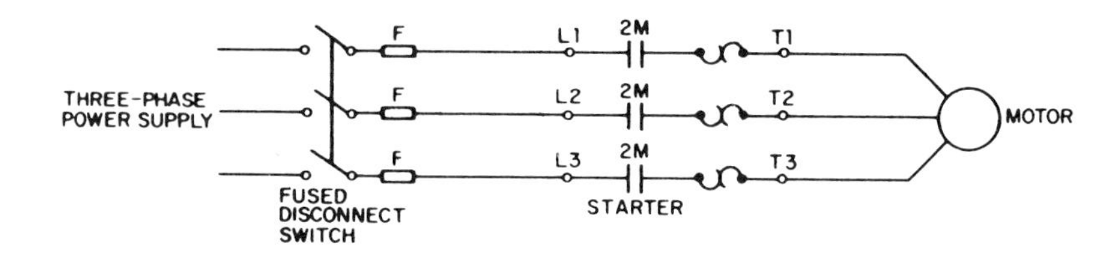

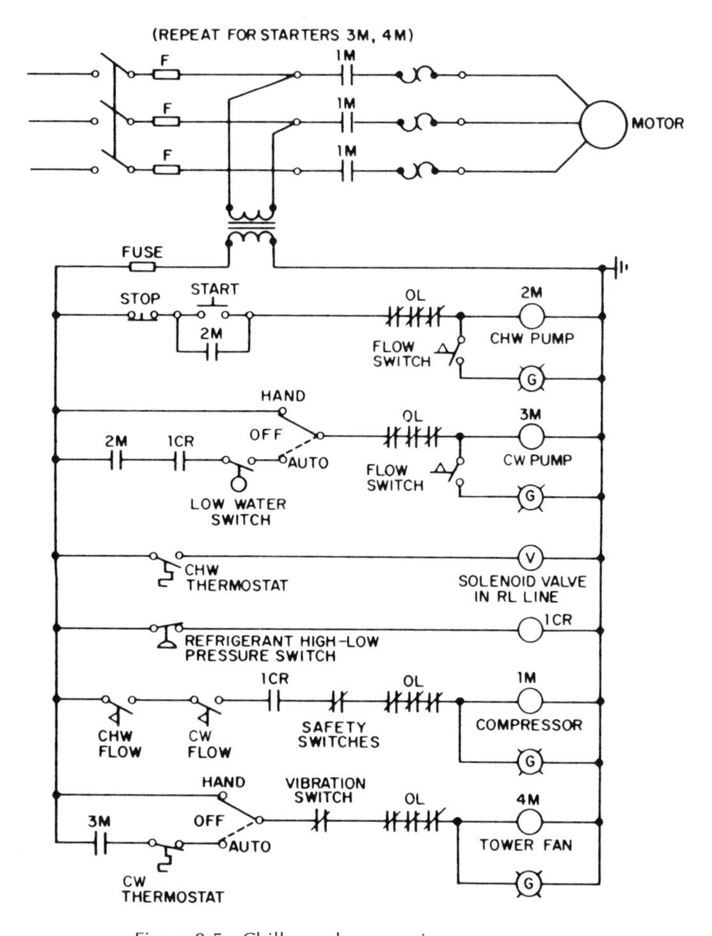

Figure 8-5 Chiller and accessories.

A float switch in the cooling tower sump will stop the condensing water pump if the water makeup system fails and the water level gets too low. A vibration switch will stop the cooling tower fan if the fan blades become damaged or get out of alignment.

The control system just described contains a number of conventional symbols for various types of operating and safety switches. These and many other standard symbols for electrical devices are in common use but not everyone uses the same standards. It is necessary to define exactly what your symbols mean.

8.4 ELECTRICAL CONTROL OF AN AIR HANDLING UNIT

Figure 8–6 shows a simple air handling unit electrical control sequence. Control power is obtained from a control transformer in the supply fan starter. The supply fan is started manually. Interlocks provide for

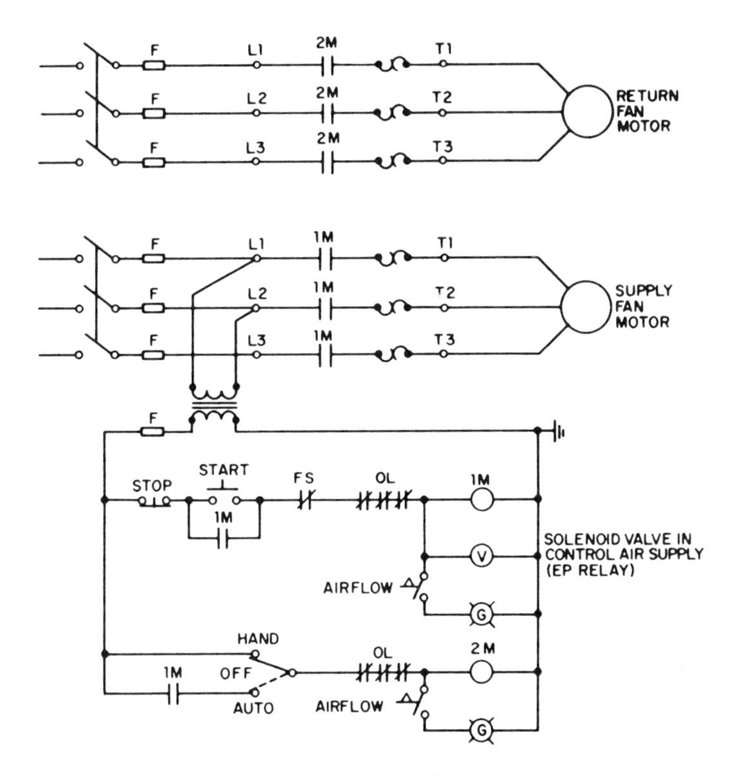

Figure 8-6 Air handling unit with return fan.

operating the return fan and supplying power to the temperature control system whenever the supply fan is running. The diagram shows an air solenoid for supplying compressed air to the temperature controls, but this could also be an electric relay for supplying power to electric or electronic temperature controls.

8.5 EXAMPLE: A TYPICAL SMALL AIR CONDITIONING SYSTEM

Consider now that we are designing the controls for a small commercial-type building air conditioning system, and we need to communicate with the electrical engineer.

The first step is reverse communication: We ask him what are the characteristics of the power system he is supplying—voltage, phase, cycles. He may answer that he is providing 120-208 V three-phase, 60-cycle four-wire, which gives us a choice for our motors of either 120 V single-phase or 208 V single- or three-phase. It is common practice in the industry to use the higher voltage and three-phase when available for all "large" motors. "Large" generally means ½ or ¾ hp and larger. The higher voltage and the use of three-phase power improves motor efficiency and thus lowers operating cost.

For this example our system will include an air handling unit with a 5 hp motor, a 2 hp return air fan, a ⅛ hp toilet exhaust fan, a 15 hp water chiller, a 5 hp air-cooled condenser, a boiler, two 1 hp pumps and a ¼ hp air compressor for the pneumatic temperature control system.

The schematic ladder diagram would appear as in Figure 8–7. We indicate the chiller safety and operating controls by the notation "See Manufacturer's Wiring Diagram," since all package chillers are factory wired and the arrangement varies from one manufacturer to another. We might even have to correct this diagram "as-built" to accommodate the actual equipment requirements. All the other equipment layouts are similar to what we have previously discussed. The air-cooled condenser fan is interlocked with and to the chiller in much the same relationship as the condensing water pump in Figure 8–5. The boiler is assumed to be a small, low-pressure type with an atmospheric burner and a built-in factory-wired control system, for which we provide 120 V power.

It is also desirable to have a schedule of motor starters and controls, such as Figure 8–8. We fill in part of this, as shown, and give it to the electrical engineer along with the schematic. (The "NO" and "NC" contacts are the auxiliary contacts required in the starter.) He can then fill in starter sizes, make other electrical decisions, and reproduce everything on the electrical drawings. And we now have assurance that the finished job will perform electrically to the requirements of our design.

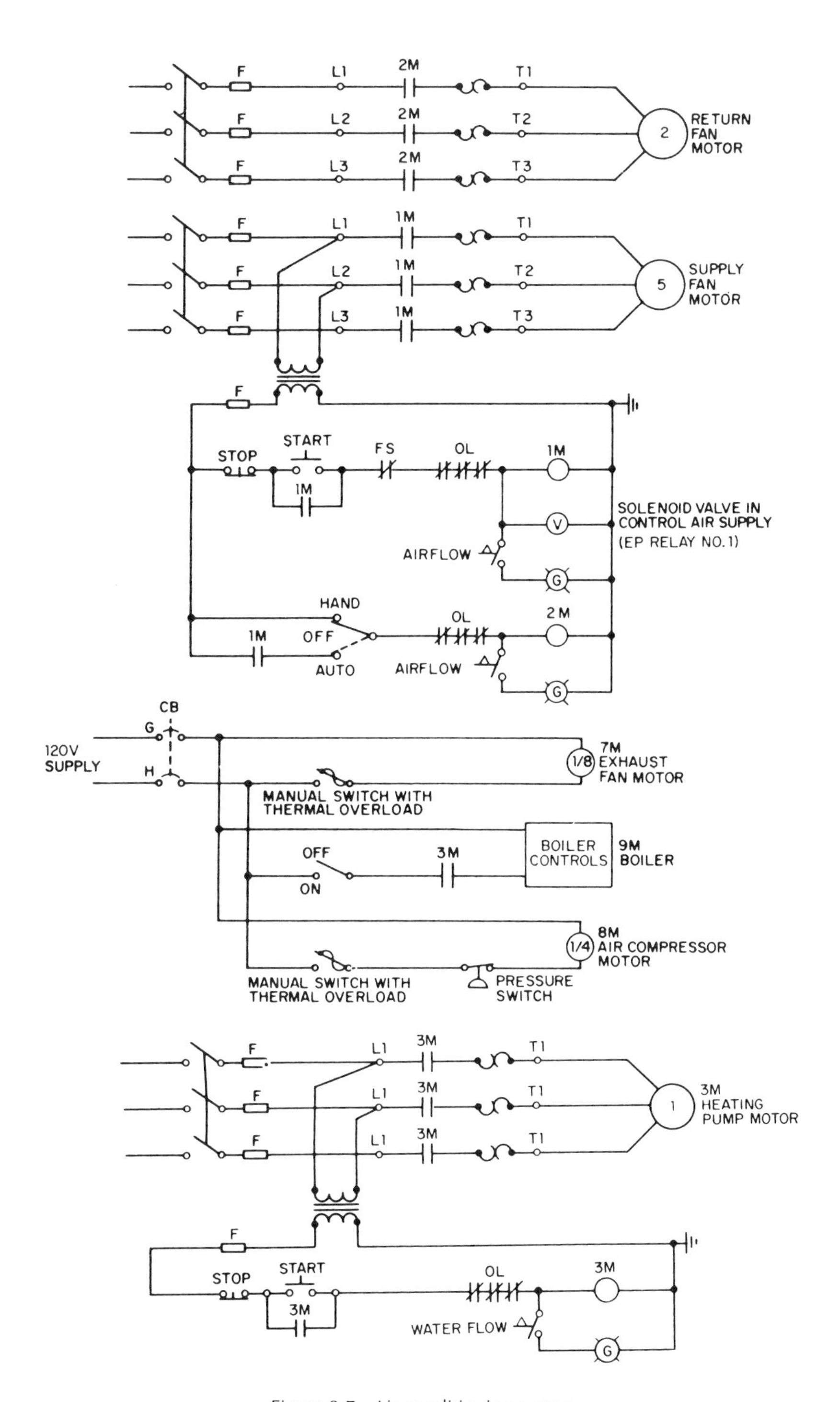

Figure 8-7 Air conditioning system.

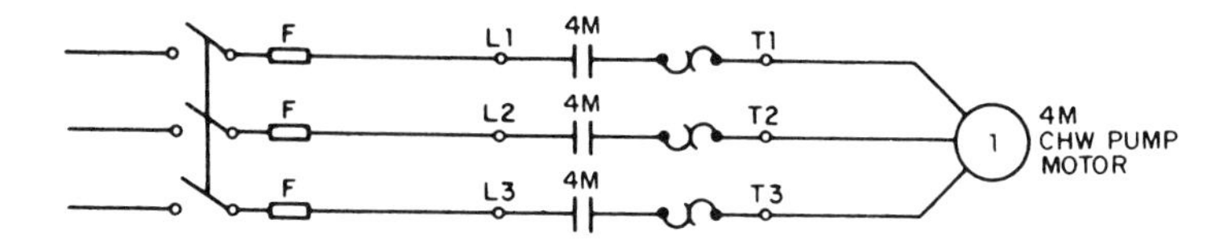

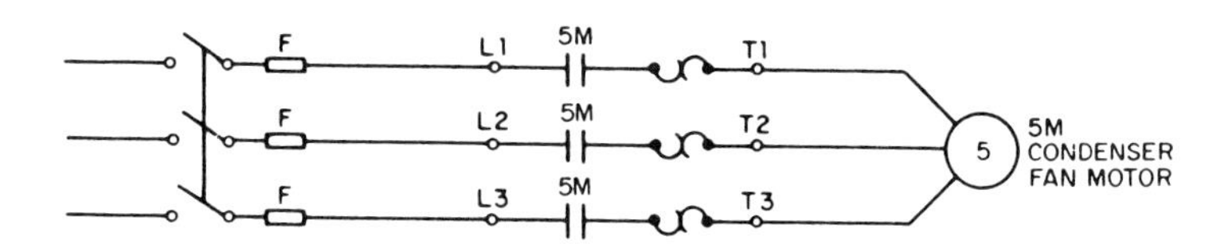

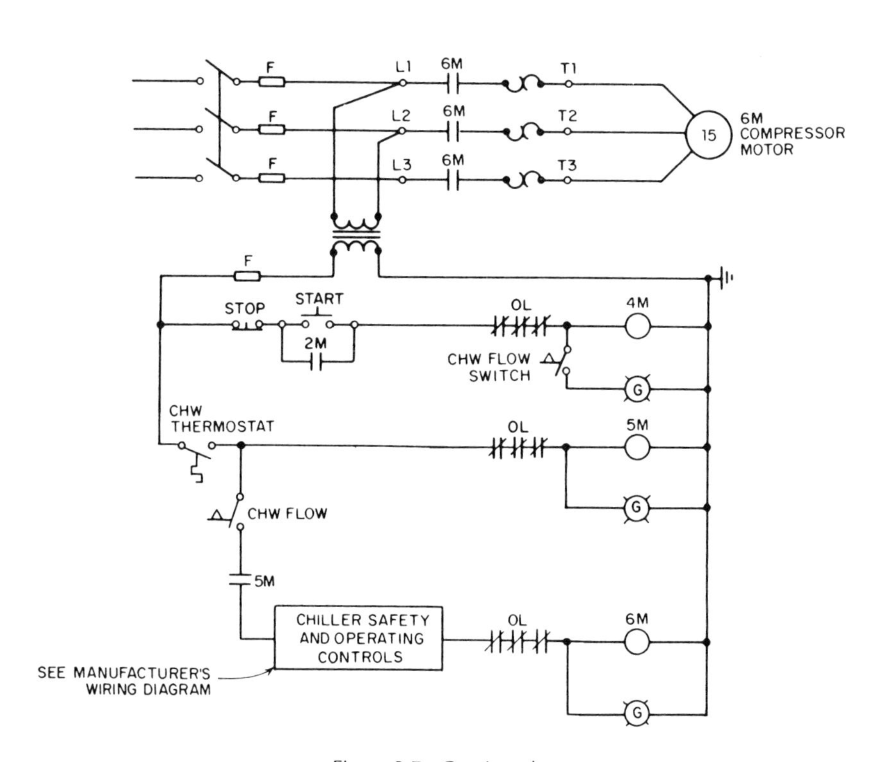

Figure 8-7 *Continued*

	Mechanical Symbol	1M	2M	3M	4M	5M	6M	7M	8M	9M
	Mechanical Equipment Served	Supply fan	Return fan	Heating pump	CHW pump	Condenser fan	Compressor (refrig.)	Exhaust fan	Air compressor	Boiler
MOTOR DATA	Horsepower	5	2	1	1	5	15	1/8	1/4	—
	Voltage	480	480	480	480	480	480	120	120	120
	Phase	3	3	3	3	3	3	1	1	1
	K.W. input									0.20
	Windings									
STARTER DATA	Type							none	none	none
	NEMA size									
	Number of speeds	1	1	1	1	1	1	1	1	—
	Control volts	120	120	120	120	120				
	Pushbutton	S-S		S-S	S-S	—				
	Selector switch		HOA			—	Controls by manufacturer			
	Pilot lamp	G	G	G	G	G				
	Normally open contact	2		2	1	1				
	Normally closed contact	—	—							
REMOTE CONTROL	Control volts						120	120	120	120
	Pushbutton									
	Selector switch							on-off	on-off	on-off
	Pilot lamp							—	— Press switch	—
RELAYS	PE relay number									
	EP relay number	1								
	Cont. relay									
INTERLOCK FROM			1M				5M CHW flow			3M

Figure 8-8 Motor starter and control schedule.

While this illustration is necessarily simple, the principles can be applied to a system of any size and complexity. Complications are limited only by the designer's ingenuity. But, as stated before, it is best to avoid complexity if possible.

If your knowledge of electric circuitry is limited, it is strongly recommended that you study a good basic text on the subject. The more common errors among inexperienced designers are placing two or more "loads" in series (none of them work properly except in special cases) or having no load in a circuit (short circuit). The ladder schematic helps to show these errors readily. Another, not so easily seen problem is a false or "feedback" circuit, which can give full or, more often, partial control power where it is not desired. These are perpetrated even by experienced designers and arise when one circuit feeds two or more loads in parallel. If there is any question about the possibility of such a problem, use a relay or relays to isolate the extra loads.

8.6 ELECTRIC HEATERS

In Chapter 6 the control of electric heaters was dealt with in very generalized terms. Figure 8–9 shows a detailed electric control and power circuit diagram for a single-phase, single-stage heater, with manual and automatic reset high-limit controls, an electric thermostat, an airflow switch, and a relay-contractor. Figure 8–10 shows controls for a three-phase, two-stage heater, with high-limit controls, two-stage thermostat and contactors. There are many possible variations to these basic arrangements, depending on the heater size and sequencing requirements.

8.7 REDUCED-VOLTAGE STARTERS

So far in this chapter only small motors and across-the-line starters have been considered. The "inrush" or starting current on any motor is three to six times the normal running current. As motors get larger, the local electric utility requirements and good engineering practice require that we take steps to reduce this inrush current. This is done by means of reduced-voltage starters. The motor size above which reduced-voltage starters are used is a function of the voltage (higher voltages mean less current for the same horsepower) and overall size of the building electrical service. The electrical engineer can advise you of the requirements for any particular job.

Reduced-voltage starters are of several types. Some require special matching motors, others work with "standard" motors. Remember that any reduction of inrush current also reduces starting torque.

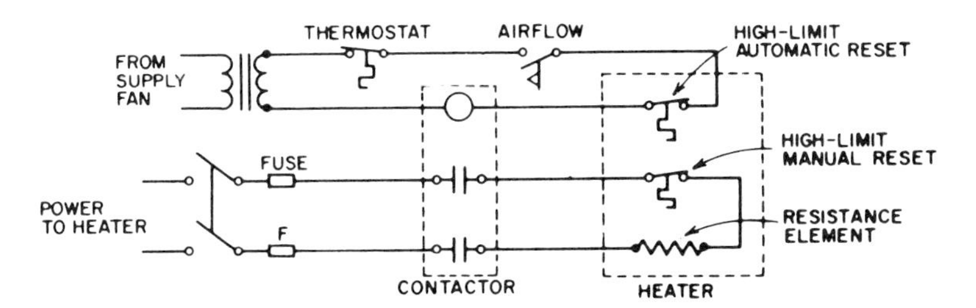

Figure 8-9 Electric heater; single-phase, single-stage.

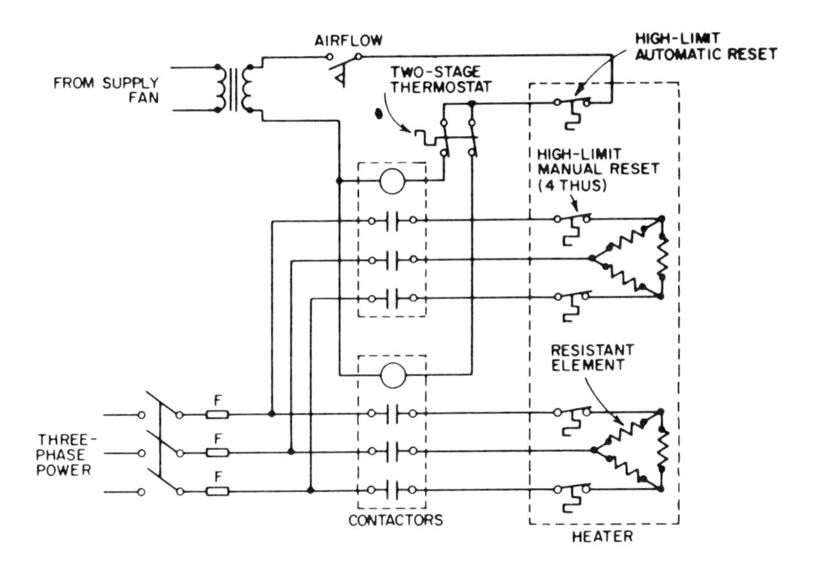

Figure 8-10 Electric heater, three-phase, two-stage.

8.7.1 Part-Winding Starter

Perhaps the simplest reduced-voltage starter is the part-winding starter. This must be used with a special type of motor—a part-winding motor. Figure 8–11 shows the circuit schematically, including the special motor winding. When the "start" button is depressed coil 1M1 is energized and the contacts serving part of the motor windings are closed. The motor starts and accelerates for a preset time period, as determined by the timing relay (1TR). At the end of this period the 1TR contact closes and all the windings are energized. Inrush current will be about 60% of across-the-line inrush. This starter also provides closed transition, which is discussed later in this chapter.

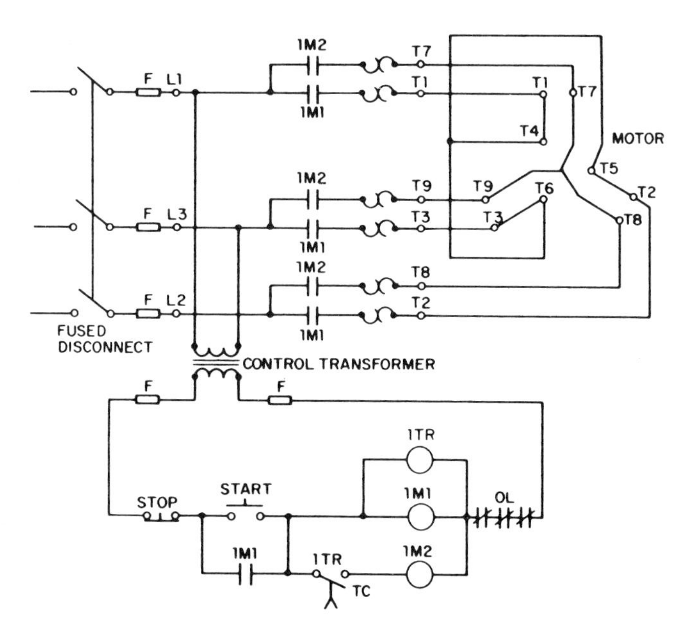

Figure 8-11 Part-winding starter.

8.7.2 Auto-Transformer and Primary Resistor Starters

These starters may be used with any standard squirrel cage motor. Inrush may be limited to as little as 50% of maximum.

Figure 8–12 shows a primary resistor starter. When coil 1M is energized, closing contact 1M, the resistors reduce the current flow during the startup and acceleration period. When 1TR times out, coil 1A is energized, closing contact 1A. This shunts out the resistors and applies full voltage to the motor. This system may be furnished with two, three or more steps of acceleration. This is also a closed-transition circuit.

Figure 8–13 shows an auto-transformer starter. This arrangement has an excellent ratio of starting torque to power input. The "start" button is depressed and energizes relay 1S, which in turn energizes relay 2S. This provides current through the auto-transformer coils in the starter, starting the motor at reduced voltage and current. After an acceleration period the 2S timed opening contact opens, deenergizing 1S, and then the 2S timed closing contact closes, energizing relay R. This shunts out the transformer coils and applies full voltage to the motor windings. 2S is deenergized, restoring those contacts to their normal position with coil R held in by a maintaining contact. Note that coils 1S and R are mechanically interlocked so that only

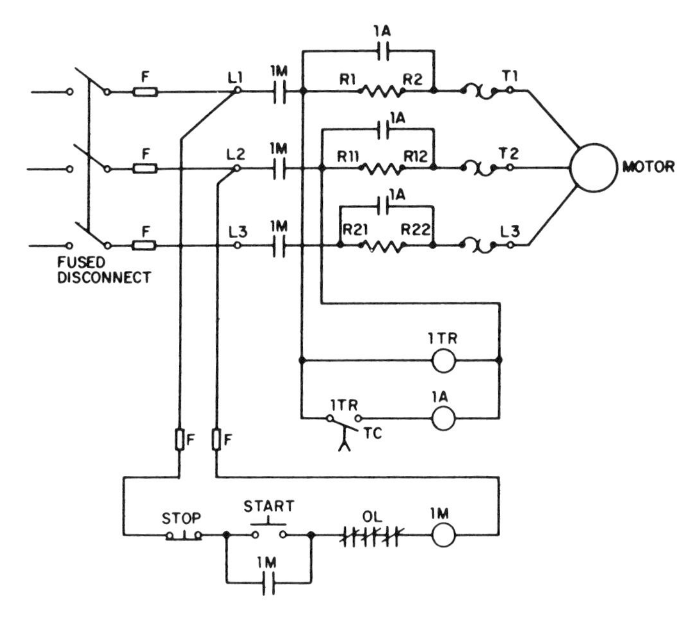

Figure 8-12 Primary resistor starter.

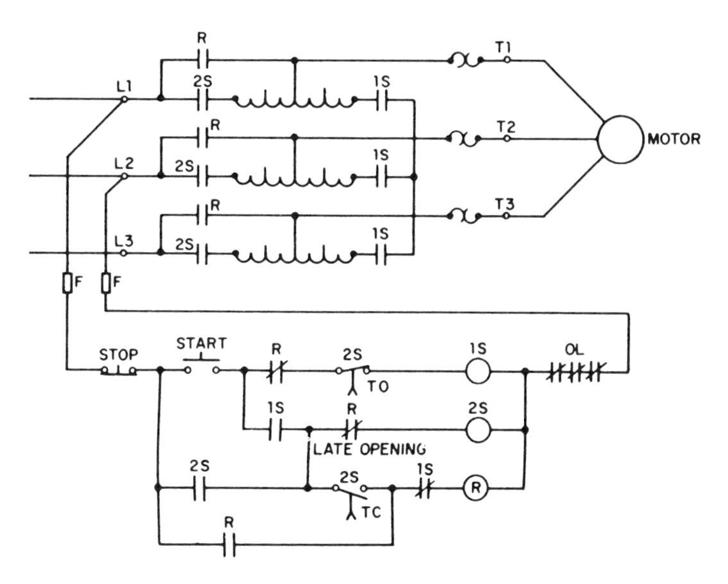

Figure 8-13 Auto-transformer starter.

one can be energized at a time. Note also that the transformer coils are tapped for various percentages of line voltage, usually in the range from 50% to 80%. The tap selected governs the inrush and starting torque.

8.7.3 Wye-Delta Starters

Wye-Delta starters require specially wound motors, but have excellent starting characteristics, especially with very large motors. They are available with both open- and closed-transition arrangements.

Figure 8–14 shows a closed-transition arrangement. A pilot control circuit (not shown) activates control relay 1CR. Contact 1CR closes and energizes relay 1S. This in turn energizes 1M1 and the motor windings are energized in a "Wye" configuration. This allows the motor to start with low current and voltage draw. After an acceleration time period determined by the setting of 1TM, contact 1TM closes, energizing relay 1A. This first shunts a portion of the current through the resistors and then the late-opening 1A contact disconnects 1S, opening the contacts in the "Wye." The NC contact in 1S recloses energizing 1M2 and the circuit is now in running configuration with motor windings connected in a "Delta" arrangement. The pilot control circuit will usually include an anti-recycle timer which prevents restarting the motor if it is stopped within 20 min of the original start. Too-frequent starting may damage the starter.

8.7.4 Solid State Starters

A new development in reduced voltage starters is the "solid state" starter. A typical arrangement is shown in Figure 8–15. The contactors found in electro-mechanical starters have been replaced by silicon controlled rectifiers (SCR's), which provide proportional control of current flow to the motor. Current is sensed by current transformers (CT) (which replace the "overloads"). Current and voltage data are fed into a controller which drives the SCR's. Current during startup can be held to a desired maximum. When the SCR's are fully "on" they offer essentially no resistance to current flow and therefore act like closed contacts. Overcurrent and low-voltage protection during operation are inherent in the control logic.

8.7.5 Open and Closed Transition

All of the starter arrangements shown in the preceding paragraphs had closed-transition arrangements; that is, there was no break in the

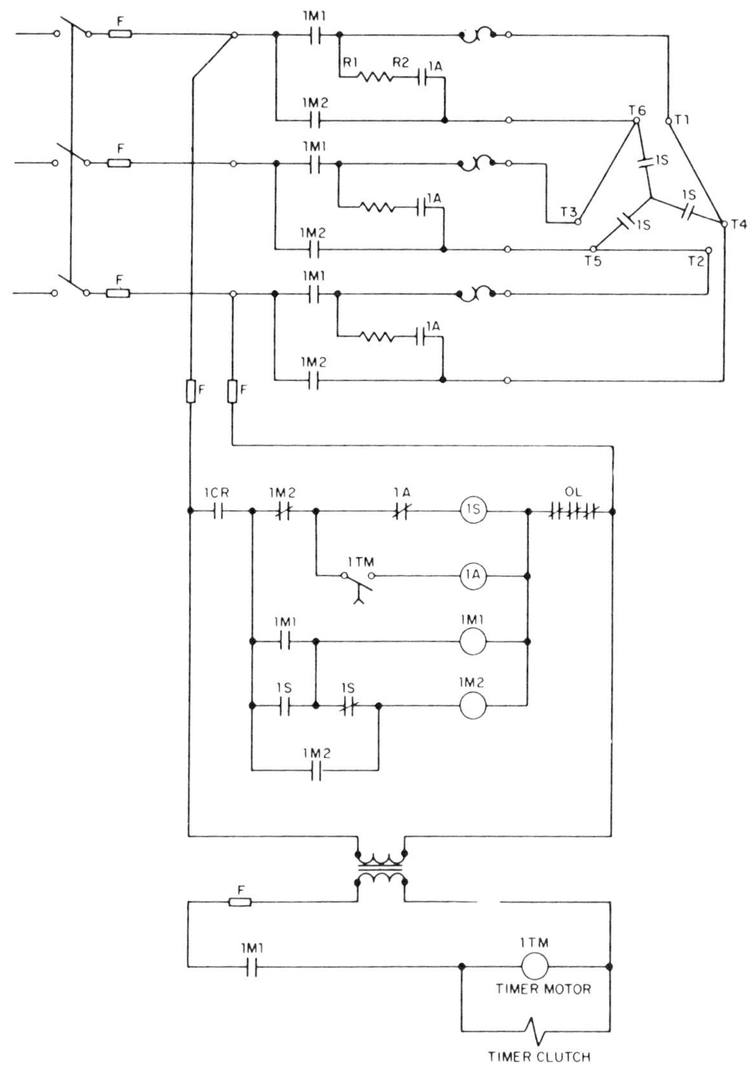

Figure 8-14 Wye-Delta starter.

current flow during the starting sequence. The switching relay contacts were arranged so that the steps overlapped.

Open-transition starters are also available. The switching is somewhat simplified because there is a momentary break in current flow at the changeover from "start" to "run" configuration. This momentary break can cause a transient current of a very high value, sometimes called a "spike." With small motors or on large electrical distribution systems, these spikes may be un-

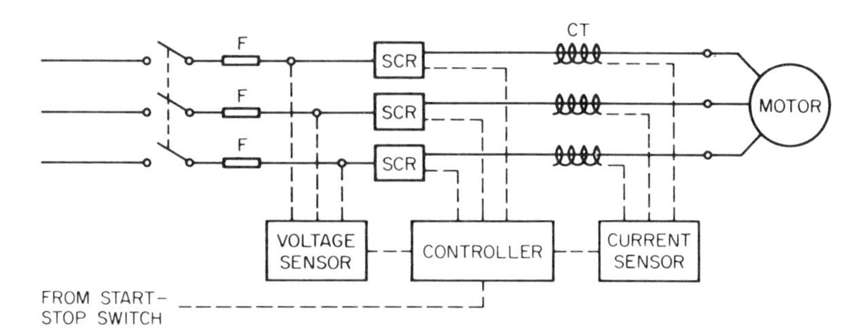

Figure 8-15 Solid state starter.

important. But many times a transient of this nature can disrupt an electrical distribution system, or cause a malfunction of a sensitive electronic device, such as a computer. Check this out with your electrical engineer.

8.8 MULTISPEED STARTERS

Two-speed fans are frequently encountered on air distribution systems or cooling towers. The motors commonly used in these applications are either variable-torque or two-winding motors. There are other types of multispeed motors, but they are seldom encountered in HVAC systems.

Figure 8–16 shows a variable-torque motor with simple two-speed control. Pushing the "slow" or "fast" pushbutton will start the motor at that speed. In addition to the electric interlock contacts, the two coils are usually interlocked mechanically to prevent damage to the motor.

Figure 8–17 shows the different starter arrangement used for a two-winding motor. The control circuit can be the same for either type of motor. The advantage of a two-separate-winding motor is that the starter circuit is a bit simpler and the motor will continue to function in an emergency on only one winding if the other is damaged. Also, two windings allow the use of any combination of speeds desired, while a single-winding variable-torque motor must operate with high speed at some fixed ratio to that of low speed, usually 2:1.

The control circuit shown in Figure 8–16 is satisfactory for small motors. However, the same starting problems occur as previously discussed under reduced-voltage starting. A "progressive" control, as shown in Figure 8–18, has the same effect as reduced-voltage starting. When the "slow" button is pushed the motor simply starts on slow speed. When the "fast" button is pushed the motor starts first on slow speed, then after a time delay for acceleration, switches to "fast." This is an open-transition starter since it is necessary to disconnect the "slow" coil before connecting the "fast" coil.

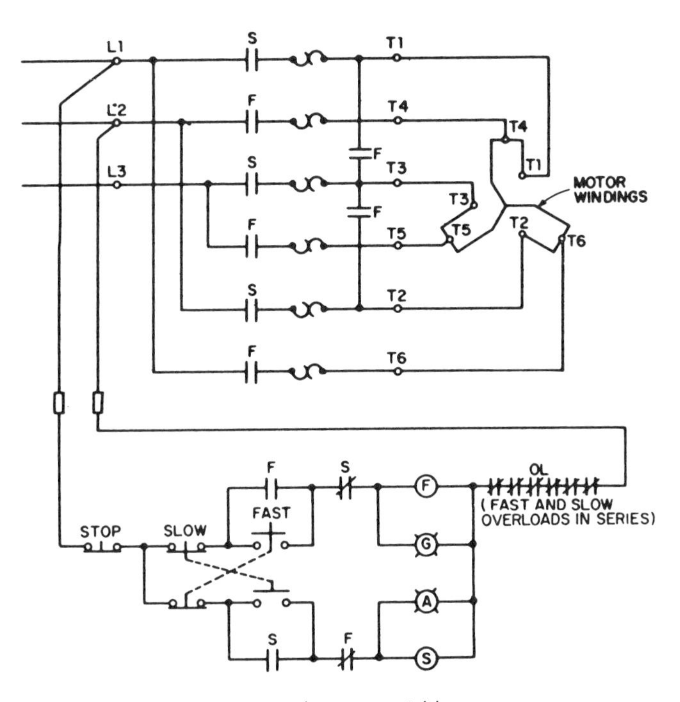

Figure 8-16 Two-speed starter; variable-torque motor.

An additional problem arises when switching from "fast" to "slow." An abrupt change will cause a dynamic braking effect, since the motor wants to run "slow" but the inertia of the fan—and this can be of great magnitude—opposes this slowdown. The resulting currents can damage the motor. It is therefore desirable to disconnect the "fast" coil and allow the system to decelerate to somewhere near "slow" speed before connecting the "slow" coil.

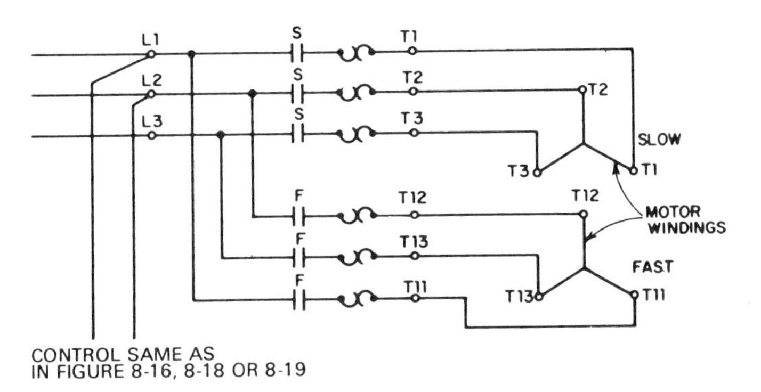

Figure 8-17 Two-speed, two-winding starter.

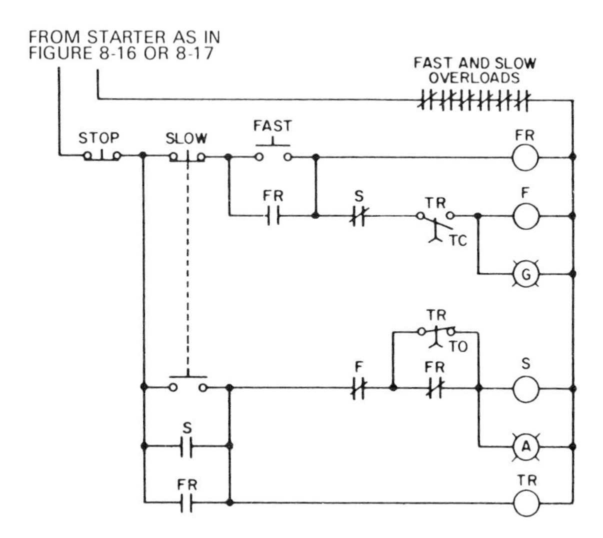

Figure 8-18 Two-speed, progressive-start.

Figure 8–19 shows a two-speed, progressive-start, timed-deceleration control circuit. When the "fast" button is pushed the motor starts on "slow" and after a time delay switches to "fast." When this occurs, relay 2TR is energized and the 2TR contact in the "slow" circuit opens. Now, when the "slow" button is pushed, the "fast" coil and 2TR are deenergized, but the 2TR contact provides a time delay to allow the motor to decelerate before the "slow" coil is energized. Relay SR provides a holding circuit during the delay period.

8.9 VARIABLE SPEED CONTROLLERS

Variable speed motors have been in use for many years. The direct current (DC) motor is inherently a variable speed device since speed is a function of current flow. However, DC motors and controllers are more expensive to purchase and install than the more common AC squirrel cage motors.

Standard AC motors may be speed-controlled by varying the input frequency to the motor. This is now done by means of "solid state" technology, employing circuits by means of which the frequency can be infinitely varied from the standard 60 Hz to any lower value. The controller responds to a varying electrical input signal and usually has an adjustable low-limit setting since damage to the motor may occur at very low speeds. The input signal

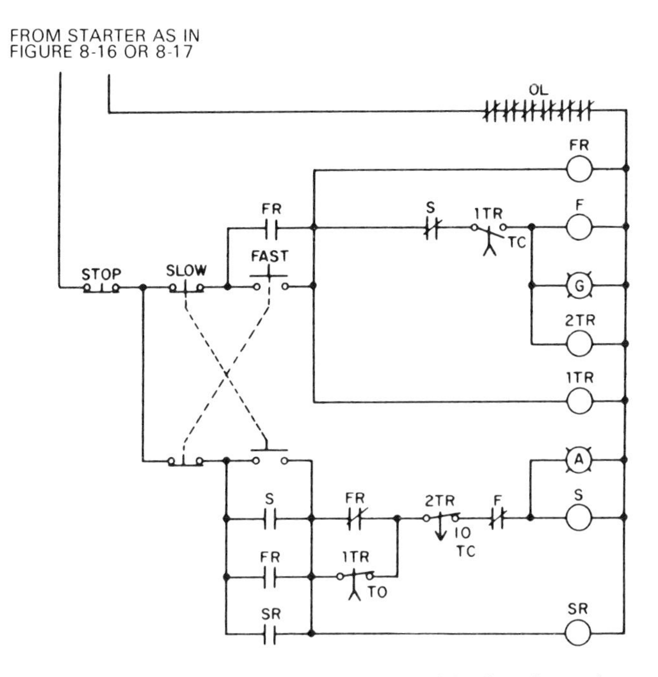

Figure 8-19 Two-speed, progressive-start, time delay from fast to slow.

may come from an electrical control system or from a pneumatic system, through a transducer as shown in Figure 8–20. This is a typical VAV system, with duct static pressure control of motor speed.

Variable speed controllers are available in capacities up to several hundred horsepower. They replace conventional starters since current flow is automatically limited to full load current and no "inrush" current is experienced.

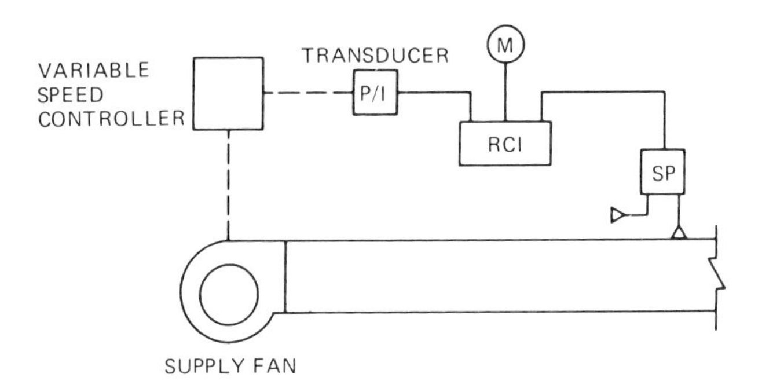

Figure 8-20 Variable speed motor controller for VAV.

8.10 SUMMARY

The discussions in this chapter have been necessarily short. It is not the intention to present a detailed treatise on electrical engineering. It is hoped, however, that you now have a better idea of the electrical problems inherent in the design of any control system and how to cope with some of them.

9 Special Control Systems

9.1 INTRODUCTION

The control systems which have been discussed up to this point might be considered "ordinary," "run-of-the-mill" or "everyday" types. They represent the majority of the work being currently designed, and necessarily, since most air conditioning systems can be adequately controlled without frills. The basic schemes discussed can be applied in many combinations to solve the great majority of control problems.

There are, however, frequent opportunities to apply our knowledge to the solution of more exotic, complex and special-purpose problems. This chapter will discuss some of these solutions in detail, to give you an idea of the possibilities in HVAC controls.

9.2 CLOSE TEMPERATURE AND/OR HUMIDITY CONTROL

Frequently encountered are criteria for extremely close control of temperature or humidity, or both. These may and do occur in a number of unrelated end uses, ranging from machine shops to hospitals to computer rooms. Not too surprisingly, these criteria are often combined with requirements for high-efficiency filtration. Let us consider some of these applications.

9.2.1 Standards Laboratory

A standards laboratory, for either primary or secondary standard comparison, always requires extremely close control of temperature in order to maintain dimensional stability in the standard gage blocks. "Close control" here means specified system differentials of 0.1° to 0.2°F. While the absolute accuracy of available temperature sensors is probably not better than plus or minus 0.5°F, the system can sometimes be designed and calibrated to control within the specified differentials.

There are many special features required to make such a system controllable. Besides a high air change rate at a low differential between room and supply air temperatures, it is necessary to maintain the interior wall, floor and ceiling surfaces at or near the room temperature. This is sometimes done by using a ventilating ceiling, with a hollow wall serving as a return plenum, all heavily insulated. One excellent method involves the use of a constant cold air supply with reheat as close to the room as possible. Reheat can be by means of hot water or electric heating coils. Steam cannot be used because of its high temperature. The hot water supply temperature cannot be greater than 80°F. A higher water temperature causes too great a rise in the supply air temperature, even though the control valve may be only cracked open. Electric heaters must be modulated by saturable core reactors, or solid-state controls such as silicon-controlled rectifiers.

Figure 9–1 shows a system for temperature control in a standards laboratory using hot water reheat. A special heat exchanger in required to provide 75° to 80°F supply water, using central hot water, steam or electricity as the heat source. This low temperature is necessary to avoid a "system gain" which would make the system uncontrollable. The temperature controller must be an electronic industrial type, to achieve the necessary sensitivity. Since there will, inevitably, be some temperature variation throughout the room, the controller must be mounted as close to the gaging work station as possible, even if this means providing a pedestal or hanging it from the ceiling. The diagram indicates zone control, for one of a group of standards rooms.

Most standards rooms require only a small amount of outside air since the room is normally unoccupied.

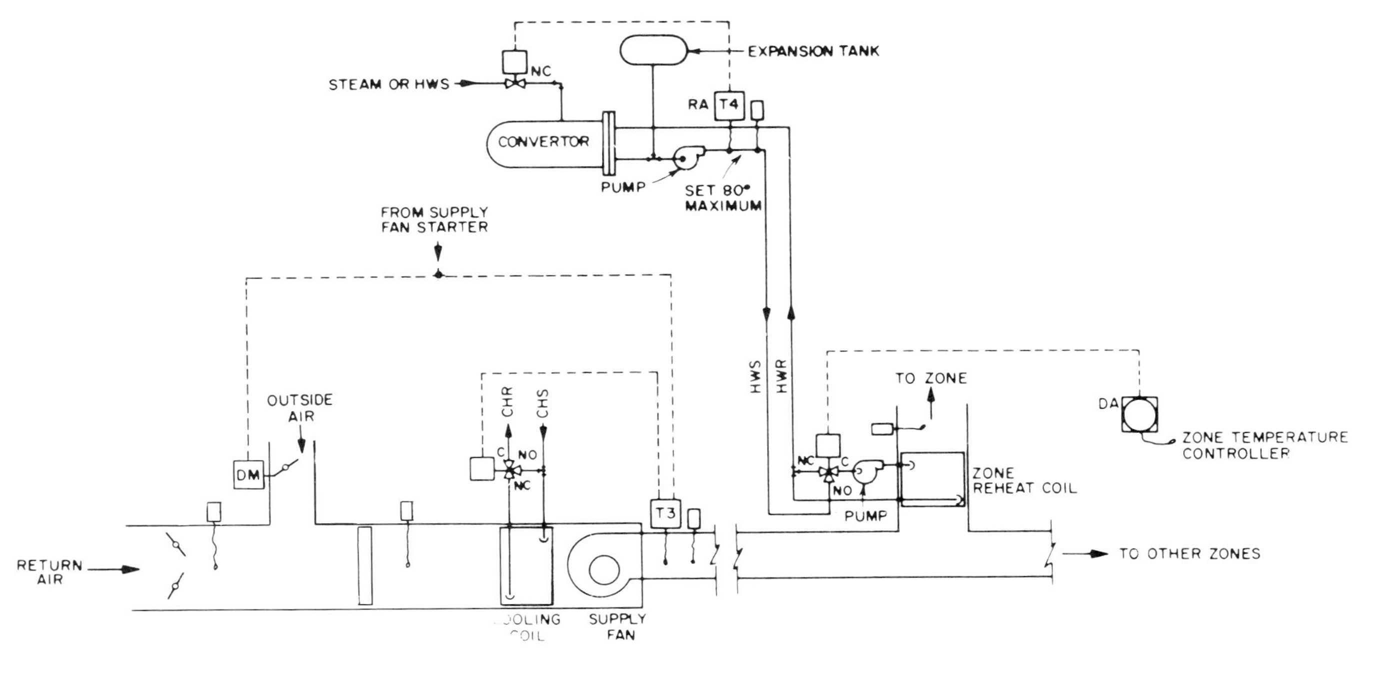

Figure 9-1 Control for standard laboratory.

9.2.2 Clean Room

Clean rooms generally require high airflow rates to maintain the clean conditions, with temperature differential requirements in the order of plus or minus one to two degrees F. Humidity, too, is often controlled. Because the airflow rate required for cleanliness is much higher than that required for temperature control, a double air handling system may be used. Figure 9–2 shows the system and control arrangement for a horizontal laminar flow clean room. A high flow rate fan and filter system circulates the air from a plenum to the room and back again. Air temperature and ventilation rate are governed by a small air handler with heating and cooling coils and a fixed minimum outside air damper. Relief dampers are provided with static pressure control. Temperature controls are conventional, with a room or return air thermostat as the principal controller. A humidistat and humidity control equipment could readily be added.

9.2.3 Hospitals

There are a number of areas in hospitals which require close control of temperature, humidity, cleanliness or all three. These include especially: surgery and delivery rooms, nurseries, intensive care units, and laboratory and research areas. Since many present codes do not allow recirculation of air from these spaces, 100% outside air must often be used. While the laboratory area should have its own separate supply and exhaust system, the other areas can be supplied by a single air system with heat reclaim, humidity control and zone reheat. Figure 9–3 shows such a system. The outside air passes through a heat reclaim coil which is transferring heat to or from a similar coil in the exhaust air. The air is further preheated as required and passed through a cooling coil with the discharge temperature corresponding to the lowest design dew point required for any of the controlled zones. Zone thermostats control reheat coils, and zone humidistats control humidifiers in the branch ducts. The psychrometric charts in Figure 9–4 show the cycles for summer and winter design conditions.

9.2.4 Computer Rooms

Computer rooms have many similarities to clean rooms, but here high airflow rates are the result of high heat gains from the computer equipment. Cleanliness is not quite so important, and humidity control is required. Temperature control should be as accurate as possible (plus or minus one

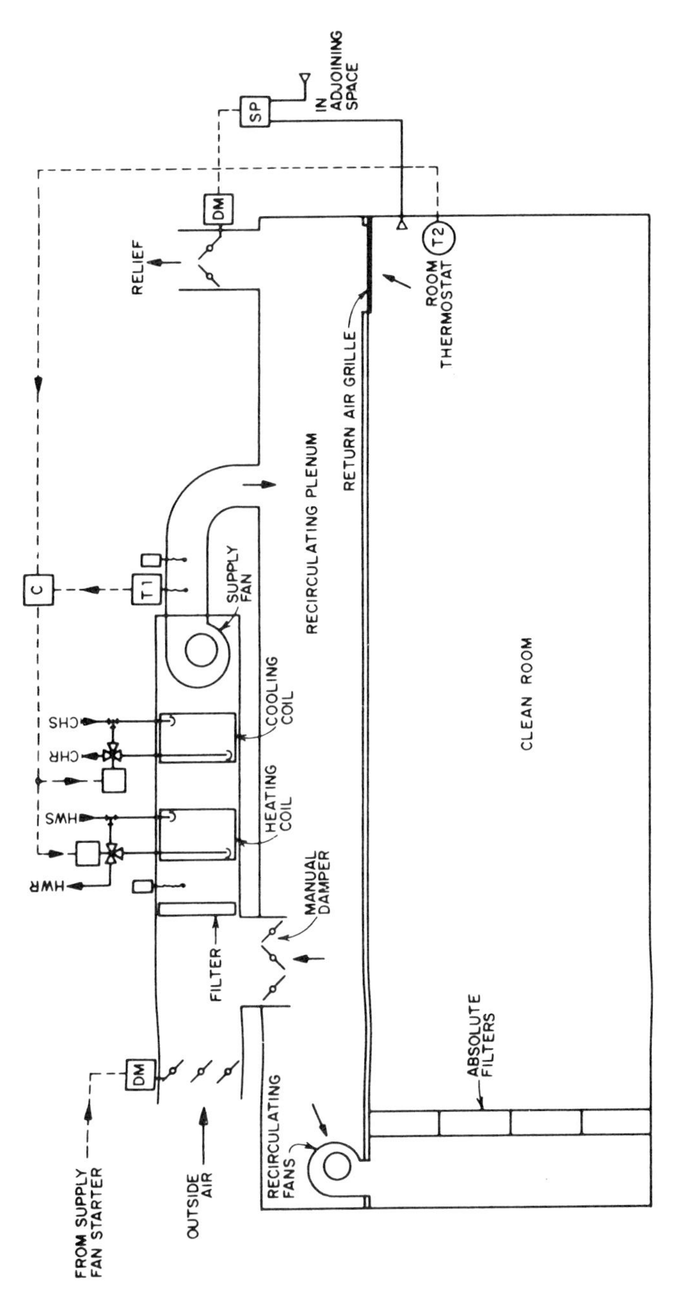

Figure 9-2 Temperature control for a clean room.

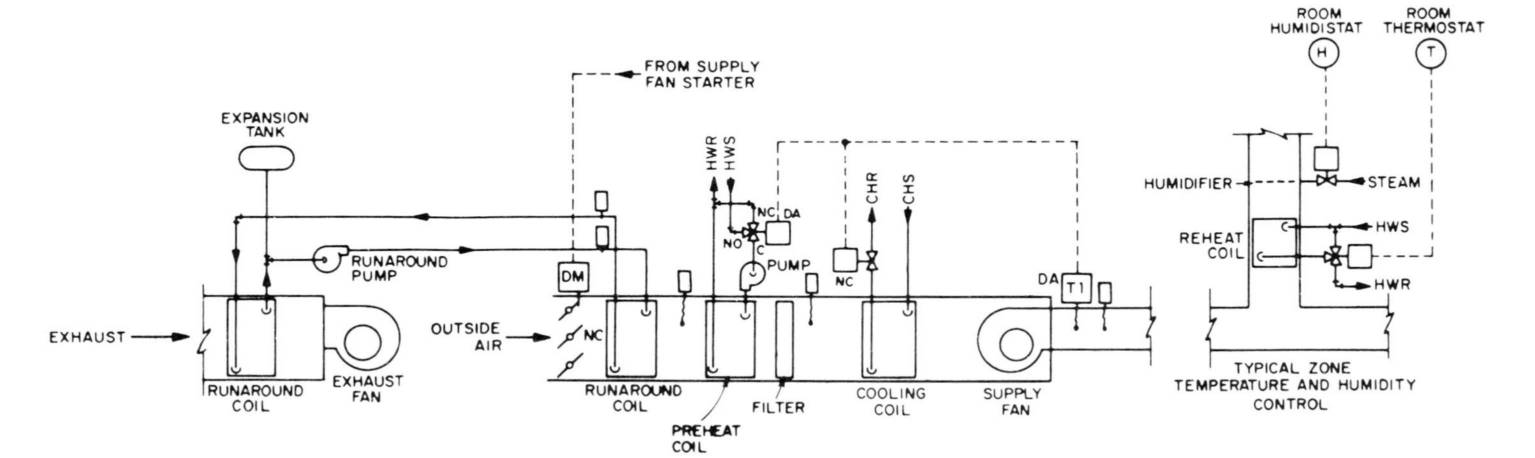

Figure 9-3 Air system with heat reclaim; 100% outside air, zone reheat.

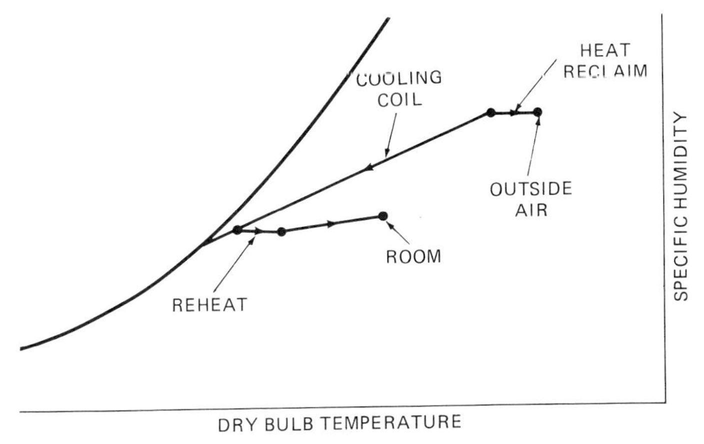

Figure 9-4A Summer cooling cycle for Figure 9-3.

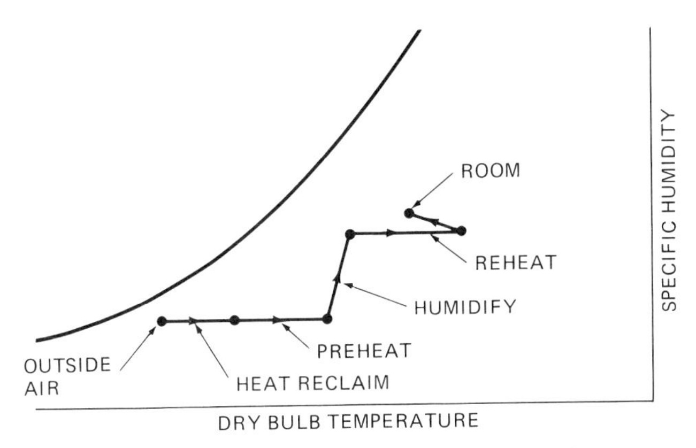

Figure 9-4B Winter heating cycle for Figure 9-3.

degree or less), since some solid-state devices are sensitive to rapid rate-of-change of temperature. Location of the room thermostat and air distribution patterns become very important, since some peripheral equipments have very high heat emission rates. Some computer equipment items require direct air supply from an underfloor plenum at a controlled temperature. Other items have a complete environmental control system within them, requiring only chilled water from an outside source. Figure 9–5 shows a hypothetical system with all of these elements included.

The best current practice is the use of packaged air-handling units—available in capacities from 2 to 20 tons—located within the computer room or rooms. Supply air is discharged downward into the space below the raised floor. This space becomes an air distribution plenum, from which air is fed

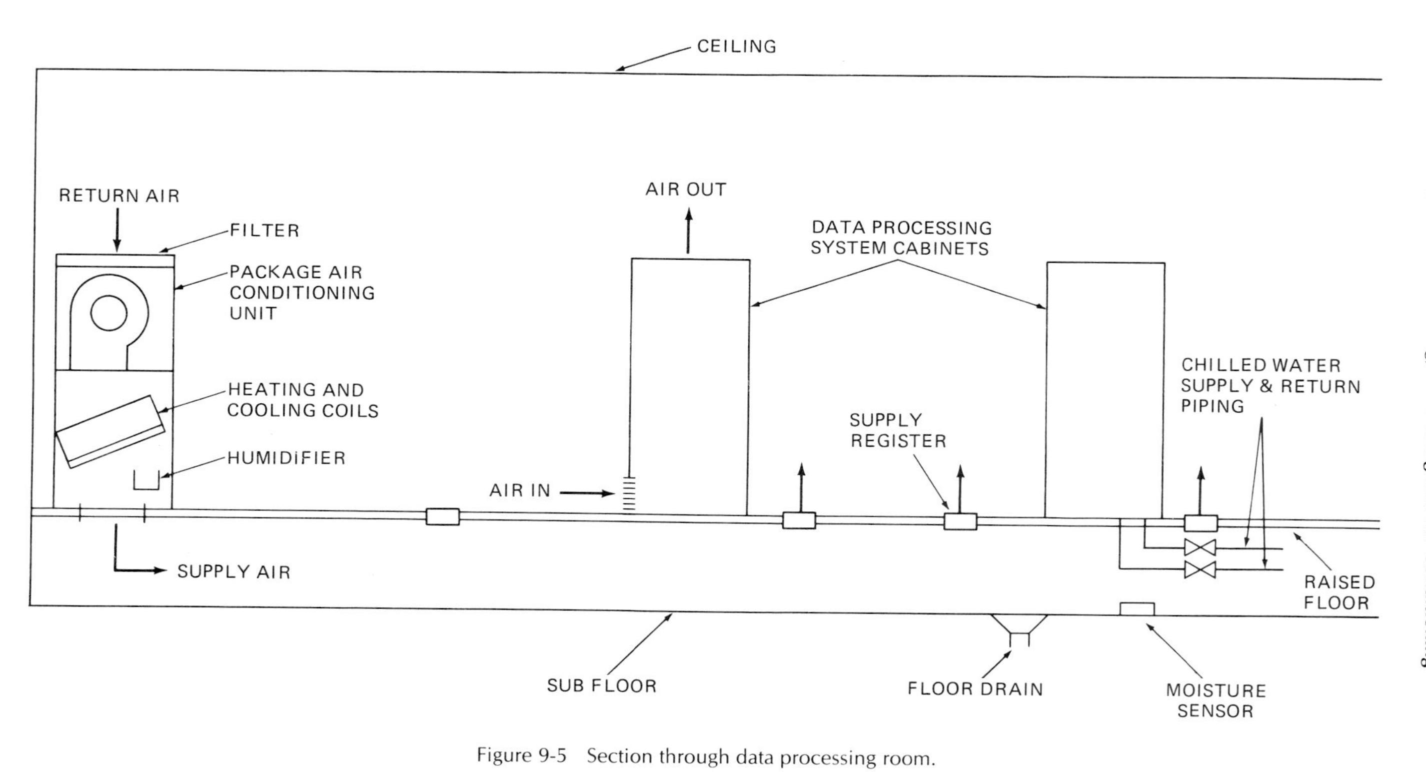

Figure 9-5 Section through data processing room.

into the room through floor registers or special perforated floor panels. Where required, computer cabinets may be supplied with air directly from the plenum. Chilled water piping is located in the underfloor space for connection to those computer elements which require it. This chilled water must be supplied at a temperature high enough to avoid condensation on the heat exchanger. For prevention of damage due to water leaks, floor drains and moisture sensors are required. Ventilation air is provided by means of an outside air connection to one or more package units or by means of a special ventilation system. The packaged units include all controls.

9.2.5 White Room

An interesting control application occurs in a machine shop "white room." Such a room, while not a "clean room," has a high degree of cleanliness. The machining operations carried on require extremely close control of temperature, and a fair degree of humidity control. Since the room must be large to accommodate machining operations, it is impossible to provide a single location for the thermostat which will adequately control the environment at all the machines. But only one or two machines are critical at any particular time. An industrial quality electronic controller may be mounted on a rolling pedestal with trailing cable connections. The controller may then be relocated anywhere in the room, as desired.

9.3 CONTROLLED ENVIRONMENT ROOMS FOR TESTING

Environmental test rooms are used in industry for testing products. Prefabricated chambers complete with controls may be purchased from several manufacturers.

Environmental test rooms for physiological testing of humans are usually custom designed as part of a laboratory complex. The air conditioning system is required to provide a wide variety of environments, for example, from 32° to 120°F and simultaneously from 20% to 80% relative humidity, with close control of differentials at any set point. Obviously, a complex system is required.

Figure 9–6 shows an air conditioning and control system which will satisfy the requirements of a physiological environmental test room. The small amount of outside air required for ventilation is mixed with return air and then flows through a cooling coil. This is a direct-expansion coil, with its own air-cooled condensing unit, and with widely spaced fins to minimize

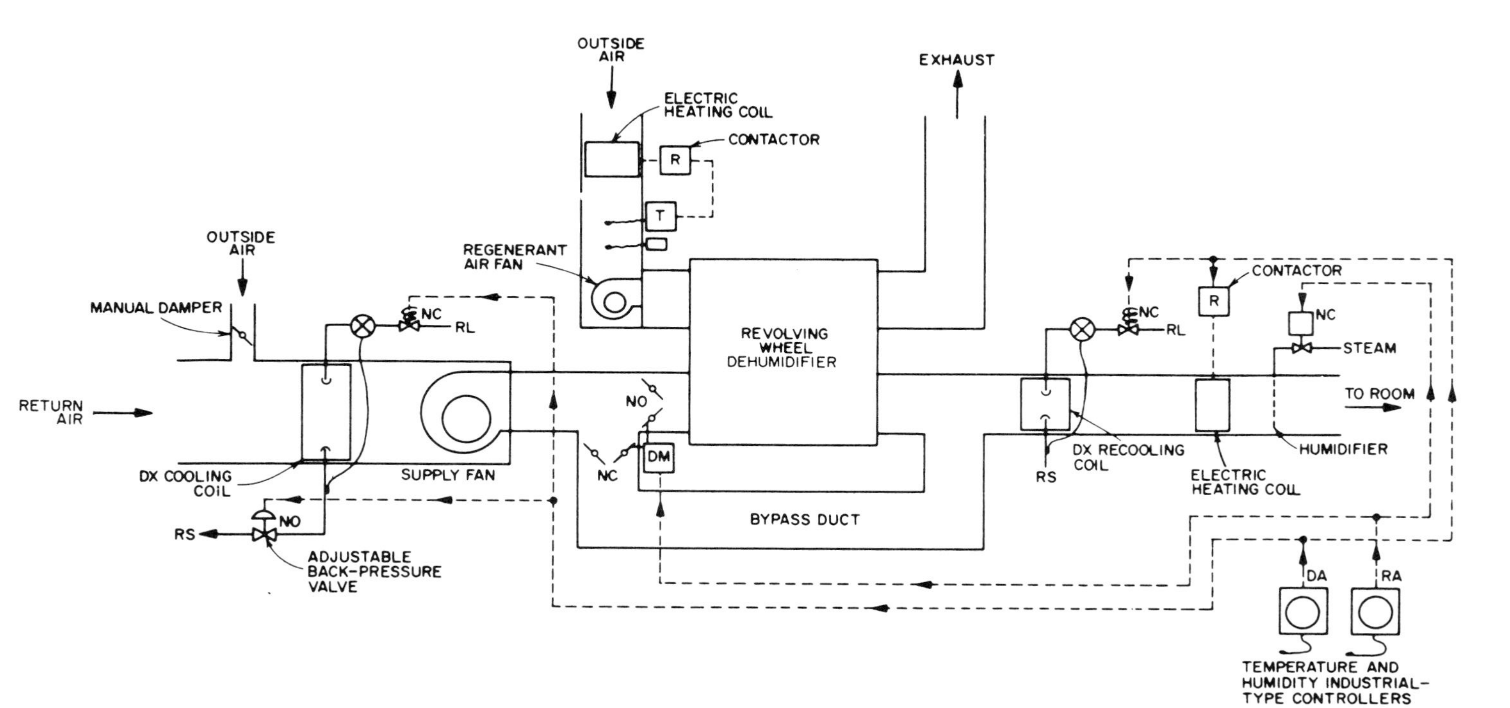

Figure 9-6 Environmental test room.

the effect of ice buildup at low temperatures. The coil is provided with an automatically adjustable back-pressure suction valve which is operated by the room thermostat to control the coil suction pressure and thus the minimum temperature of the air leaving the coil.

The air then passes through the supply fan and to a chemical dehumidifier which is provided with a bypass and a regeneration air duct and fan. The dehumidifier includes circular "trays" of a chemical absorbent which rotate through the supply air and regenerant air in series. The supply air is dried, and the heated regenerant air removes the moisture and carries it away. An electric heater and duct thermostat are provided for controlling the temperature of the regenerant air.

The humidistat controls the proportions of air flowing through the dehumidifier and bypass since direct control of the dehumidifier is not easy to accomplish. The supply air leaving the dehumidifier is hot, due to the heat given up by the hot absorbent as well as the heat due to the chemical reactions. It must, therefore, be recooled, and a second cooling coil and condensing unit are provided for this purpose. An electric reheat coil provides final temperature control when high temperatures are needed. A steam humidifier provides high-level humidification.

The basic thermostat and humidistat are recording-type industrial electronic instruments with remote sensors, so that remote monitoring and adjustment of room conditions are available.

9.4 SOME EXAMPLES FROM PRACTICE

The next paragraphs will describe some rather complex HVAC and control systems taken from the archives of Bridgers and Paxton, Consulting Engineers, Inc. and used with their permission. An exception is the system shown in Figure 9–9, which is described in detail in bibliography reference 33. The layouts have been somewhat simplified but all of the essential elements are retained. These examples illustrate some of the problems encountered in practice and their practical solution. You will note that, complex as these systems are, the basic elements are the same as those that have been discussed throughout the book. These elements are simply combined in new and ingenious ways to solve specific problems.

9.4.1 Office Building with Water Source Heat Pump

The Simms Building (Figure 9–7) is a 12-story office building, with the long axis east-west and large glass areas on the north and south faces. The thermal system is a heat pump with water from two wells as both

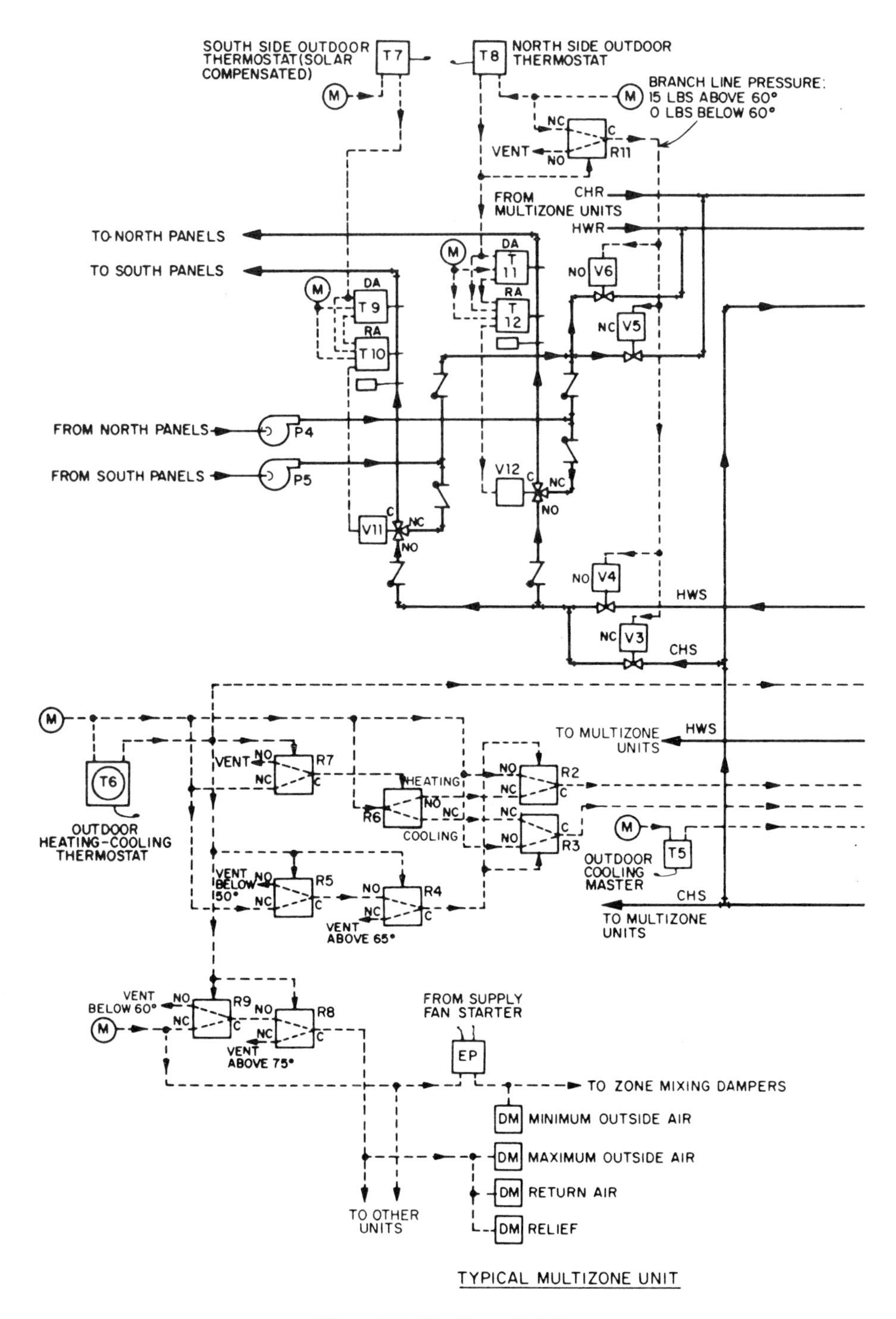

Figure 9-7 The Simms building.

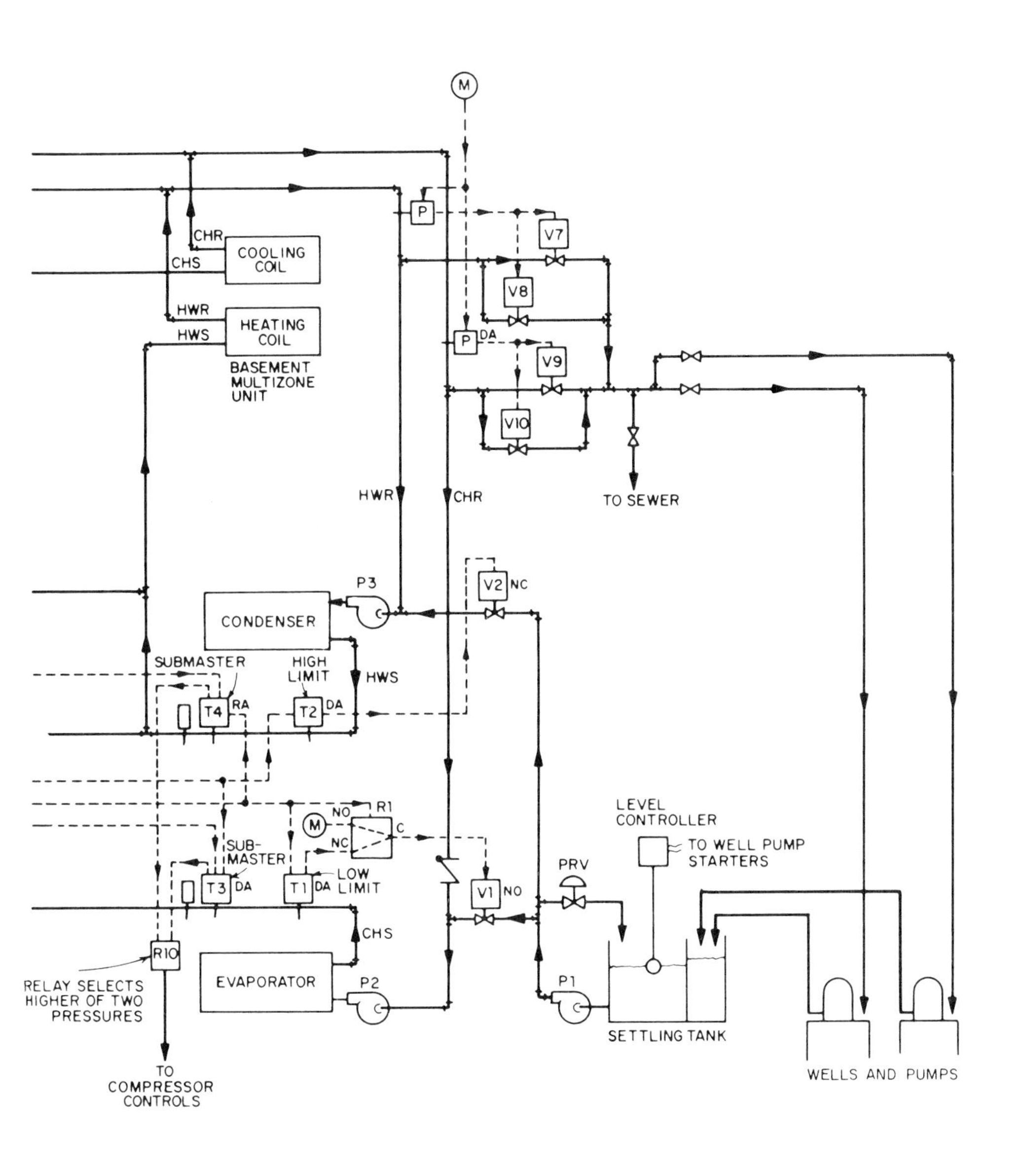

Figure 9-7 *Continued*

source and sink. The south-facing exterior zones have such high solar loads that cooling is required even on cold days (as low as 19°F outside). The basic air conditioning is provided by multizone units, with supplemental heating and/or cooling provided by wall and ceiling panels at the north and south exteriors. Because of the low-temperature heating water available from the chiller condensers (115°F maximum), heating coils are much larger than normal and hot water flow, as well as chilled water flow, to the multizone unit coils is uncontrolled. Panel water supply temperatures are carefully controlled on a schedule set by an outdoor master thermostat. The well pumps are operated manually so that the shallow well is used for heating supply and the deep well for cooling supply. Valves in return water to wells are manually positioned so that return water goes to the well which is not pumping.

Pump P1 takes the well water from the settling basin and pumps it to the evaporator and/or the condenser, depending on the position of valves V1 and V2. V2 is controlled by a high-limit thermostat T2 in the HWS leaving the condenser, and opens and makes up cooler well water to the condenser to avoid excessively high condensing temperatures. V1 is controlled by a low-limit thermostat T1 in the chilled water leaving the evaporator and makes up "warmer" well water to avoid freezeup. Submaster cooling and heating thermostats T3 and T4 control the chiller compressor through relay R10, to maintain design hot and chilled water temperatures. The set points of these two thermostats are reset by outdoor master thermostats T5 and T6. Thermostat T6 also controls a complex series of logic relays to provide control air to the heating or cooling controls as required by the season. Relays R4 and R5 operate to limit well water makup to that required by thermostat T2 when outdoor temperatures are between 50° and 65°. (V1 is closed by means of R3 and R1.)

Pumps P2 and P3 are the primary pumps for the chilled water and hot water circulating systems, respectively. Hot and chilled water circulate through the multizone unit coils without control at a preset design flow rate.

Pumps P4 and P5 are secondary pumps for the panel water circulating systems. Valves V3, V4, V5 and V6 are positioned by outdoor thermostat T8 (through relay R11) so that chilled water is supplied to the panel system when outdoor temperature is above 60° and hot water is supplied when the outdoor temperature is below 60°. Three-way modulating valves V11 and V12 are controlled by submaster thermostats in the panel supply water, with supply temperature set points reset by outdoor thermostats. The south panel system outdoor thermostat is "solar compensated," to take account of the solar load.

Any well water which is added to the system must be returned to the other well, and this is accomplished by means of valves V7 and V8 for chilled water and V9 and V10 for hot water. These valves are opened in sequence by pressure controllers.

The multizone air handling units have conventional mixing damper controls with zone thermostats. Minimum outside air dampers open when the supply fans run. Maximum outside air and relief dampers open only between 60° and 75° outdoor temperature. The economy cycle is not used because at lower temperatures the system needs to utilize generated heat to minimize the use of well water. At ideal conditions the entire system is self-balancing, that is, no well water is needed and heating and cooling loads are in balance.

The diagram has been greatly simplified (as compared to the original) to make it more easily understood. There are actually two pumps each at P1, P2 and P3. There are two chillers in the heat pump system. The compressor control system, not shown, consists of a set of sequenced switches to start and stop compressors and load or unload cylinders in response to demand. Service and bypass valves are provided extensively throughout the system.

9.4.2 Life Science Building

A college laboratory and office building for the study of "life sciences," that is, biology, zoology, etc., presents a completely different problem. Here there are high volumes of exhaust, some from fume hoods with high-pressure drop in the hood, and the rest as general exhaust with low-pressure drop through a grille. Static pressure controls must be provided to keep fumes and odors from traveling from the laboratories to adjoining corridors and offices. Each laboratory must have its own temperature and, in some cases, humidity control.

Figure 9–8 shows a solution to these problems. A single, large variable-volume-with-reheat system serves the entire building. A single exhaust system serves all the exhaust requirements. A booster exhaust fan is provided at each fume hood to overcome the hood pressure loss. A runaround heat reclaim system is provided. This serves to partially preheat the 100% outside air supply. A preheat coil with circulating pump provides the rest of the preheating to 53°. In summer the runaround coil system provides some pre-cooling and the cooling coil provides additional cooling to 53°. Total air supply varies according to demand and is controlled by a static pressure controller which senses "end-of-main" pressure and modulates inlet vane dampers on the supply fans to maintain this at +0.25 in. of water with respect to outdoors . Modulation of the variable-volume supply dampers will cause a change in the "end-of-main" pressure. Each room supply is provided with a variable-volume damper and reheat coil. These are controlled by the

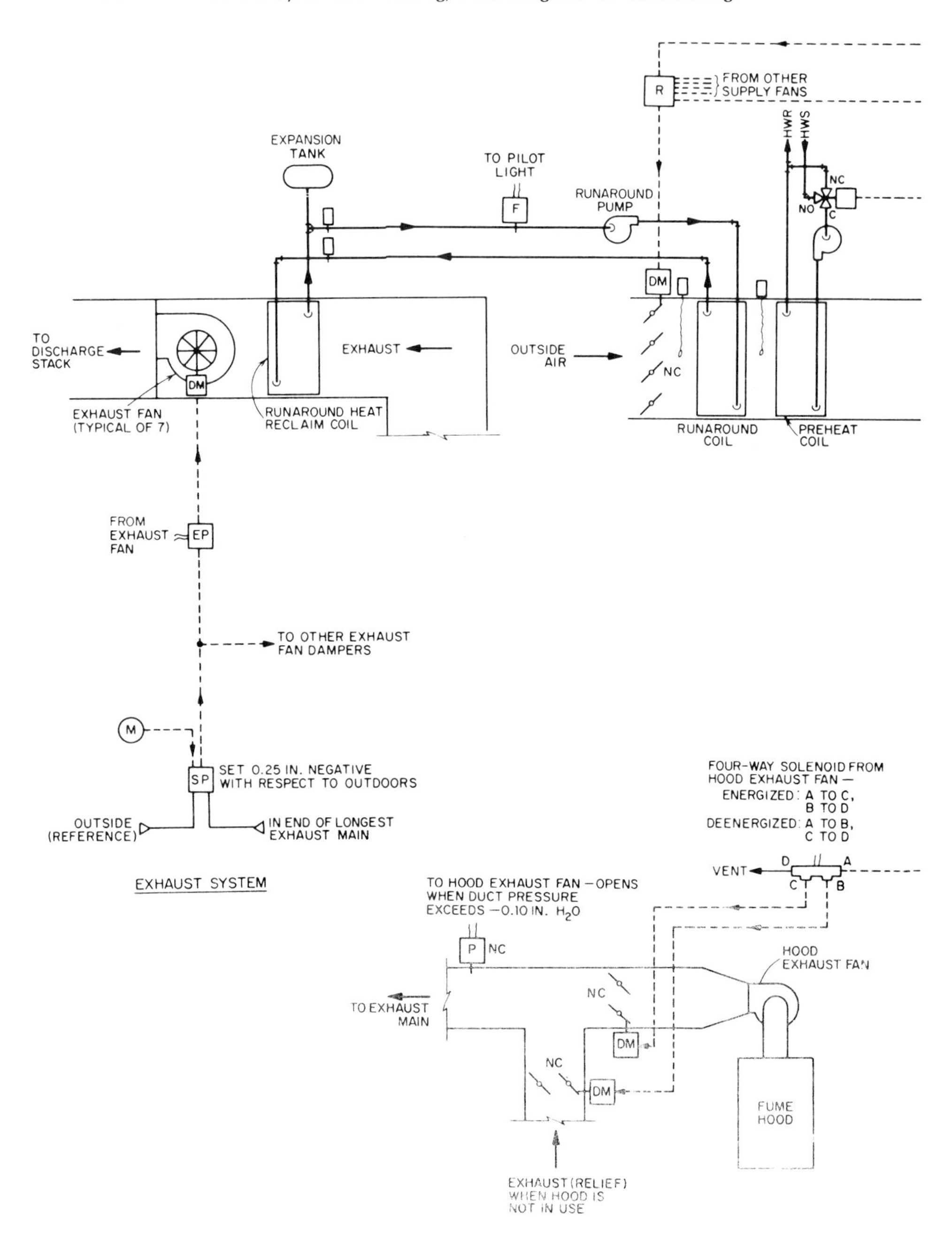

Figure 9-8 Life science building.

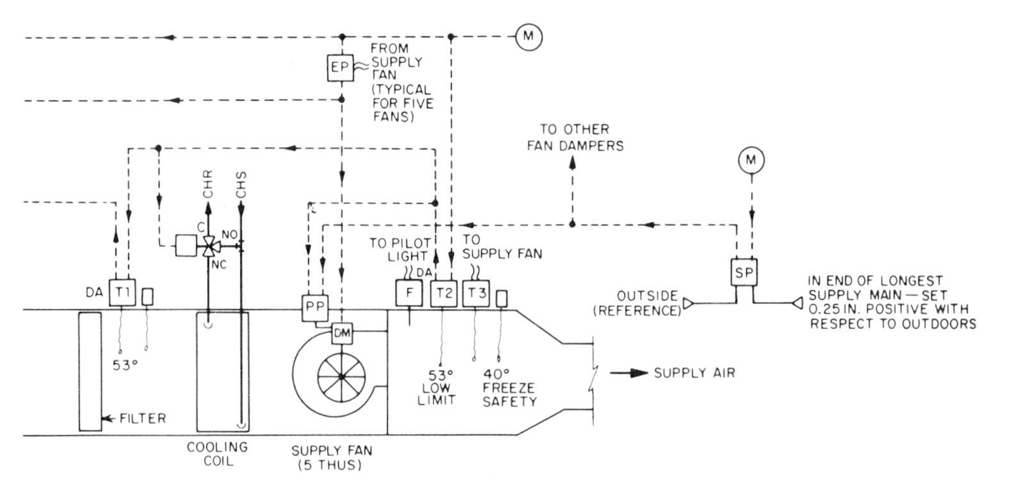

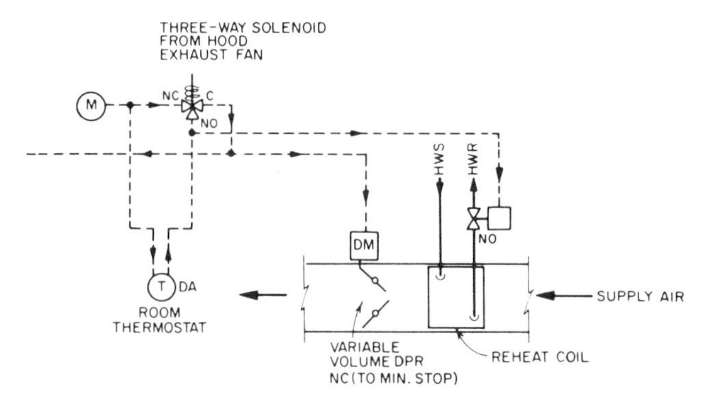

LABORATORY TEMPERATURE CONTROL
(ROOMS WITH FUME HOODS)

Figure 9-8 *Continued*

room thermostat as described in paragraph 7.5.1, except that when the fume hood exhaust fan is started (manually) a solenoid relay causes the supply damper for that room to go full open. Another solenoid relay opens the damper in the hood exhaust duct and closes a damper in the room "general exhaust" duct. A pressure sensor in the exhaust branch main to the hood will stop the fume hood fan if the branch main pressure becomes greater than −0.10 in. of water. The exhaust system is maintained at an end-of-main negative pressure of −0.25 in. of water with respect to outdoors by a static pressure controller which modulates inlet vane dampers on the main exhaust fans. This combination of pressure sensors maintains a pressure gradient from supply to exhaust which effectively prevents exfiltration from the laboratory areas.

Some rooms require humidity control, and this is provided by steam humidifiers in the room supply ducts, with control by room humidistats.

Environmental rooms in which temperatures may be varied from 60° to 90°F are provided with high-capacity reheat coils and unit coolers.

9.4.3 Laboratory Building

The system described in the previous paragraph is an ideal solution to the pressurization problem and will work well when everything is properly set up and calibrated. In practice it is difficult to establish or maintain proper calibration and adjustment of the static pressure sensors and controllers. The many interactions among the building elements tend to upset the system. A simpler control system is shown in Figure 9–9. This system provides stepwise adjustment of supply air flow rates to match changes in exhaust air flow rates. The rates are pre-balanced so that proper pressure *relationships* are always maintained though the actual *differential* pressure may vary.

The system includes a large, variable-volume exhaust system. (A separate fan for each hood could be used, together with a variable general exhaust system.) A matching variable-volume supply air system is provided. The VAV box (or boxes) for each room is sized to provide slightly less than the total exhaust from the fume hoods in the room, since the lab room pressure should always be negative with respect to its surroundings. When the hoods are not in use (manual switches in "off" position) the general exhaust damper is open and the VAV box is controlled by the room thermostat. When it is desired to use a hood the two-position manual switch is turned to the "in-use" position. This opens the hood exhaust damper and simultaneously opens the VAV box to a pre-set position which will provide matching supply air. The general exhaust damper is closed. If more than one hood is in use,

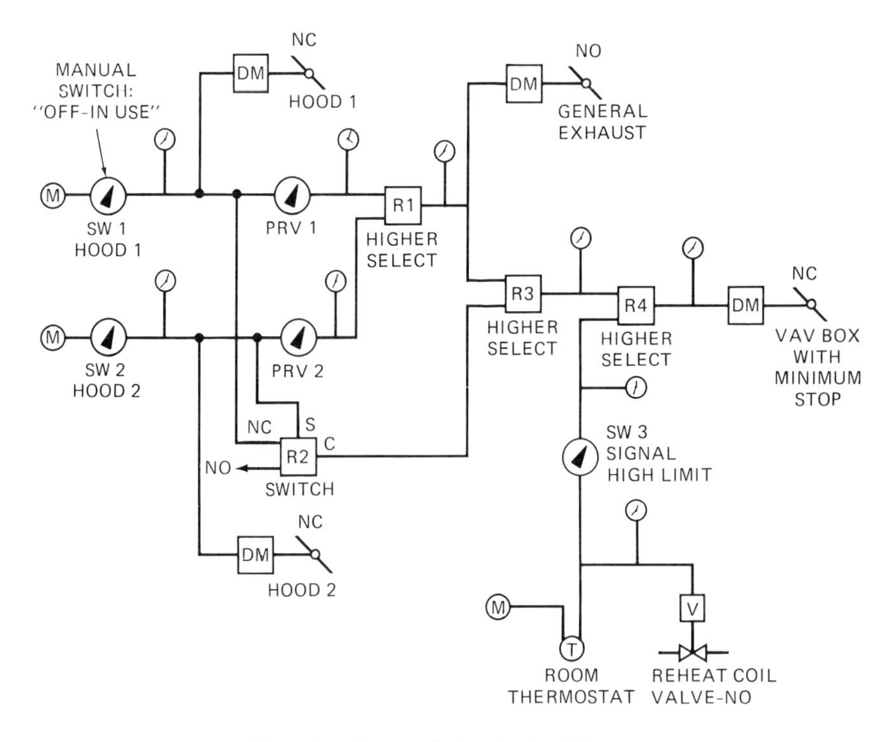

Figure 9-9 Pressure balancing in a laboratory.

the VAV box will open further (or another VAV box may be used). The complex of relays is used to match either or both hoods in any sequence. More than two hoods in a room can be accommodated, though the control complexity increases geometrically. When a hood or hoods are in use reheat may be required to maintain the room temperature. The central supply and exhaust systems could incorporate heat reclaim.

9.4.4 Office Building with Internal Source Heat Pump

An office building for an electric utility company had as a basic criterion that electricity should be the only energy source. Studies indicated that an internal heat source pump would be the most economical solution. Figures 9–10 and 9–11 show the highlights of the resulting system.

The air system, Figure 9–10, consists of two main systems. One of these provides cooling only and serves the interior zones, which, of course, require no heating. This system includes return-relief fans and economy cycle

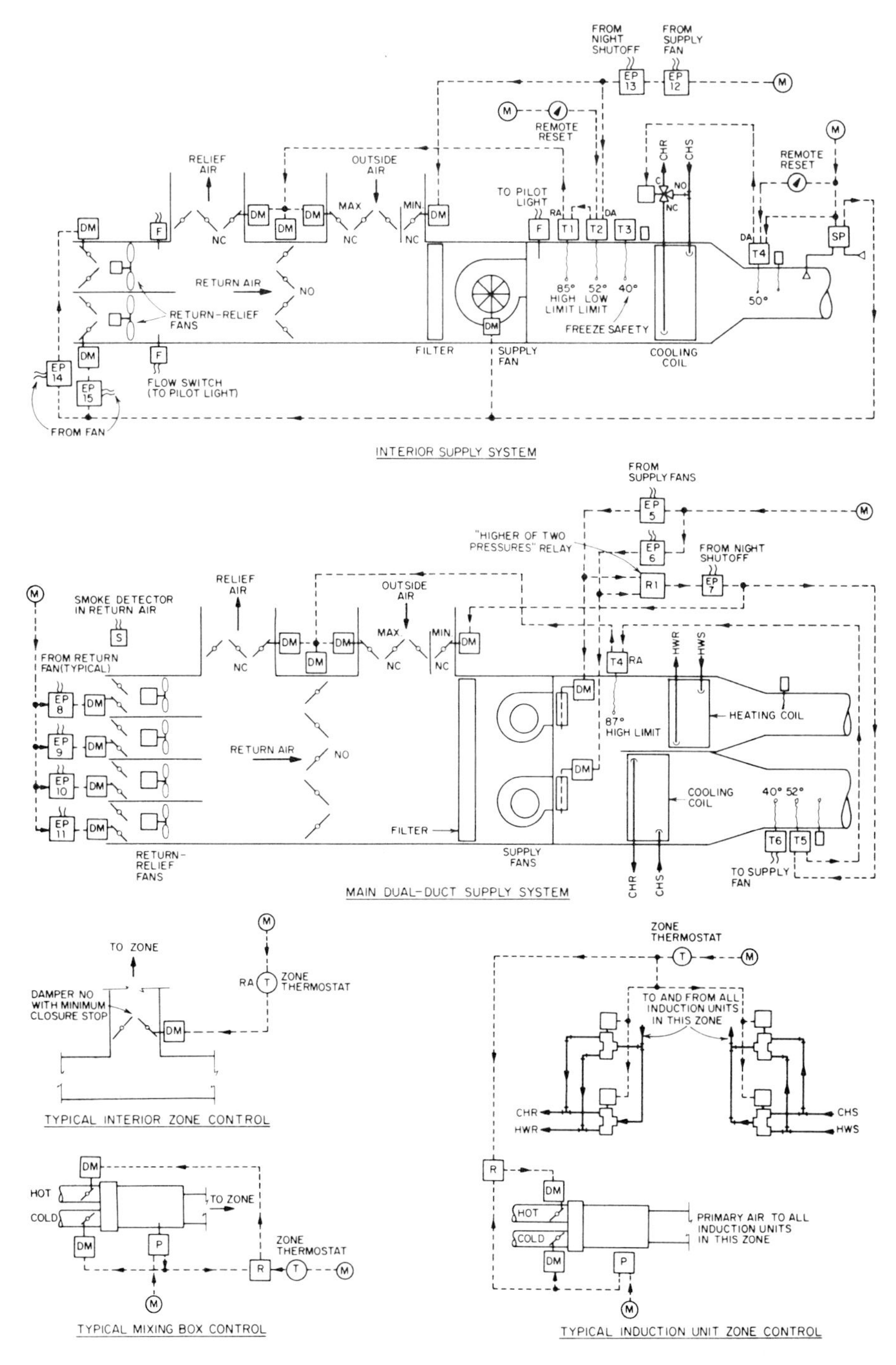

Figure 9-10 Office building with internal source heat pump—air systems.

outside air control, with the added feature of night-time shutoff of outside air. The cooling coil valve is controlled by a discharge thermostat to provide 50°F supply air. Individual rooms or zones are provided with variable-volume dampers. A static pressure controller in the supply main senses the pressure changes as the variable-volume dampers modulate and adjusts the supply air volume accordingly by means of inlet vane dampers on the supply fan.

The other air system is a dual-duct, high-velocity type, again with return-relief fans and special outside air control. Notice that the low-limit "mixed-air" thermostat is in the cold plenum and senses the temperature leaving the cooling coils, rather than the true mixed-air temperature. Since the cooling coils are uncontrolled, this has the effect of limiting the outside air quantity in mild weather, allowing the use of internal heat for heating exterior zones, rather than dissipating it in relief air. Night shutoff of outside air is also provided. The supply and relief fans are provided with dampers which close when the fan is off, thus preventing short-circuiting of air when not all the fans are running.

Perimeter walls are provided with a continuous series of induction units and these are served by zone air systems supplied through dual-duct mixing units. The induction unit coils are controlled by means of the zone thermostat and four-pipe control valves. The mixing unit pressure controller is set to provide a relatively high outlet pressure, to insure proper operation of the induction units.

Mixing units with low outlet pressures are provided for the overhead air distribution systems which supply most of the air to the exterior and public zones.

Figure 9–11 shows the system for providing hot and chilled water. Three package water chillers are arranged in "cascade" fashion, that is, the chilled water and condensing water flow through them in series and with counter flow. Thus, each chiller operates on about the same temperature difference between condenser and evaporator. The cascade system makes possible greater overall temperature changes in the hot and chilled water, allowing smaller water quantities to be used. At light loads one or two of the chiller units may be shut down manually, but normally all units are running and capacity is controlled by relay R12, which selects the greater of the two pressures fed to it by the reverse-acting hot water and direct-acting chilled water temperature controllers, T3 and T4. Output pressure from these controllers flows through relays R10 or R11, which are positioned by the manual selector switch. Below 45° outside, "heating" is selected and the hot water controller controls. Above 60° outside, "cooling" is selected and the chilled water controller controls. Between these two temperatures, "auto" is selected, and both controllers are effective. The one with the greater demand

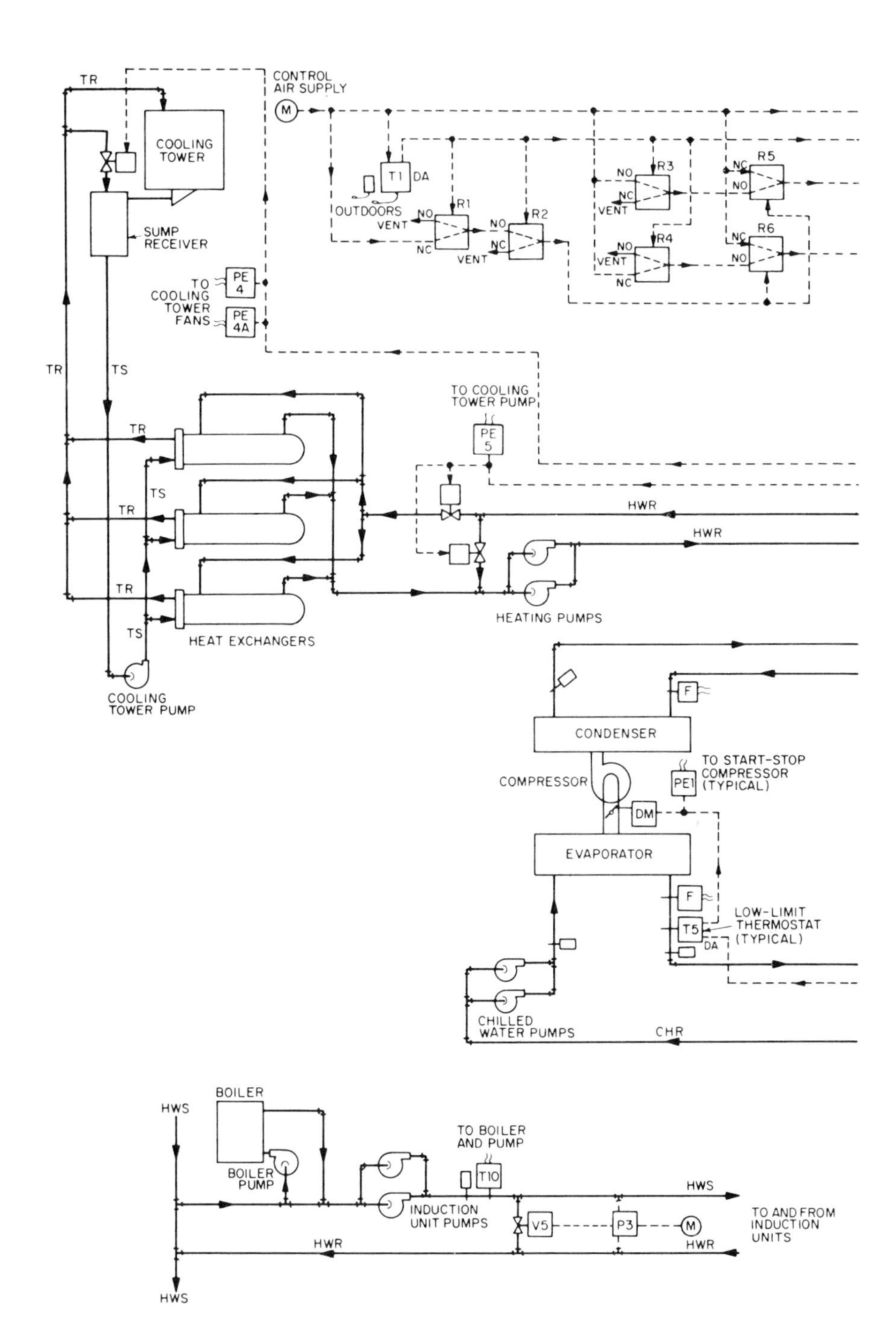

Figure 9-11 Office building with internal heat pump—water system.

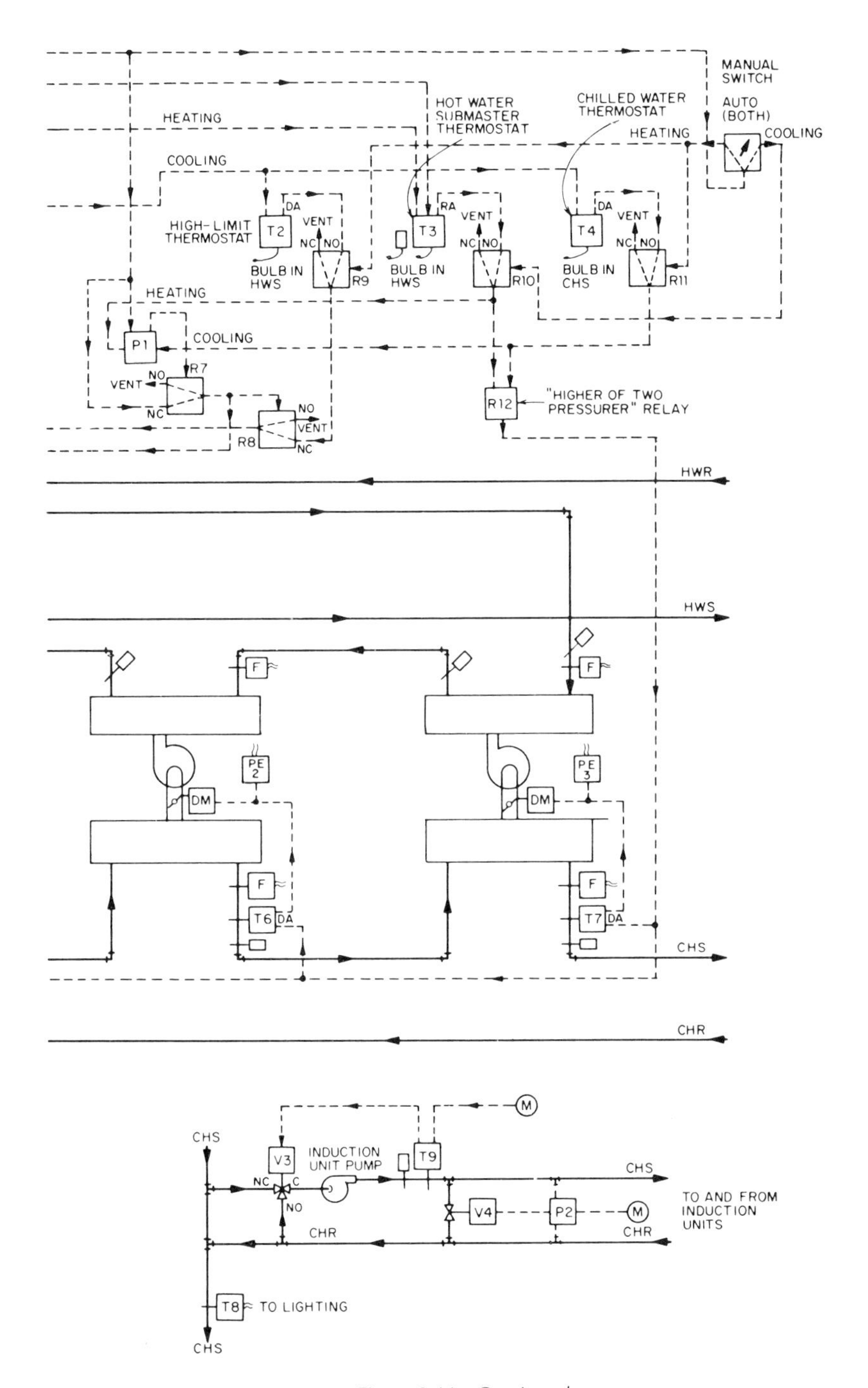

Figure 9-11 *Continued*

controls the chiller capacity. Also, in the "auto" position, changeover may be automatic as described below.

In addition, these two controllers operate through differential pressure control P1 as follows: If the "cooling" branch pressure is greater than "heating," the output of P1 is equal to the difference in pressures. If the "heating" branch pressure is the greater, the output is 0 psi. A 7 psi output pressure changes relay R4 to provide a 15 psi output which starts the cooling tower pump (through PE5) and positions relay R8 to allow the hot water high-limit controller T2 to operate (provided the manual switch is on "cooling" and R9 is in the normal position). Also, valves V1 are positioned to circulate the hot water through the heat exchangers and thus dissipate excess heat to the cooling tower. Thermostat T2 acts to start and stop the cooling tower fans and position the tower water bypass valve, V2, in accordance with demand.

When the output of P1 is 0 to 3 psi, the cooling tower and pump are inoperative and valves V1 are positioned to recirculate the hot water and bypass the heat exchangers. T2 is also inoperative since it is assumed the heating load is sufficient to keep the hot water cool enough to prevent excessive condensing temperatures.

The hot water temperature controller T3 is reset by an outdoor master thermostat, which also operates through relays R1 through R6 to provide control air supply to T2, T3 and T4, thus: When the outdoor temperature is below 45° the branch pressure from T1 is low, all relays are in the normal position and heating control only is provided. Control air is supplied through R3 and R5 to the hot water submaster controller T3, which is reset to some high control point. As the outdoor temperature (and T1 branch pressure) increase, relay R1 switches, switching R5 and R6 which provide control air for both heating and cooling. On a further increase in outdoor temperature, R3 and R4 are switched, but outputs of these relays are blocked by R5 and R6. At about 65° relay R2 is switched, returning R5 and R6 to normal and providing control air for cooling only through R4 and R6. As the outdoor temperature falls, a reverse sequence takes place.

The induction unit chilled water pump is manually started. Thermostat T9 in the induction chilled water supply positions three-way valve V3 to prevent too low a temperature of the supply water. Differential pressure control P2 modulates valve V4 to maintain a maximum differential pressure from supply to return. This is effectively a bypass valve, to maintain flow if the induction unit control valves are closed. A similar control is provided in the induction unit hot water system.

The induction unit hot water pumps are started manually. No control is normally provided in the hot water supply, but a low-limit thermostat T10 may bring on an electric boiler and circulating pump if the temperature falls too low at night.

A low-limit thermostat T8 in the chilled water supply operates through relay R13 to turn on portions of the building lights, thus creating a false load and maintaining a minimum chilled water temperature.

9.5 SUMMARY

It is obvious that the systems described above required a great deal of thought and experience to design. In any case, however, the overall system is built up from relatively simple subsystems, and it is their interplay which creates the complexity. Careful study of these designs will be rewarding and educational. Some of these arrangements may appear to violate the rule of simplicity, and perhaps you can find a simpler way to accomplish the same program.

10 Supervisory Control Systems

10.1 INTRODUCTION

The increasing size of modern buildings and building complexes, and the difficulty of obtaining competent operating personnel have led to an increasing use of central supervisory control systems. These systems allow one person, at a central location, to monitor and control the operation of up to several thousand elements of the heating, ventilating and air conditioning systems in the building or complex. The addition of alarm circuits and audio or audio-visual communications improves security as well as simplifying and improving the degree of control which can be maintained. The additional first cost of such a system can usually be amortized in a few years by more efficient maintenance and operation. Where numerous tenants are involved, the increased tenant satisfaction may be sufficient justification for the added cost.

10.2 HARD-WIRED SYSTEMS

"Hard-wired" central systems were the first used. As the name implies, this is simply the extension of the conventional individual control wires to a central point. Each element in the control system, whether start-stop control of a motor, indication of a temperature, reset of a control point or malfunction indication requires one or more wires running all the way to the control panel, plus a separate device on the panel. Obviously, a building need not be very large before this becomes too cumbersome to be useful. Many such early systems were built, though few are still in use.

The transmission of signals from distant air handling units and other equipment to the central control point poses some problems. Two-position or digital signals are simple. These are used for starting and stopping motors and indicating on-off or out-of-control conditions. Analog signals for measuring flow, pressure, temperature and the like are more difficult. These signals may be transmitted as low-voltage direct-current signals, or as air pressure signals which can be translated and read at the receiving end by means of a transducer. In either case, line losses due to distance become important, and careful calibration and compensation is necessary.

10.3 MULTIPLEXING SYSTEMS

A complex involving a great many similar control systems such as air handling units can be more efficiently and easily handled by the use of multiplexing relays. A multiplexing system provides one set of channels to cover the required control functions which are common to all units. A relay at each individual air handling unit allows that unit to be selected for monitor and control through the common channels. When the routine for one unit is completed, another unit can be selected, with the same common channels used. The relays used are called multiplexing relays. The control and monitor wiring is greatly simplified. After basic wiring is provided, we need only one additional wire for each additional station. Figure 10–1 shows the elements of this system in simplified form.

When the selector switch is positioned for a particular system, coil no. 1 in the multiplexing relay is energized, closing contacts no. 1 and putting the system "on-line" to the control elements at the central panel. Then the motor may be stopped or started from the central panel, with pilot light indication provided. In the wiring shown here, local start-stop pushbuttons and pilot light are also provided.

Analog functions, such as temperature indication and reset, can be added by using a different type of relay.

For additional information, a projection screen may be provided. Later systems of this type use coding signals to energize the relay. This requires only a few wires to address a large number of relays, with an automatic projector and slides to give the operator a visual picture of the system with which he is dealing.

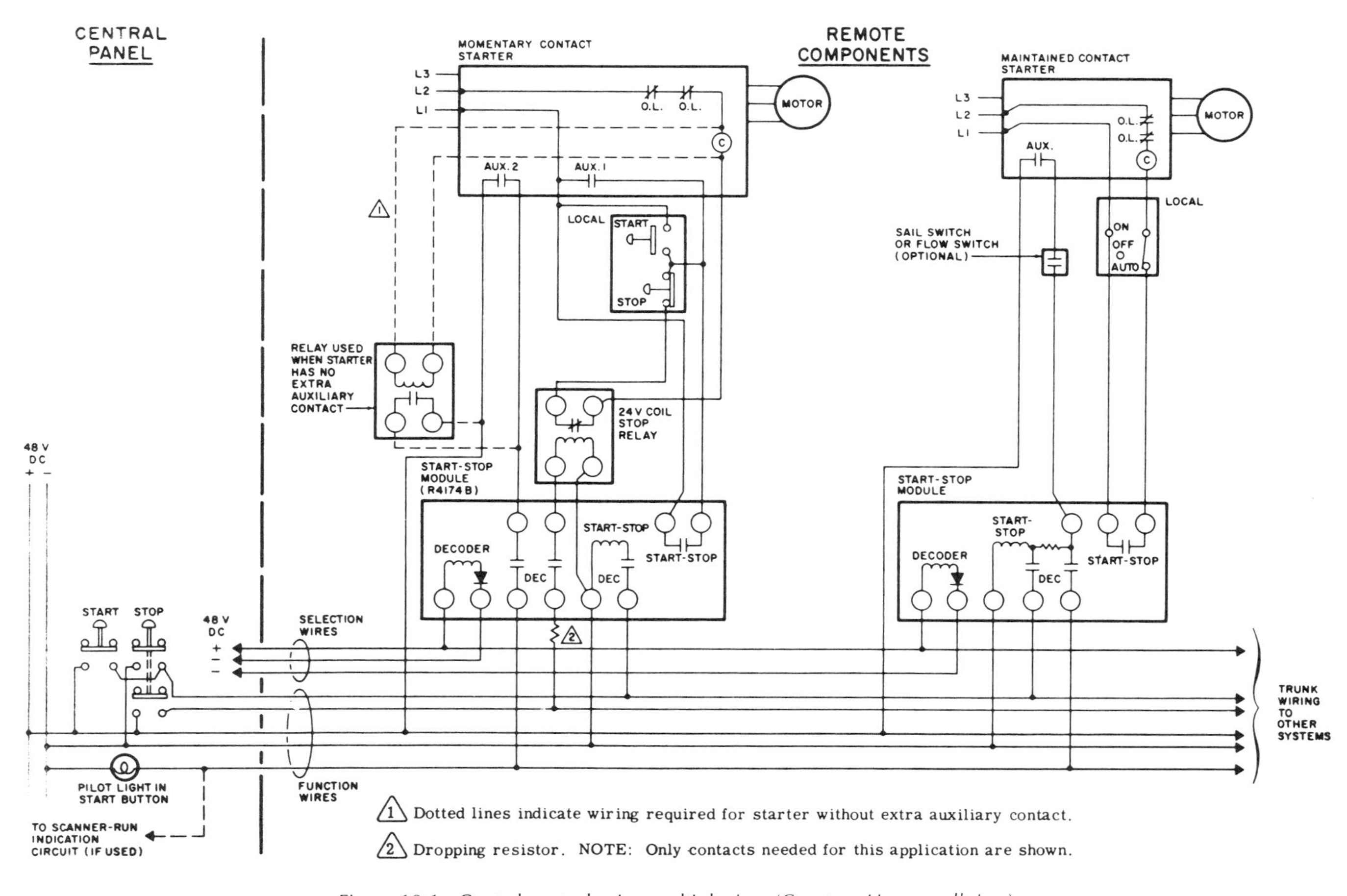

1. Dotted lines indicate wiring required for starter without extra auxiliary contact.
2. Dropping resistor. NOTE: Only contacts needed for this application are shown.

Figure 10-1 Central control using multiplexing. *(Courtesy Honeywell, Inc.)*

Alarms for system malfunction must be hard-wired, or they can be superimposed on the basic system wiring using various proprietary devices based on solid-state generated high-frequency signals.

Where the equipment rooms are remote from the central control panel, as is usually the case in a large complex, the control system may include an audio or audio-visual communication system. With this available, the actions of a maintenance man doing trouble-shooting on the equipment may be guided by the supervisor at the central location. Closed-circuit TV can also improve security by providing visual observation of unoccupied equipment rooms. Though a few such systems are still in use, they are obsolete today, having been replaced by computer-based systems.

10.4 COMPUTER-BASED SYSTEMS FOR MONITORING AND CONTROL

With each new edition of this book it has been necessary to rewrite this section extensively. Technological change is so rapid that control designers and users find it difficult to keep up. Because of the emphasis on new equipment and techniques it is easy to lose touch with fundamentals. The mistake that many people make is that of expecting the computer to solve control problems. It won't. The computer is simply a (possibly) better tool for use in control systems. Treated as a "solution" it may even aggravate the problem.

The term "computer-based" refers to a control system which utilizes some type of programmable digital computer. The computer is often referred to as "intelligent" because it can be programmed to make decisions. Some of the most promising new developments are in the area of *adaptive control,* a branch of *artificial intelligence.* A computer with adaptive control capabilities will learn from experience which decisions are "best" and modify its operating programs accordingly.

The once clear-cut distinctions among various types and sizes of computers have become blurred. Increasingly, small devices have greater and greater capabilities. Some local loop controllers utilize programmed "chips" to increase their sophistication and provide better communication to supervisory computers. All of this tends to result in a hierarchical system in which *intelligence* is found at all levels. The term "distributed intelligence" is often used.

These computer-based systems are called by various names: MCS (Monitoring and Control System), BAS (Building Automation System), EMCS (Energy Management and Control System), and others. Functionally they are all equivalent.

10.4.1 Functions of a Computer-Based System

The computer-based system can provide *monitoring, intervention control*, or *direct control* of local loop elements. It can also provide historical data summaries, data analysis and maintenance scheduling.

Monitoring means more or less continually looking at the status of the various points connected to the computer. It usually includes comparing the status with some norm and providing an audible, visual and/or printed display of off-normal (alarm) conditions. Monitoring is essential if any form of control is provided.

Intervention control is used when the computer has no direct control of the local loop but may provide reset of set points, start/stop commands for motors or open/close commands for dampers or valves. Intervention control is used primarily for energy conservation and efficiency in HVAC system operation. This is the most common mode of supervisory computer operation.

Direct control of the local loop means that the computer provides all of the control functions, interfacing directly with the sensors and controlled devices. This is usually called Direct Digital Control (DDC). DDC performed by a dedicated computer contiguous with the HVAC system provides good, reliable control (see paragraph 10.4.5 below). When a remote supervisory computer is used to perform DDC, reliability and accuracy may be poor. Reliability decreases because of the possibility that failure of the computer or the communication system would lead to loss of control. Accuracy decreases as a function of the number of points interfaced to the computer. As the number of points increases, the frequency with which the computer can scan (look at) any specific point decreases. At some frequency this leads to loss of accuracy and finally to loss of control. In the opinion of this author, about 1000 points is the largest number with which the typical supervisory computer can deal satisfactorily on a real-time basis. This mode is not recommended.

It should be noted that the so-called EMCS (Energy Management and Control System) is simply a subset of the more general MCS (Monitoring and Control System). The two systems are essentially identical, differing only in the emphasis placed on certain activities.

10.4.2 Elements of a Computer-Based System

The elements of a computer-based system are shown in Figure 10–2. It will be noted that they form a hierarchy. Specific functions are allocated to each level but levels may sometimes overlap or be combined. Various types of communication links are used between levels.

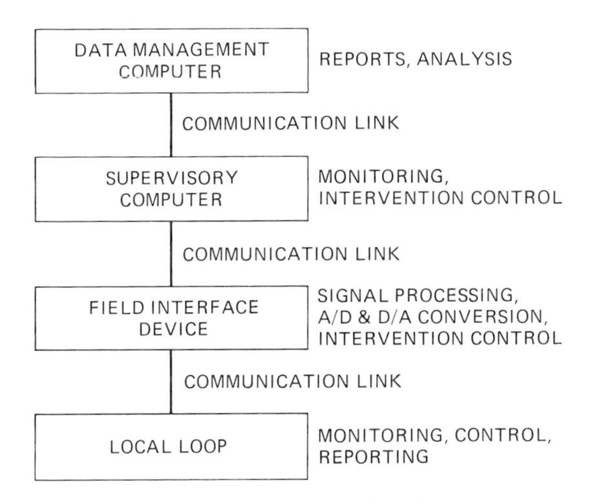

Figure 10-2 Elements of a computer-based control system.

10.4.2.1 Local Loop Control The lowest level is the local loop. This includes sensors, controllers, controlled devices and other devices as needed. If the controller is replaced by a computer the result is called direct digital control (DDC). (See paragraph 10.4.5, below.) In the ideal system if the supervisory computer or communication systems fail the local loop will continue to function. This arrangement is made to improve reliability. In addition to the sensors required by the local loop, other sensors will usually be provided for use by the supervisory computer. Intervention control at the local loop level will often be performed by the supervisory computer or the FID. This includes such functions as start/stop, open/close, reset of set points and similar operations.

Figure 10–3 shows a single-zone AHU similar to that shown in Figure 7–3 but with interface to the supervisory system. Each of the numbered rectangles represents an individual *point*. The supervisory system uses many monitoring and control points not normally provided in the local loop, e.g., the flow switch in the supply duct is used to prove fan operation. If desired, other points can be added for sensing damper and valve positions.

10.4.2.2 Input/Output This portion of the system is required for interfacing between the various forms of signals to and from the HVAC equipment and the digital signals used by the computer. In a typical MCS the element is called a Field Interface Device (FID), though some manufacturers use other names. Most state-of-the-art FIDs are *intelligent*; they include a micro-computer and can perform some supervisory and intervention control functions. The typical FID can interface to several local loops, as well as

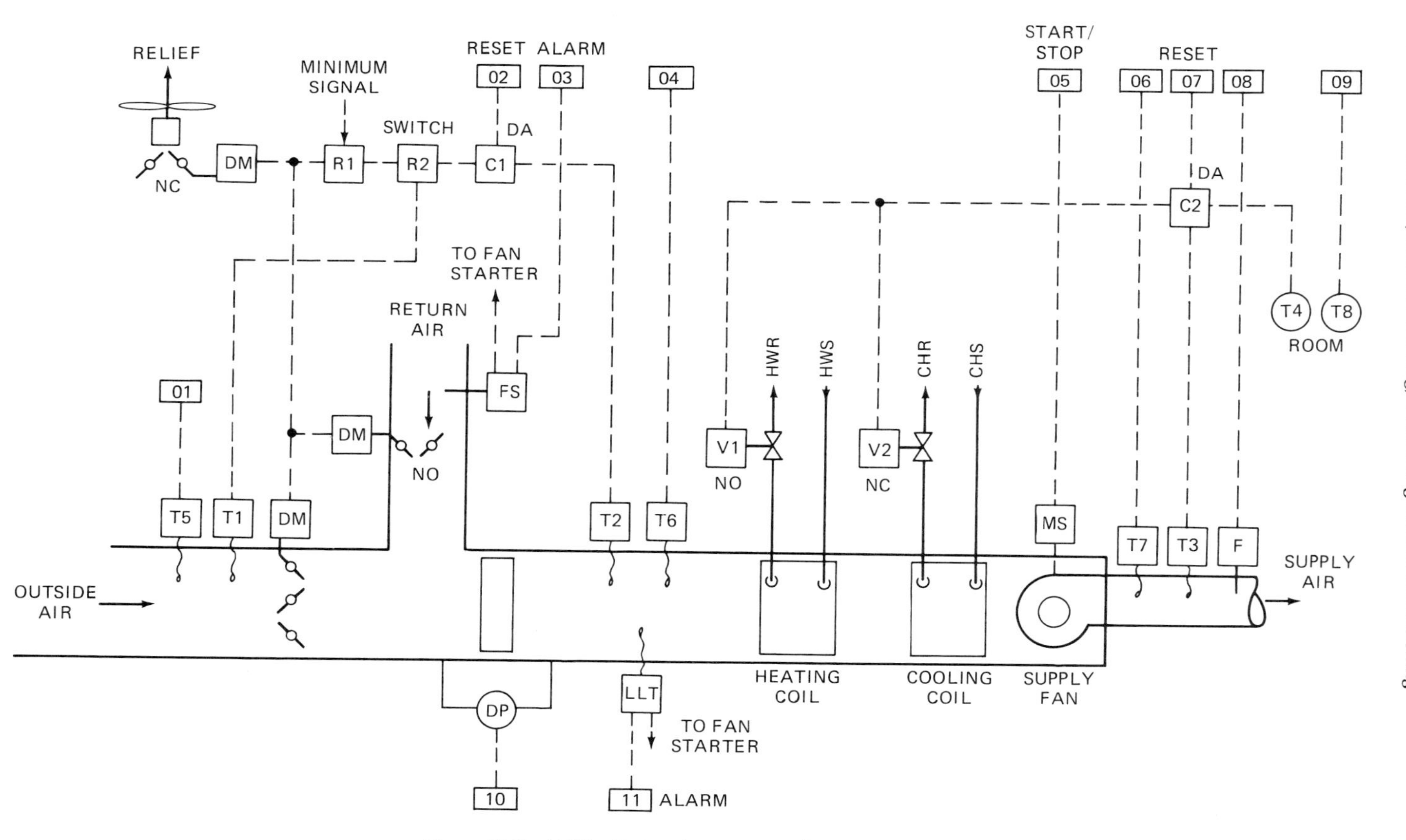

Figure 10-3 AHU with interface to supervisory computer.

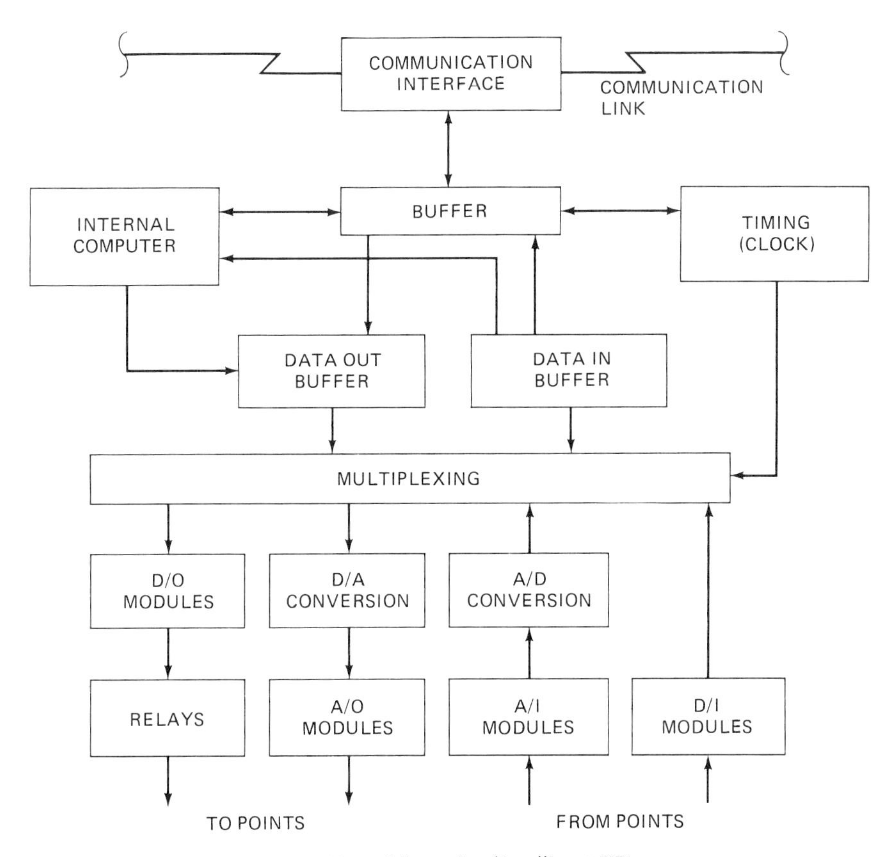

Figure 10-4 Schematic of intelligent FID.

to additional monitoring and control points. All point connections are hard-wired and each point has a unique *address*.

Figure 10–4 is a schematic representation of an intelligent FID. There are four kinds of points. Analog In (AI) signals are from sensors for temperature, pressure, humidity, etc. These analog signals can vary infinitely over the range of the sensor. Analog Out (AO) signals are analog commands, such as reset of a set point. Digital In (DI) signals are contact closures or openings, showing status or alarm conditions in a two-position mode (on/off, open/closed). Digital Out (DO) signals are two-position commands (start/stop, open/close).

Analog In signals must be *conditioned*, a process of evaluation and interpretation, and converted from analog to digital form (A/D conversion) for use by the digital computer and transmission over the communication link. Analog Out signals are generated by the computer in digital form and must be converted to analog form (D/A conversion) for transmission to the point.

Digital In signals require no conditioning. Digital Out signals usually require an amplifying relay to handle the power requirements of the controlled device, e.g., a motor starter. The amplifying relay uses a low-voltage, low-current solenoid to open and close contacts which can handle higher currents and voltages.

The FID addresses many points but the computer can only deal with one point at a time, therefore a multiplexer is needed. The multiplexer controls traffic, allowing the computer to address each point in sequence or on demand. For the communication link a similar procedure is followed. *Buffers* are provided for temporary storage of data.

The *communication interface* connects the FID to the communication link. Messages on the link include the address of the point to which the message relates. The FID can ignore all messages except those relating to its connected points. When the FID sends a message to the supervisory computer the point address is included.

An intelligent FID will *scan* its connected points on a regular basis (every few seconds). The status data are stored and compared with data from previous scans and with limit conditions stated in the programs. If a significant change takes place the information is transmitted to the supervisory computer and some action may be initiated by the FID.

The FID and local loops deal with events in *real time*, that is, as the events actually happen.

10.4.2.3 Supervisory Computer The term "supervisory computer" is used here to describe what is more often known as the *Central Processor* or *Central Console* or *Control Console*. It includes the elements shown in Figure 10–5. The computer has the major analysis and decision-making capability for the system and can interface to a large number of FIDs. A large amount of memory is required and this is usually provided by means of up to a megabyte or more of internal RAM (random-access-memory chips) and by a *hard disk*, which stores information on a electromagnetic track. A typical hard disk can store 20 to 40 megabytes of information (1 *byte* equals 8 *bits*. (A bit is a single element of data, in binary usage having a value of either 1 or 0—on or off.) In addition, most computers use *floppy disks*, small flexible disks with storage capacities of 360, 720, or 1200 kilobytes. Floppy disks are used for inputting pre-assembled software, whereas hard disks, with tape backup, are in general use for long-term data storage.

The computer used to be a standard mini-computer or a special-purpose processor. Because of the increasing capabilities of smaller machines, present day systems more often use standard micro-computers or even personal computers (PC). Any computer requires some special software to adapt it for use in the supervisory monitoring and control situation. Dealing efficiently with real-time input/output is one of the more difficult software problems.

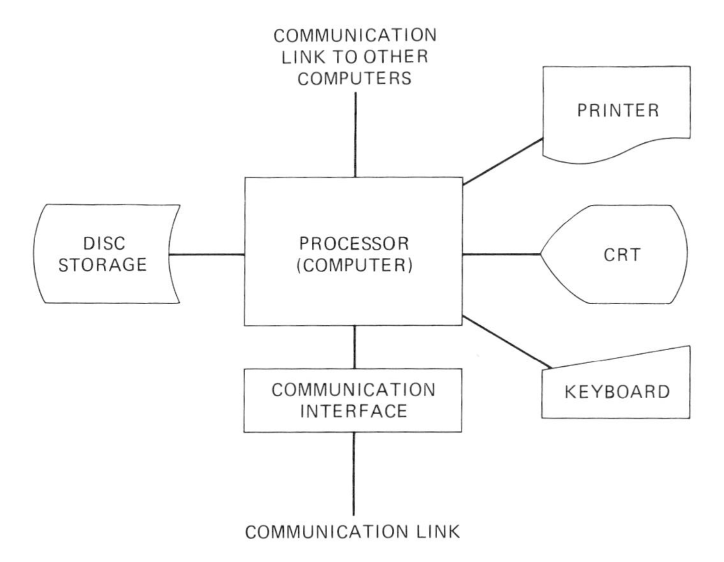

Figure 10-5 Elements of the supervisory computer.

Figure 10-6 MCS console. (*Courtesy Johnson Controls*)

The other part of the Central Console is the *operator-machine interface* (OMI). By means of the OMI the system operator can obtain information, issue commands, and change or add to programs. The elements of the OMI are the printer (or printers), the cathode ray rube (CRT), and the keyboard.

The printer provides a hard-copy record of system operation—whatever the program or operator asks for. This may include status reports, trend logs, operator transactions, program changes, summary reports, maintenance reports and much else. Restraint is needed to avoid being swamped with too much paper. One should print only those things which will be used or are essential history.

The CRT is the operator's window to the system. Here are displayed status information and sometimes graphic schematic diagrams of the HVAC system elements. Keyboard inputs are displayed, together with program information. Interactive programs will display menus and "prompts" which must be answered by the operator.

10.4.2.4 Communication Links There are two principal modes of communication used in the computer-based system. The FID is connected to each of its related points by means of *hard-wiring*—a separate, permanent two-wire link. Signals may be either analog or digital (binary) and are usually low-voltage—24 volts or less. The length of these links should not exceed about 1000 feet, since analog signals tend to attenuate over distance.

Connections between the FID and the upper levels of the hierarchy are made using a *coaxial cable, twisted pair* (Figure 10–7) or *glass fiber optics*. The most common of these is the twisted pair—a pair of wires, usually shielded and/or grounded to minimize interference. Fiber optics have a great deal of promise and should be used more often in the future as costs decrease. Fiber optic systems are virtually interference-free and can handle very high transmission speeds.

Transmission at this level is digital, a continuous stream of bits which are grouped into *words* which follow one another along the wire or cable (sometimes called a *data highway*). Each word consists of 16, or sometimes 32, bits and includes an address, a message and parity bits which are used for checking accuracy. Sometimes accuracy is checked by retransmitting the word two or three times and comparing the transmissions. Speed of transmission is measured in *baud rate*, equivalent to bits per second. Typical hard-wire network baud rates are 1200, or sometimes less, up to a maximum of 9600.

A *modem* (modulator-demodulator) is a device which allows the communication link to interface with a telephone circuit for transmission over long distances. The modem converts the digital signal into an analog signal, or vice versa. Typical modem speeds are 300 to 1200 baud. Higher rates can be used if specially conditioned phone lines are provided. The telephone

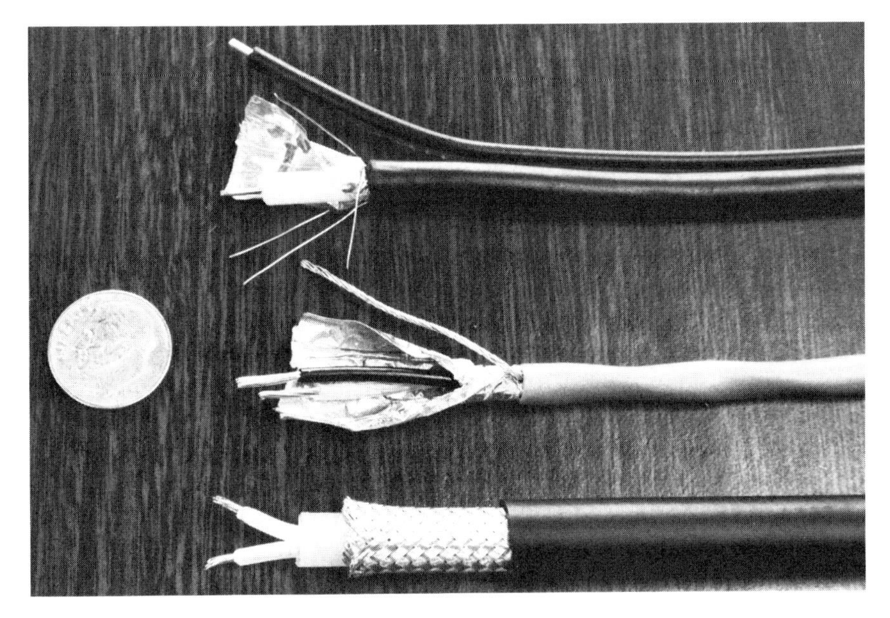

Figure 10-7 Communication cable. Top to bottom: 1 coaxial cable, 2 twisted pair with ground wire, 3 twisted pair with shield. *(Courtesy Powers Regulator Co.)*

circuit may be a *leased line*, dedicated to the system, or a *dial-up* modem may be used. This device automatically dials the number of the destination modem whenever data are to be transmitted. This is less costly in terms of the telephone bill and can be used when real-time monitoring and control are needed only occasionally, for example, when intelligent FIDs are used.

An alternative communication method is the *power-line carrier* system. This is most often used in lieu of hard-wiring when fairly long distances need to be covered. The signals to be conveyed are combined with a high-frequency "carrier" signal which is superimposed on the existing power wiring within a building. With the proper interface equipment signals can be sent and received from any location within the building. Power-line carrier signals will not cross a transformer, so special provisions must usually be made for communication between buildings.

10.4.2.5 Data Management Computer This higher-level element is used only in very large systems with several supervisory computers. Its function is to acquire data from the supervisory computers and to analyze and assemble the data into management reports and summaries. This will allow upper-level managers to evaluate the system operation and quickly determine the results of changes in operating policies. At this level there is no need for a real-time interface.

10.4.2.6 Intercom Some computer-based systems include an intercom system to allow the console operator to "listen-in" on equipment rooms or communicate with mechanics in the field. This requires a completely separate set of equipment and analog communication links, with no direct relationship to the computer system.

10.4.3 Security and Fire Reporting

Security and fire reporting functions are fundamentally the same monitor-control-alarm functions used in HVAC supervision. The systems which are used for HVAC control can therefore easily be adapted to these additional functions. In some localities code authorities require the use of "proprietary" and UL labeled systems for fire reporting. The basic HVAC system with a few special features will meet this need.

Fire reporting systems, particularly in high-rise buildings, require a special control panel for fire department use. The computer-type system is well fitted for this arrangement, with the special panel protected against unauthorized use by a key lock or special code. Many of the HVAC control manufacturers have a version of their HVAC supervisory control adapted especially for fire and/or security use.

Security systems get into many areas, such as door locks activated by magnetic cards, intrusion alarms, and proximity sensors. Closed-circuit TV can also be used, though this is not a part of the basic supervisory system.

10.4.4 Software

The function of the computer (or processor) is to supervise the operation of the system. It provides the necessary programs, timing, addressing, response to data, analysis and interpretation of operator input, and control of the communication system. These functions are performed by means of *software*, that is, the *programs* which the computer follows. A program is a step-by-step set of instructions which direct the computer in each of its actions. These instructions must be carefully written in a special language (which is different for each make of computer). For example, pressing a function key on the operator's console will trigger a lengthy instruction set, each step of which is performed in sequence with the desired end result obtained only if each step is correctly written and performed. The standard programs available from each control manufacturer have been carefully written and thoroughly tested for errors. This testing process is called "debugging" and can sometimes be very frustrating to the programmers and system analysts who are responsible for software development. "Customized" programs, written especially for a particular project can become very expensive. "Standard" software which has already been written and debugged is comparatively much less costly.

Program languages include such standard languages as FORTRAN and BASIC, but more often are special purpose languages developed for use with the system. Process controllers provide for input of program logic directly from the keyboard, sometimes using a special language or special terms. Operator commands, using the keyboard, are not "programs" but are used to override automatic program functions (as in "start-stop") or to modify the Data Base (as in changing set points).

Programs fall into several classifications. *Operating* programs perform the various functions for which the system is designed (control, monitoring, etc.). The *Data base* is a list of all points in the system, together with the address and data pertinent to each point. *Diagnostic* programs continually survey the computer system elements, diagnosing and reporting hardware problems. *Executive routines* are the basic programs which tell the computer how to perform each of the operations necessary for proper functioning.

Most control systems manufacturers now offer "modular software." These programs are written to cover specific functions such as timed start-stop of equipment, data acquisition and analysis, and "optimization" of various kinds. The programs are designed to operate on any system by accepting the data base for that system. This kind of software is comparatively inexpensive and allows the operator to take advantage of the computer's flexibility in ways which are beyond the capability of the simpler supervisory systems.

For many process control systems a small cassette tape is used to input programs. The programs are compiled on a large computer, then stored on the cassette tape, for play back into the process control computer (Figure 10–8). Major program changes are handled in this way, though most systems have programming capability from the operator's console. The term "keyboard programmable" usually refers to the ability to adjust limits, add or delete items in the data base, and similar functions which do not change the basic programs.

One of the major difficulties with computer-based systems has been, and still is, a lack of compatibility among the various systems available. Systems and devices made by one manufacturer cannot be used with those of another manufacturer because of differences in hardware and software design. The result is that once a particular manufacturer's equipment has been installed the owner must look only to that manufacturer for future changes and additions. This is primarily a software problem and relates to two areas: language and message protocol. Even such common languages as BASIC and FORTRAN may have slight differences from one computer to another. Many control systems require the use of special languages, completely incompatible with any other. Message protocol—the way in which messages are struc-

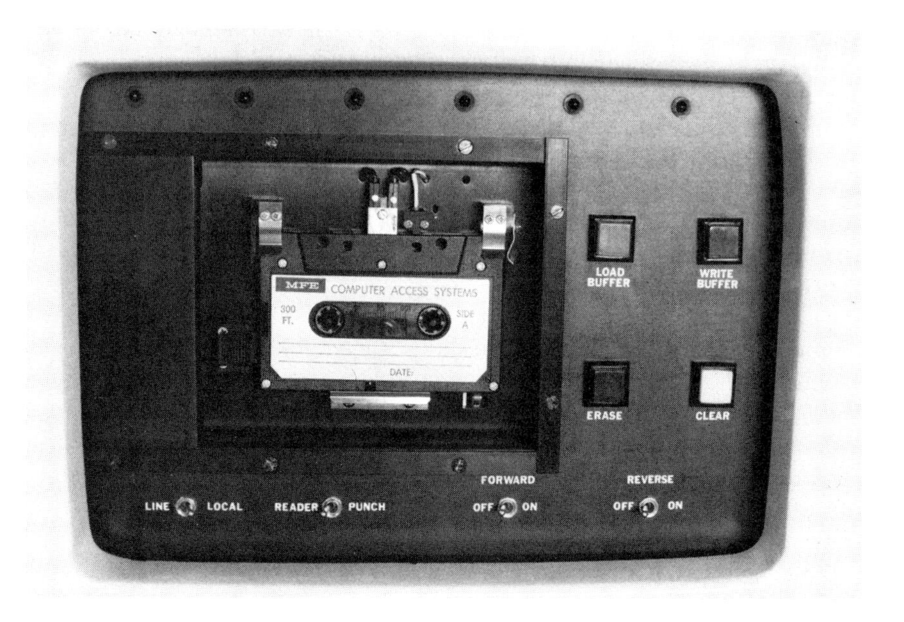

Figure 10-8 Cassette tape for program input. *(Courtesy Powers Regulator Co.)*

tured, scheduled and checked for accuracy—varies greatly among the various software suppliers.

There is, currently, a trend toward greater compatibility. A number of independent software suppliers are offering complete operating systems which can accept many different kinds of hardware. These suppliers can also provide custom software systems for equipment not on their regular lists.

10.4.5 Direct Digital Control

Direct Digital Control (DDC) means that a digital computer is substituted for the "regular" local loop controller. The DD controller will usually include most of the switching, selecting and discriminating functions as well. "Discrimination" in DDC can be very effective, since the DD controller can be programmed to reject any obviously erroneous signals. Controllers specifically for DDC are manufactured by many suppliers. In addition, Programmable Process Controllers (PPC) have been used for many years in the process control industry and are readily adaptable to DDC use in HVAC applications; HVAC is a process, too. A standard PC may also be used though it requires special input/output modules and software for real-time interface.

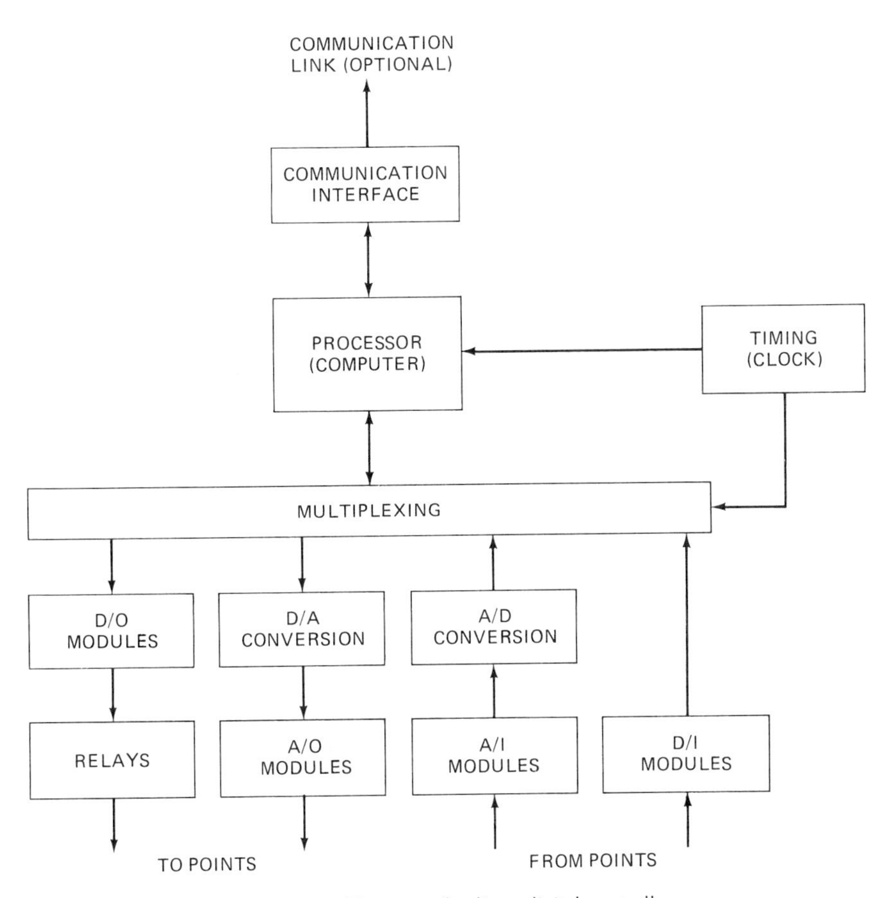

Figure 10-9 Elements of a direct digital controller.

The elements of a DDC system are functionally similar to those in the supervisory computer system. See Figure 10–9.

10.4.5.1 Local Loop Devices for DDC Local loop sensors and controlled devices are the same for DDC as for "ordinary" control. It is preferable to use electronic sensing since these signals can be used directly by the controller. For modulating output the simplicity, reliability and low cost of pneumatic operators make them preferable to electric or electronic operators, though a source of high-quality compressed air is required. Transducers are used to convert the controller output signal from electronic to pneumatic.

10.4.5.2 Input/Output Input/output modules are included with all controllers manufactured specifically for DDC. External interface modules are available for PPCs and PCs. Functionally they all operate exactly as described for the FID (paragraph 10.4.2.2 above) except that the DDC usually includes more sophisticated and complex control strategies. The number of points which can be addressed by a DD controller is a function of the controller design and internal memory size (no external memory is used). A typical controller can handle a minimum of 32 points, while some systems are much larger.

10.4.5.4 Operator Interface Most dedicated DD controllers have a small LED display, together with a special keyboard. By means of the keyboard the operator may ask for a status display of any point, may change a set point or limits, or enter new points in the data base. In some systems the keyboard/display element is separate and portable and is plugged into a controller at the operator's discretion. Some PPCs and PCs may include CRT displays, standard computer keyboards and even printers. Most DD controllers have the capability of interfacing with a CRT, standard keyboard and printer.

Figure 10-10 Direct digital controller. (*Courtesy Johnson Controls*)

10.4.5.5 The DDC as an FID A comparison of Figures 10–4 and 10–8 will show that there are many similarities between an FID and a DD controller. It follows that the advantages of DDC and supervisory monitoring and control may readily be combined. The ability to communicate with a supervisory computer is standard with most, if not all, of the hardware discussed in this section. There is a definite trend in the HVAC control industry toward this concept.

10.4.5.6 Software Software for DDC has all of the elements discussed above in paragraph 10.4.4. Because of limited memory, DDC software capability is somewhat less than that of the supervisory system. As noted above, a supervisory computer can be added if more capability is needed. Then it becomes quite easy to make changes and download programs from the supervisory computer to the DDC. Because the DD controller has greater capabilities than the FID, the supervisory computer has less to do and can deal with a greater number of points when combined with DDC.

10.5 THE ECONOMICS OF SUPERVISORY SYSTEMS

Almost the entire justification for supervisory systems is the potential savings in energy and personnel. Originally, labor-saving was more important, especially in large campus-type installations where the ability to monitor and trouble-shoot from one central location saves thousands of man-hours annually. However, it must be noted that the addition of a computer-based system includes a requirement for special training of maintenance and operating personnel. The advent of the *energy crisis* made *energy conservation* (a better term than *energy saving*) the more important factor. The so-called *optimization* programs were developed for the express purpose of maximizing efficiency and minimizing energy consumption.

The potential savings in any given situation may be estimated by comparing unsupervised operation to a carefully supervised and controlled operation. The savings must then be compared to the amortization cost of the supervisory system for justification.

In a large campus-type situation with many buildings, the economics of the supervisory system can be readily demonstrated, especially when the system includes extensive special programs for optimization and maintenance. But what of smaller installations? It is this author's belief that some type of supervisory control can be justified for any commercial building. The simple systems described earlier in this chapter can be installed at very low cost, but even something as simple as a central hardwired panel with a time clock or two for automatic start-up and shut-down can save operating cost.

The simpler special-purpose computer-type systems are useful in even fairly large installations.

The designer must therefore ask "What does the owner want, what does the owner need, and what can the owner afford?" The system that fits these criteria should be recommended and used.

10.6 BENEFITS OF THE COMPUTER SYSTEM

The computerized system has some interesting and valuable side benefits (for only slight additional costs).

10.6.1 Optimization

One of these side benefits is a program which will continuously analyze performance of the various systems, compare actual performance with some "ideal" standard and make adjustments to approach this standard and thus improve efficiency.

10.6.2 Data Management

Since the computer has already acquired a great mass of data about operating conditions, it is a simple matter to have this digested and printed out on any schedule and in any form desired. Some of that data may be very valuable for evaluating the existing design and improving future designs.

10.6.3 Maintenance Schedules

The regular preventive maintenance programs can be stored in the computer memory and printed out on schedule to remind the operator that they must be performed. As a further check, he must notify the computer of proper performance on schedule, or a follow-up notice will be printed by the computer.

Maintenance sensors can be made a part of the monitoring equipment. Such sensors will detect overheating of motors or bearings, excessive vibration, loud and extraneous noises and numerous other malfunction indicators.

10.6.4 Energy Consumption

Where a central chilled water or heating system serves a number of tenants, it may be necessary to meter the services supplied to each tenant. Special meters with outputs which can be read by the computer may provide an automatic measuring and billing service. Further, any sudden change from normal in a tenant's consumption would be instantly detected, and cause a warning message to be printed for investigation by operating personnel.

10.7 TRAINING FOR MAINTENANCE AND OPERATION

One of the most important considerations with a computer-based control system is the need for properly trained and motivated personnel for operation and maintenance. The basic rule is the same for any situation: "No system is better than the people who operate it." The operating people should preferably be involved during design and construction of the system, with training furnished by the contractor. Training should continue after the system is installed and working.

10.8 SUMMARY

The principles of supervisory control described in this chapter are relatively constant. The details of implementation are changing almost daily, due to the highly competitive computer industry. The control system designer must determine the latest state-of-the-art and then apply it, using fundamental principles, to satisfy his needs.

11 Psychrometrics

11.1 INTRODUCTION

Psychrometrics is the particular branch of thermodynamics devoted to the study of air and water vapor mixtures, commonly referred to as moist air. Since moist air is, in most cases, the final transport medium used in the air conditioning process, psychrometrics is of great interest to the HVAC system designer and operator.

This chapter will present a simple discussion of psychrometric charts and their uses with a minimum of theory. For a detailed theoretical discussion and complete psychrometric tables, see the ASHRAE HANDBOOK, FUNDAMENTALS, current edition.

11.2 PSYCHROMETRIC PROPERTIES

The properties of the air/water vapor mixtures which are used in HVAC design include dry bulb, wet bulb and dew point temperatures; relative humidity; humidity ratio; enthalpy; density and atmospheric pressure.

11.2.1 Temperature

Dry bulb temperature (DB) is measured with an ordinary thermometer. "Temperature", when not otherwise defined, means dry bulb temperature. *Wet bulb temperature* (WB) is measured using a thermometer with a wet cloth sock wrapped around the bulb. The air being measured is blown across the sock (or the thermometer is moved through the air) allowing moisture to evaporate. Evaporation has a cooling effect which is directly related to the moisture content of the air. Wet bulb temperature will thus be lower than dry bulb temperatures unless the air is *saturated* (100 percent relative humidity). The difference between the two temperatures is called the *wet bulb depression. Dew point temperature* (DP) is the temperature to which a given sample of air must be cooled so that moisture will start condensing out of it. When the air is saturated the dry bulb, wet bulb and dew point temperatures will all be equal.

11.2.2 Relative Humidity

Relative humidity (RH) expresses the relationship of the amount of moisture in the air to the amount the air would hold if saturated at that dry bulb temperature. It is defined as the ratio of the partial pressures of the water vapor at the two conditions. (See the ASHRAE HANDBOOK.) *Percent humidity* is not the same as relative humidity, and is not used in HVAC design.

11.2.3 Humidity Ratio

Humidity Ratio, sometimes called specific humidity, designated by the symbol "w," is the amount of water in the air expressed as a ratio: pounds of water per pound of dry air. In some places grains of water per pound of dry air is used. 7000 grains equal one pound.

11.2.4 Enthalpy

Enthalpy, designated by the symbol "h," refers to the heat content of the moist air, in Btu per pound of dry air. As used in psychrometrics it is not an absolute value, and relates to an arbitrary zero, usually at zero F. For this reason, differences between two values of *h* are valid, but ratios are not.

11.2.5 Density and Volume

Density refers to the weight of the moist air, with units in this text of pounds of dry air per cubic foot. *Volume* is the reciprocal of density.

11.2.6 Atmospheric Pressure

Variations in atmospheric pressure due to elevation above or below sea level have an important effect on the values of the various properties. This is because the total pressure of the mixture varies with atmospheric pressure while the partial pressure of the water vapor in the mixture is a function only of dry bulb temperature. High altitude tables and charts are available for elevations to about 7500 feet above sea level, while the Bureau of Mines provides a low altitude chart for depths to about 10,000 feet below sea level.

11.3 PSYCHROMETRIC TABLES

Tables of psychrometric properties are available from several sources, including the ASHRAE HANDBOOK OF FUNDAMENTALS. For elevations to about 2000 feet above or below sea level the "standard pressure" tables may be used. For higher and lower elevations, new tables can be calculated using basic psychrometric equations.

11.4 PSYCHROMETRIC CHARTS

A basic tool used in HVAC design is the psychrometric chart. This chart is simply a graphical representation of the properties described above. The ASHRAE chart, Figure 11–1, used for illustration in this text, is a Mollier-type chart and is very useful for illustrating HVAC cycles and finding state points at various stages of each cycle.

11.4.1 State Points

Any point on the chart can be a *state point*. It can be defined and located by the values of any two properties. Once the point is located, the values of all other properties can be read (Figure 11–2).

Notice the relationship of the various lines on the chart. The basic coordinates on which the chart is drawn are enthalpy (h) (lines sloping down from left to right) and humidity ratio (w) (horizontal lines). Dry bulb temperature lines are nearly vertical and not quite parallel, but are uniformly spaced. Wet bulb temperature lines slope down from left to right and are not parallel to the enthalpy lines or to each other. They are not uniformly spaced; intervals between lines increase with temperature.

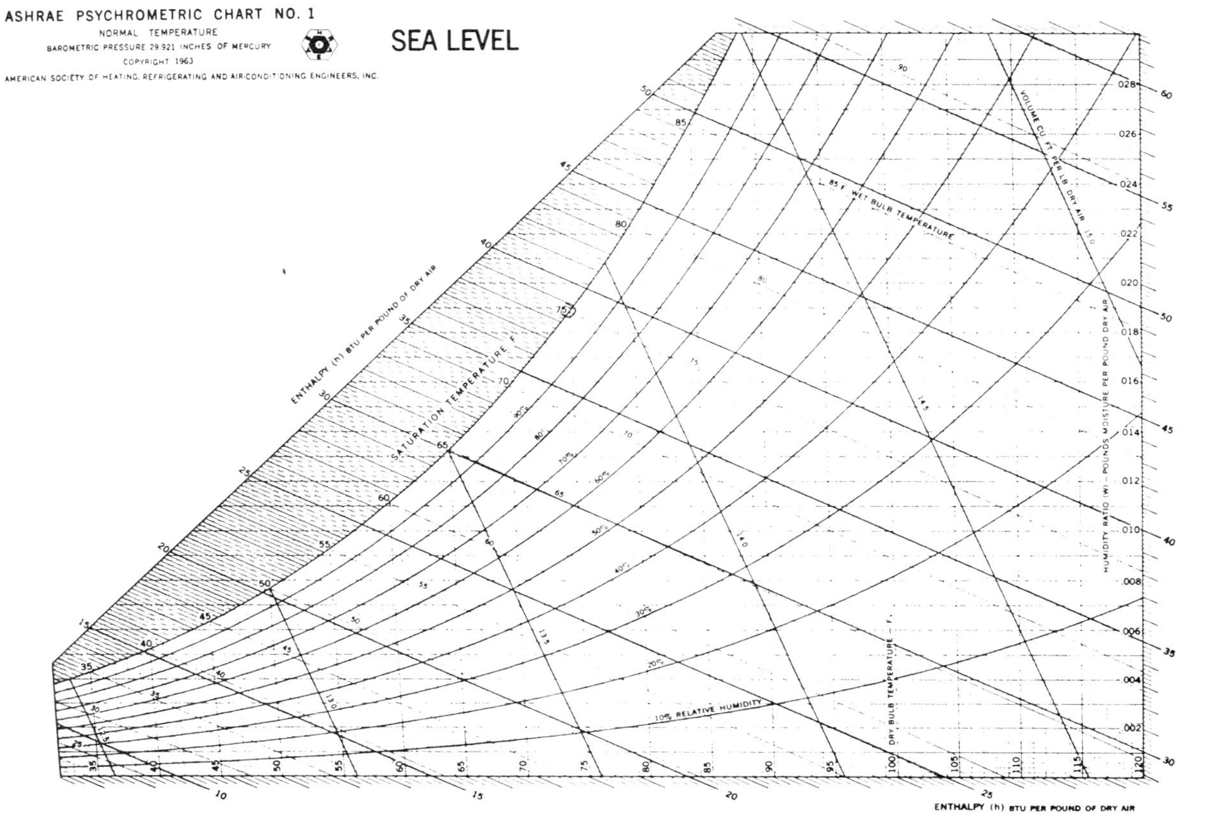

Figure 11-1 ASHRAE psychrometric chart. (Reproduced by Permission.)

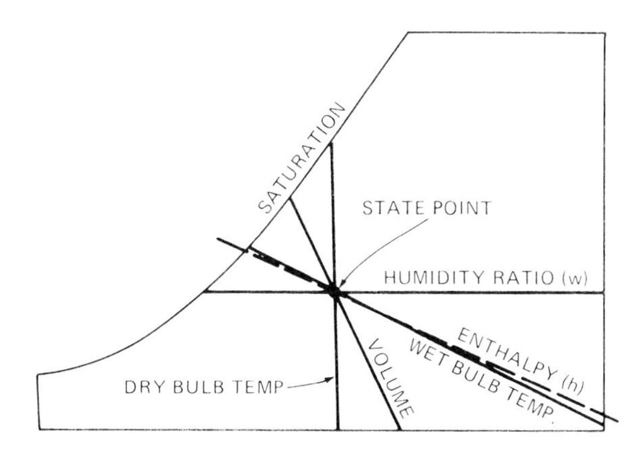

Figure 11-2 Relationships of lines of properties on the psychrometric chart.

A series of curving lines represent relative humidity, with the saturation curve equal to 100 percent RH. Saturation means that the air is holding all the moisture it can at that temperature and pressure. RH lines are almost but not quite uniformly spaced.

Lines of volume slope down from left to right and are parallel and uniformly spaced.

As Figure 11–2 shows, any point represents an intersection of all six properties, and values can be read by interpolation. For clarity, most of the enthalpy lines are shown only at the edge of the chart and may be read in the body of the chart by means of a straightedge.

11.5 PROCESSES ON THE PSYCHROMETRIC CHART

Because the chart is a Mollier-type the various HVAC processes can be readily represented as straight lines connecting two or three state points.

11.5.1 Mixing

Mixing of two air streams can be shown on the chart as a straight line connecting the state points of the individual air streams. The state point of the mixture will fall on the line, dividing in into two segments which are proportional to the volumes of the two air streams. For example, Figure 11–3 shows mixing of return air and outside air, with proportions of 80 percent

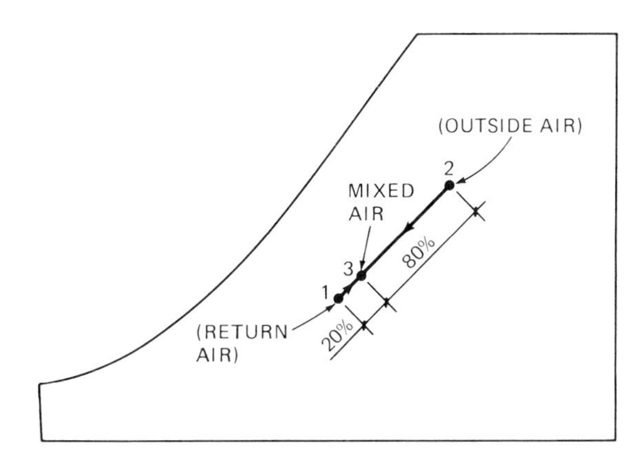

Figure 11-3 Mixing of two air streams. (Example).

return air and 20 percent outside air. Note that the mixture point is closer to the state point representing the larger air quantity.

Mixtures of 3 or more air streams require that any two points be used first, with the resulting mixture combined with the third point, etc.

11.5.2 Sensible Heating and Cooling

The term *sensible* applied to heating or cooling means that no moisture is added or subtracted as the air temperature is increased or decreased. These processes are therefore represented on the chart as horizontal lines with a constant value of specific humidity (w). (See Figure 11–4.)

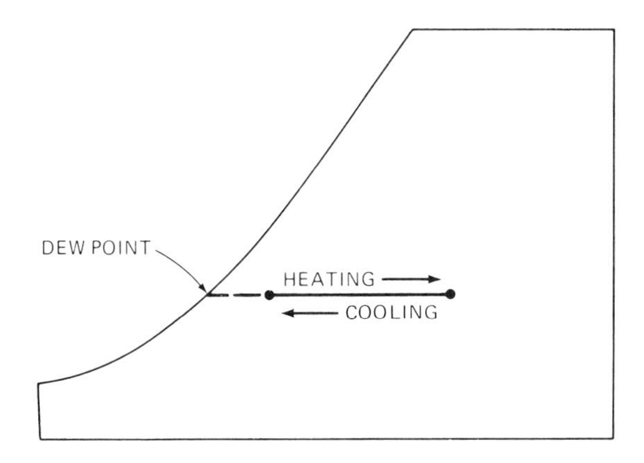

Figure 11-4 Sensible heating and cooling.

11.5.3 Cooling and Dehumidifying

If the process of sensible cooling is continued (in Figure 11–4) until the process line intersects the saturation curve, the air is said to have been cooled to its dew point temperature. Any further cooling requires the removal of moisture. This process would follow the saturation curve down and to the left.

The typical extended-fin-and-tube cooling coil does not have the capability to cool all of the air passing through it to saturation and beyond. Some percentage of the air stream will pass through the coil without any contact with tubes or fins. This is known as the coil bypass factor and is typically 5 to 10 percent, with smaller factors resulting from increased numbers of rows and closer fin spacing.

The result is actually a mixture, with some air cooled to the *Apparatus Dew Point* and some unchanged. This is shown graphically in Figure 11–5.

11.5.4 Chemical Dehumidifying

Chemical dehumidification is used to obtain very low humidities. It is usually an adsorption process, using silica gel or some similar moisture adsorbent. The process is an indeterminate curve on the chart, with the final state point determined from the equipment manufacturer's data. The air temperature always increases. (See Figure 11–6.)

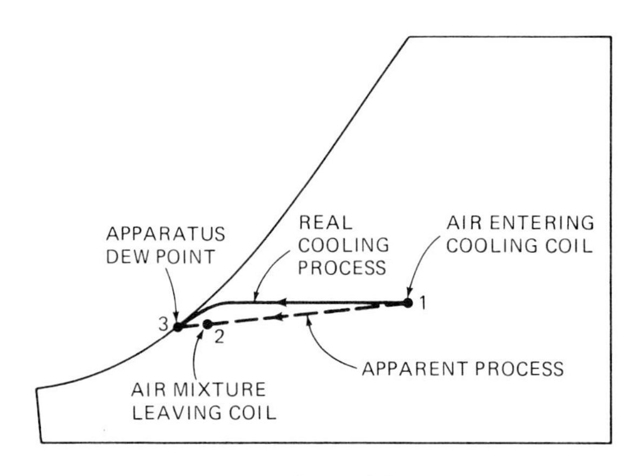

Figure 11-5 Cooling and dehumidifying.

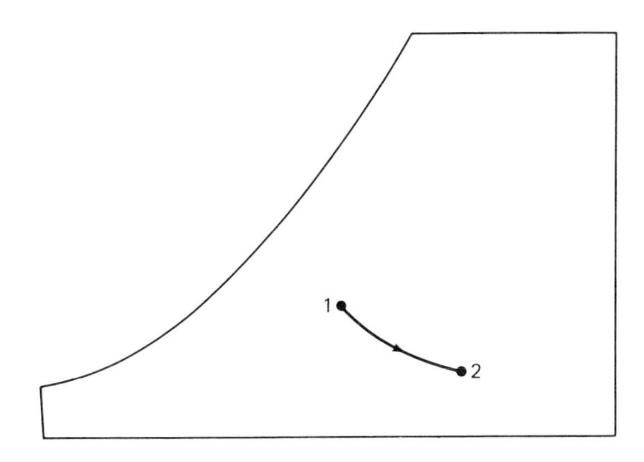

Figure 11-6 Chemical dehumidification.

11.5.5 Humidifying

Humidity may be added to the air stream in various ways, including evaporative cooling as described in paragraph 11.5.6. If humidity is added by means of a steam grid or a heated pan the process may be shown on the chart as in Figure 11–7. The process line slopes up (humidity increase) and to the right (heat added).

Adding humidity be means of unheated evaporator pans or atomizing sprays is equivalent to evaporative cooling.

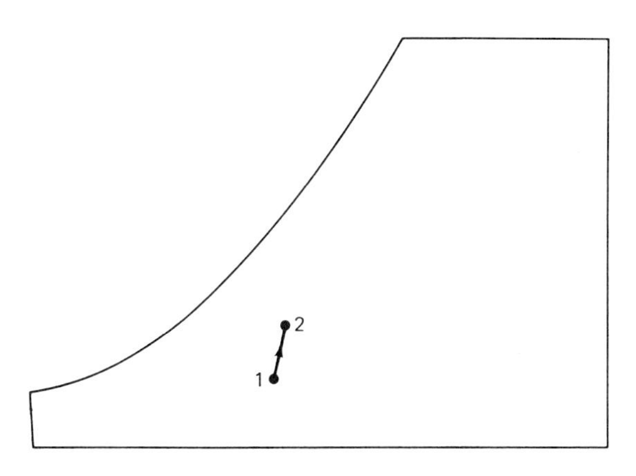

Figure 11-7 Humidification.

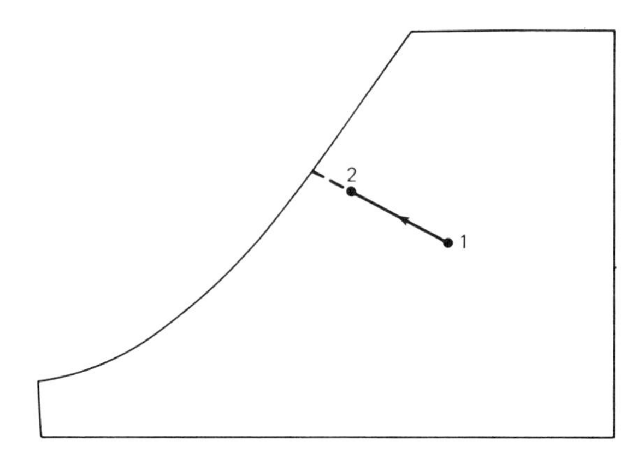

Figure 11-8 Evaporative cooling.

11.5.6 Evaporative Cooling

Evaporative cooling is a process in which water vapor is added to the air stream by adiabatic evaporation. That is, no heat is added to or subtracted from the system. The heat required to evaporate the water (latent heat) is obtained by cooling the air. On the chart, Figure 11–8, this shows as a constant wet bulb temperature process, sloping up from right to left (increasing specific humidity, decreasing dry bulb temperature). If the evaporative cooling system was 100 percent efficient the final state point would be on the saturation curve. In practice, efficiencies range from 60 to 95 percent.

11.6 HVAC CYCLES ON THE CHART

A complete HVAC cycle on the chart will include several of the processes described above. A typical cooling cycle is shown in Figure 11–9. Room and outside air conditions are determined from design criteria. For simplicity, return air is assumed to be the same as room air. The mixture condition is based on the minimum outside air required by the design. Point E represents the condition leaving the cooling coil and is determined from the desired room condition, the design CFM and the design latent and sensible cooling loads. Some reheat may be required or may occur in the distribution duct system and this is shown as the line from points E to F. Point F is the condition of the air entering the room.

A typical heating cycle with humidity control is shown in Figure 11–10. At winter outside air design, humidity must be added.

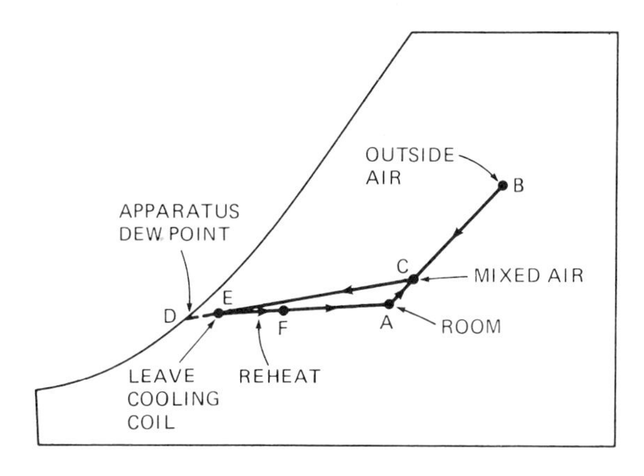

Figure 11-9 Complete HVAC cycle (cooling).

The chart is also useful for investigating the cycle at intermediate outside conditions. Figure 11–11 shows a cycle based on outside conditions of about 60°F with high relative humidity. This shows that the mixed air is nearly 100 percent outside air under economy cycle control, but some cooling is necessary to provide dehumidification, and there is considerable reheat.

11.7 IMPOSSIBLE PROCESSES

Sometimes it is not possible to go directly from one state point to another. That is, while a line can be drawn on the chart, the process is thermodynamically impossible. Drawing the cycle on the chart will clearly show

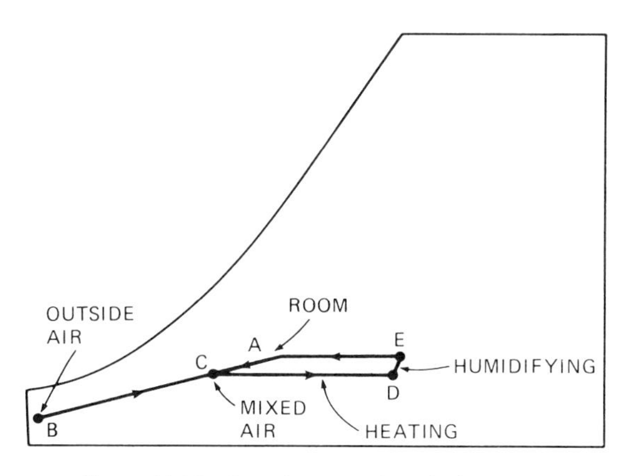

Figure 11-10 Complete HVAC cycle (heating).

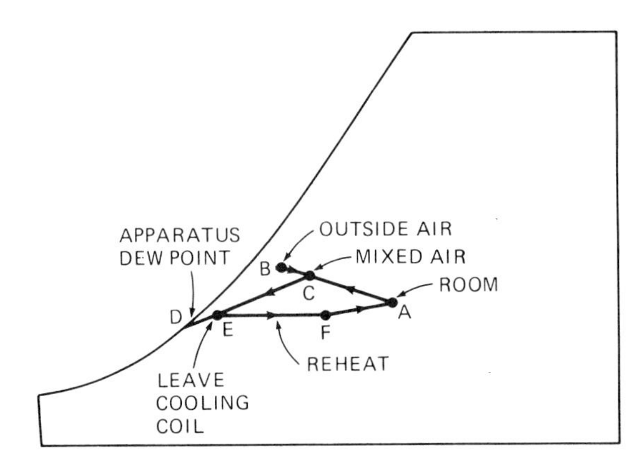

Figure 11-11 Complete HVAC cycle (intermediate).

the impossibility and suggest alternative ways to accomplish the desired result.

The most common problem is the "missing ADP" (apparatus dew point). The process of cooling air through a cooling coil requires that there be an ADP which can, in fact, exist. If the specified coil entering and leaving air conditions result in a process line which does not intersect the saturation curve, as in Figure 11–12, then the process is physically impossible. The required result can be obtained by moving the leaving coil condition to the left until an ADP is possible, and then adding a reheat process to get to the final state point (Figure 11–13). Of course, the refrigerant or chilled water in the cooling coil must have a temperature below the ADP.

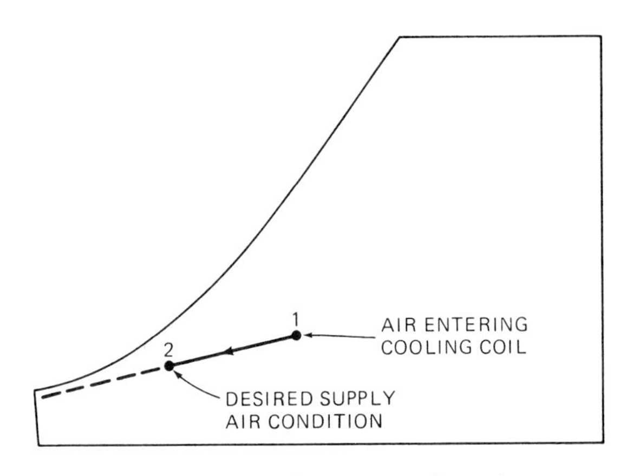

Figure 11-12 Missing apparatus dew point.

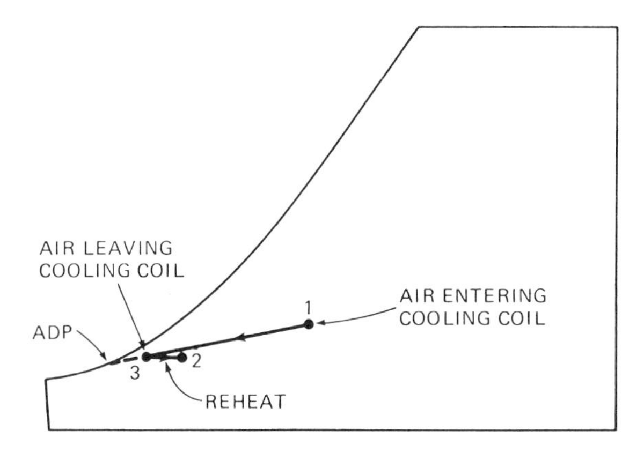

Figure 11-13 Correcting for missing ADP.

11.8 EFFECTS OF ALTITUDE

Many locations where air conditioning is used are at altitudes of several thousand feet above sea level. As noted above, the difference in pressure due to altitude becomes significant above about 2000 feet above sea level. Data obtained from a standard pressure chart may be in error at higher altitudes. High altitude charts are available and should be used when appropriate.

The effect of decreasing atmospheric pressure is to "expand" the chart

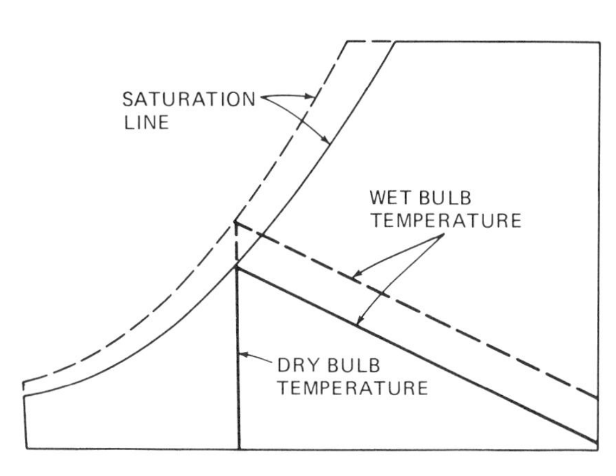

Figure 11-14 Effect of altitude on the psychrometric chart. Atmospheric pressure decreases with elevation above sea level. Solid lines—lower altitude (higher pressure). Dotted lines—higher altitude (lower pressure).

(Figure 11–14). That is, given an unchanged coordinate grid of enthalpy and humidity ratio, as total pressure decreases the chart is affected as follows:

a. Dry bulb lines are unchanged.
b. The saturation curve and RH curves move upward and farther apart.
c. Wet bulb lines move farther apart.
d. Volume lines move to the right and upward.

11.9 SUMMARY

This discussion of psychrometrics is simplified but comprehensive. Psychrometric chart applications to control design are used extensively in this text and the author considers psychrometrics an essential tool in the design of both HVAC systems and their controls.

12 Central Plant Pumping and Distribution Systems

12.1 INTRODUCTION

A "central plant" refers to a grouping of one or more chillers and/or boilers supplying chilled and hot water (or steam) to HVAC units located at various points in a building or complex of buildings. The plant is not necessarily centrally located with respect to the HVAC units. Some plants may have only one or two elements while others may contain a dozen or more chillers and boilers. Some large plants, which have grown over a period of years as a campus developed, have been spread out in two or three locations as the originally assigned spaces became too small.

The discussions in this chapter are concerned with plants of two or more chillers and boilers circulating normal temperature hot and chilled water to a scattered group of HVAC units. Steam and high temperature hot water are excluded. The typical plant consists of chillers, boilers, circulating pumps, expansion/pressurization equipment, makeup equipment, chemical feeders, controls and accessories. Chilled water supply temperature is between 40° and 45° with return water design temperature 12 to 15 degrees higher. Hot water system supply temperatures are usually 160° to 200°F with return temperatures 40 degrees lower. The hydraulic principles and energy conservation procedures are the same regardless of system size. *Diversity* becomes more of a factor as system size increases.

12.2 DIVERSITY

Diversity is very important in the design and operation of a central plant. It refers to two factors:

1. An HVAC unit seldom or never operates at peak design conditions.
2. In a group of HVAC units, no two units reach peak load at the same time.

The result of these effects is that the central plant capacity need not be equal to the sum of the HVAC unit design loads. The amount of allowable reduction is known as the *diversity factor* and is expressed as a decimal determined by the ratio of the "connected load" to the central plant capacity. Connected load is the sum of the design loads of all the HVAC units served by the central plant. For example, typical diversity factors for a college campus are 0.60 to 0.70.

12.3 CONSTANT FLOW SYSTEMS

Almost all small central plants and many larger ones are operated as constant flow systems (Figure 12–1). The same quantity of water is circulated at all times, regardless of load. At light loads the difference between supply and return water temperatures may be only one or two degrees. Chillers and boilers may be turned off but water continues to circulate through all equipment in order to maintain flow.

This situation occurs because of the use of three-way valves on HVAC units. Flow control simply consists of turning pumps off and on. Attempts to

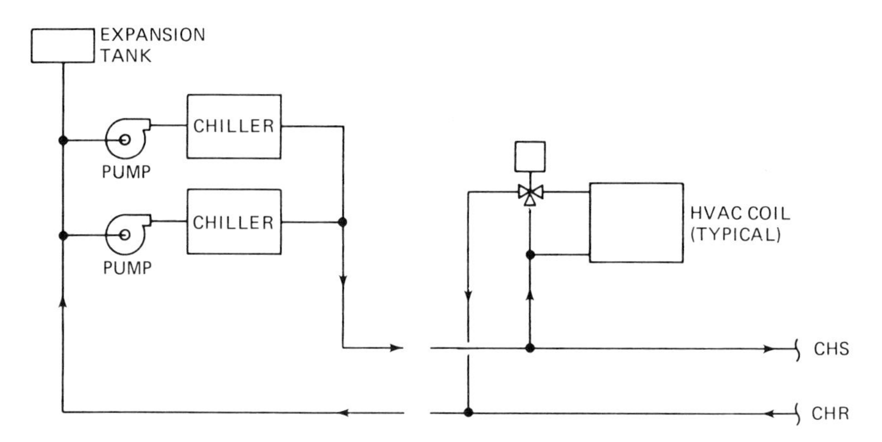

Figure 12-1 Central plant; constant flow arrangement.

save energy by turning pumps off will usually result in inadequate flow and/or pressure at the "end-of-the-line" HVAC units. Turning off a chiller (or boiler) while continuing to circulate water through its results in "mixture" supply temperature which is higher (or lower) than design.

Controls alone cannot improve the performance and decrease the energy consumption of such systems. System redesign is required to arrive at one of the variable flow methods described below.

12.4 VARIABLE FLOW SYSTEMS

One reason for using constant flow systems is the need to maintain a constant water flow rate through any on-line boiler to chiller to prevent damage and maintain efficiency. A variable flow system is designed to do this while allowing the flow in the distribution system to vary with load. This also will maintain the system design Delta T (temperature difference between supply and return) which allows more efficient operation of chillers and boilers.

Two general arrangements are used.

12.4.1 Bypass Control

The bypass arrangement in Figure 12–2 utilizes a pressure operated bypass valve and straight-through (two-way) valves at the HVAC unit

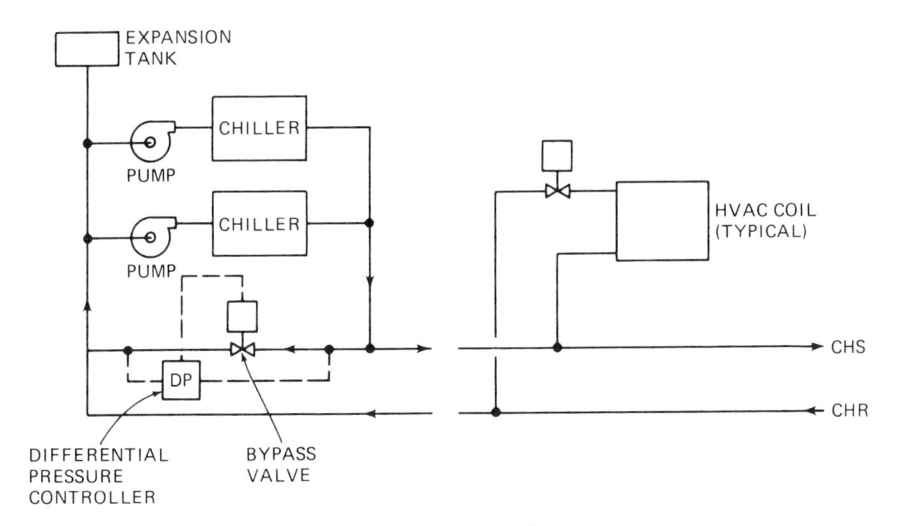

Figure 12-2 Central plant; bypass flow arrangement.

coils. As the HVAC unit control valves modulate, the distribution system flow will vary. This results in a change in differential pressure between supply and return mains which is sensed by the pressure differential controller. The controller then modulates the bypass valve to compensate. Flow through the chillers remains essentially constant.

The bypass valve should be sized for the flow rate of one chiller. Limit switches on the bypass valve should be used to indicate 10 percent and 90 percent open. If the plant is started at light load with one chiller and its pump running, the bypass will be in some modulated position. As HVAC load increases the bypass will modulate toward the closed position. Limit switch closure can be used to automatically start a second chiller and pump, or alert the operator to do so. Further load increases may bring on more chillers. As the load decreases, the opposite sequence takes place.

12.4.2 Chiller Pumps plus System Pumps

Figure 12–3 shows what may be considered an "optimum" central plant arrangement. Here the chiller pumps circulate at a constant rate through a closed loop which includes only the chillers and loop piping. Any

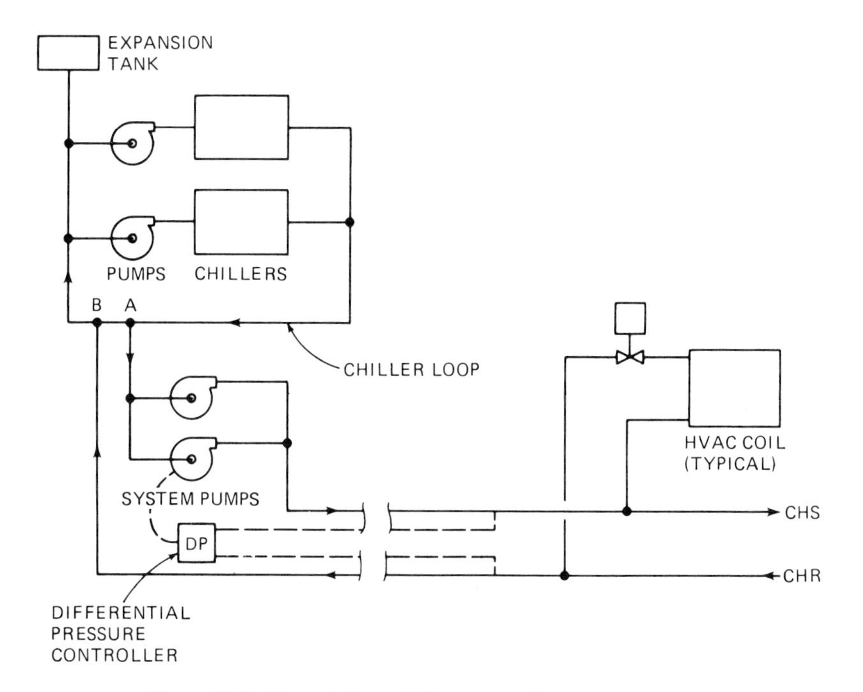

Figure 12-3 Central plant; chiller pumps plus system pumps.

number of chillers can be on- or off-line depending on the load. "System" pumps are used for the distribution with system supply and return connections to the chilled water loop located close together (points A and B). Distribution flow will vary with load. This short section of piping, common to both pumping circuits, is known as a *hydraulic isolator*. It must have a low resistance to flow compared to the whole system (keep it short and full-line size) and have no restrictions or valves. When the two circuits are hydraulically isolated, changes in flow in one circuit have no effect on flow rates in the other circuit. System pumps may be sequenced to maintain the desired minimum pressure differential between supply and return mains. Ideally, one of the system pumps will be a variable speed typ so that a constant differential pressure may be maintained.

12.5 DISTRIBUTION SYSTEMS

Piping distribution systems connect the central plant to the HVAC units. There are three arrangements in use: out-and-back, reverse return and loop.

12.5.1 Out-and-Back Distribution

This arrangement, shown in Figure 12–4, is the oldest and most common system. It has the virtue of simplicity since supply and return mains decrease in size equally as branch takeoffs for HVAC units occur. The pumps must be sized to overcome the pressure drop in the entire line plus the losses in the HVAC unit at the end. Thus, pressure differential from supply to return mains may be excessive at HVAC units near the central plant, making them more difficult to control. To clarify this, refer to Figures 12–5 and 12–6. Figure 12–5 shows the hydraulic profile at design water flow rate with balancing valves set properly. In Figure 12–6 at 50 percent of design flow rate

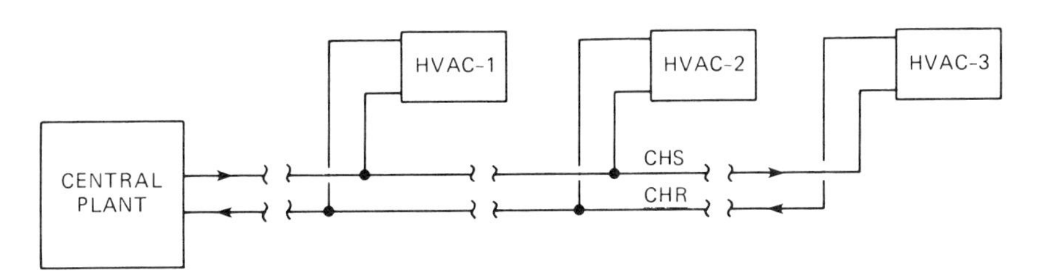

Figure 12-4 'Out-and-back" distribution.

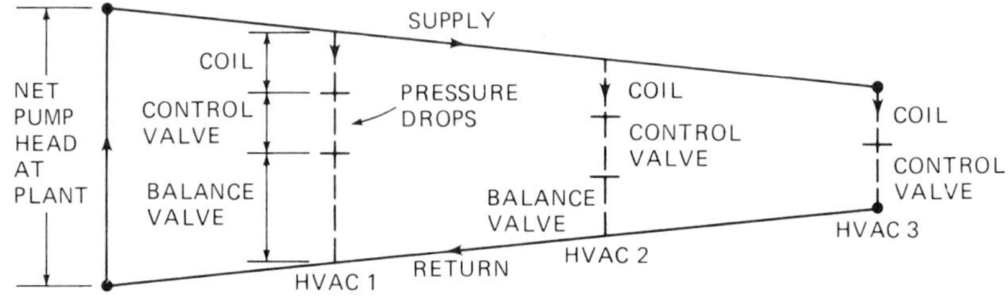

Figure 12-5 "Out-and-back" hydraulic profile; design flow rate.

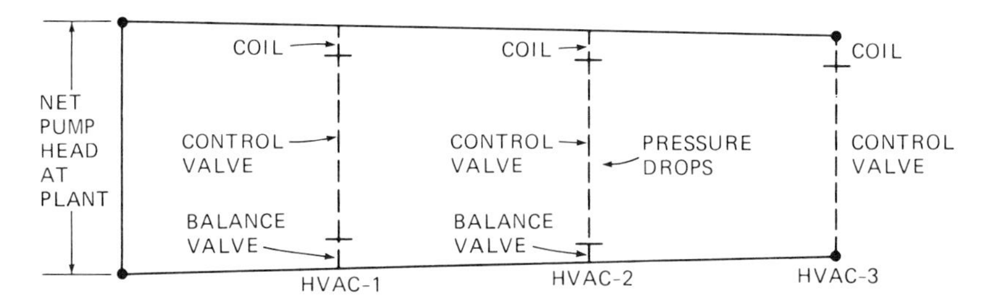

Figure 12-6 "Out-and-back" hydraulic profile; 50% of design flow.

the pressure drops through coils and balancing valves are 25 percent of design, and the control valve must make up the difference.

It is very difficult to add another HVAC unit on this system. Some main piping may become too small and the entire system must be rebalanced.

12.5.2 Reverse-Return Distribution

The reverse-return system is designed to equalize distribution pressure drops throughout the system. As shown in Figure 12–7 the return

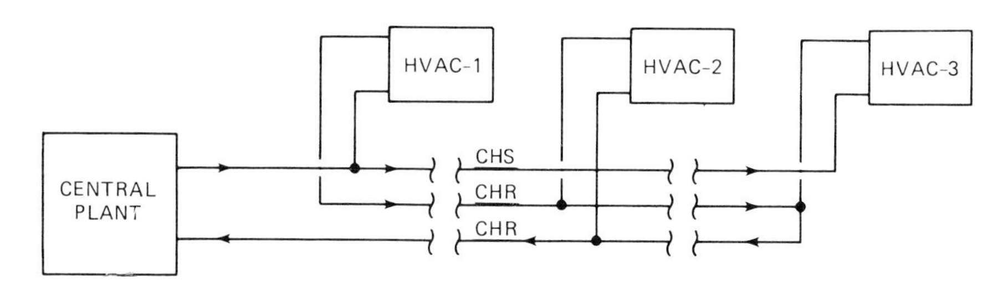

Figure 12-7 Reverse-return distribution.

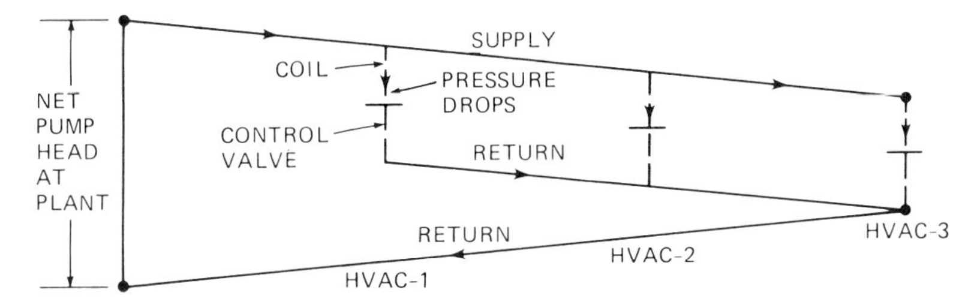

Figure 12-8 Reverse-return hydraulic profile; design flow rate

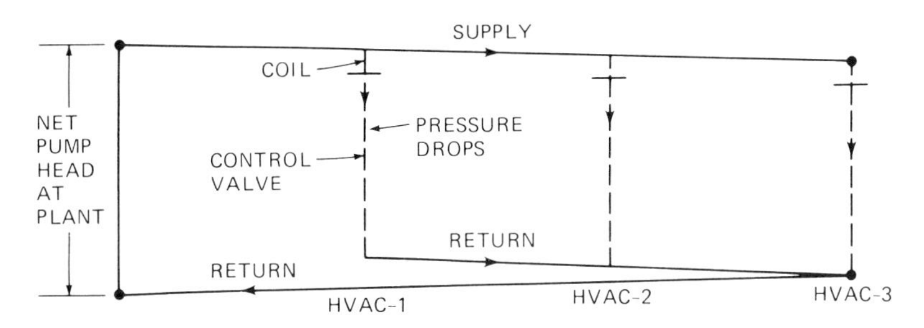

Figure 12-9 Reverse-return hydraulic profile; 50% of design flow.

line starts with the first HVAC unit on the supply line and flows parallel to the supply line until end of supply is reached. Then the return line reverses and runs back to the central plant. Thus the total length of distribution piping (and pressure drop) is approximately equal for all HVAC units. This system is more expensive than the out-and-back system but is easier to balance and control. The hydraulic profiles at design flow and 50 percent of flow are shown in Figures 12–8 and 12–9. Again, it is difficult to add HVAC units to this system.

12.5.3 Loop Distribution

For a large campus, or even a large building, and especially where changes and additions to the HVAC units are expected, a loop distribution system is preferred. The basic scheme is shown in Figure 12–10. The loop mains are of one uniform size throughout. This size is approximately 40 percent of the initial main size in a comparable out-and-back system for the same design pressure drop. Each loop is hydraulically self-balancing. Flow takes place in both directions, with pressure drop resulting, and at some point there is an equal pressure from both sides resulting in zero flow at that point. The point of zero flow/balanced pressure will automatically adjust to

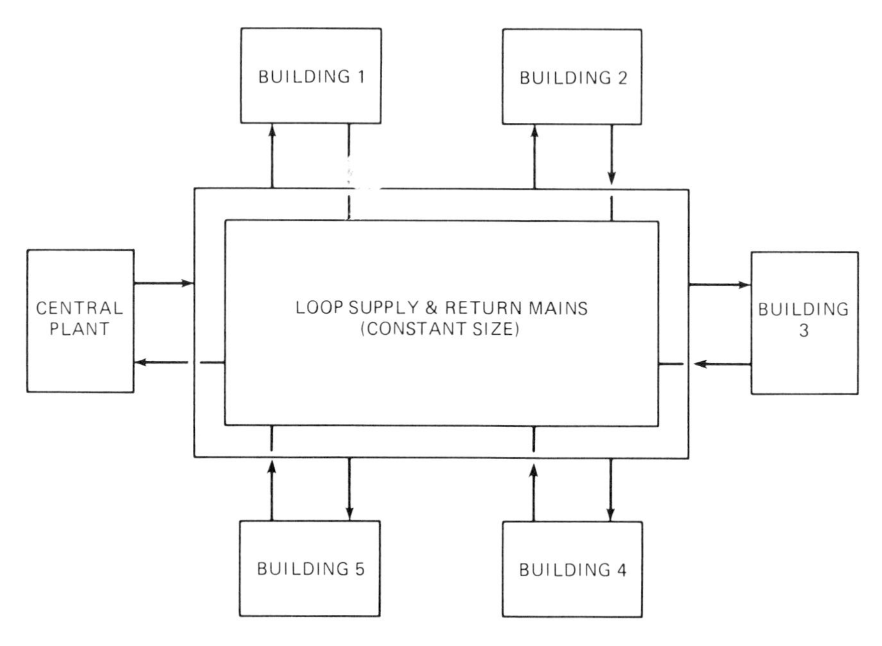

Figure 12-10 Loop distribution.

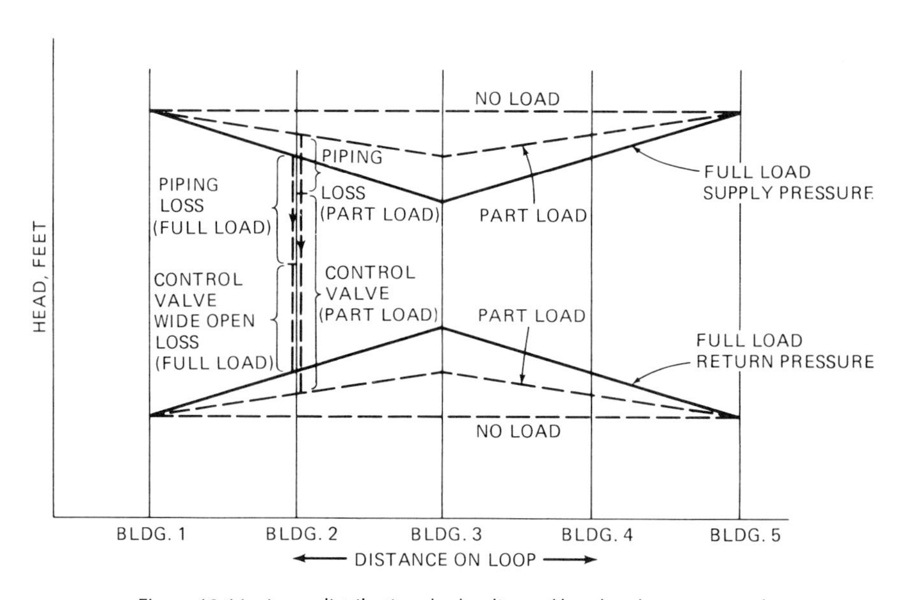

Figure 12-11 Loop distribution; hydraulic profile, plant bypass control.

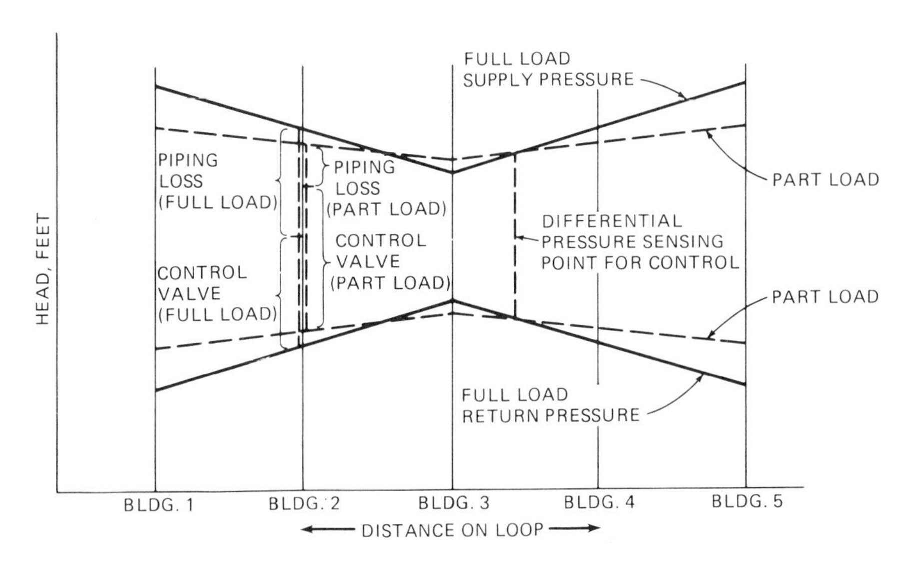

Figure 12-12 Loop distribution; hydraulic profile, variable speed pumping.

any changes in load including addition and deletion of HVAC units or buildings.

The hydraulic profile in Figure 12–11 assumes that a fixed differential pressure is maintained at the chiller plant, as would be the case with Figure 12–2, bypass control. If the system pump arrangement shown in Figure 12–3 is used, with variable speed pumping and a differential control point out in the system away from the central plant, then system pump head will be decreased at part load, saving energy (See Figure 12–12.)

12.6 BUILDING INTERFACES

When a central plant serves several buildings, those buildings may vary greatly in size, load and internal pressure losses. If the system pumps are required to provide sufficient pressure to overcome the building losses, some severe energy penalties will result. It is typical, therefore, to provide secondary pumps, at least at the larger buildings, to avoid this penalty. The important criterion when connecting secondary pumps is to provide hydraulic separation. That is, the operation of the secondary pump should not influence the pressure differential in the primary distribution system. Additionally, it should be possible to control the return water temperature from the building to maximize the water temperature differential in the primary distribution.

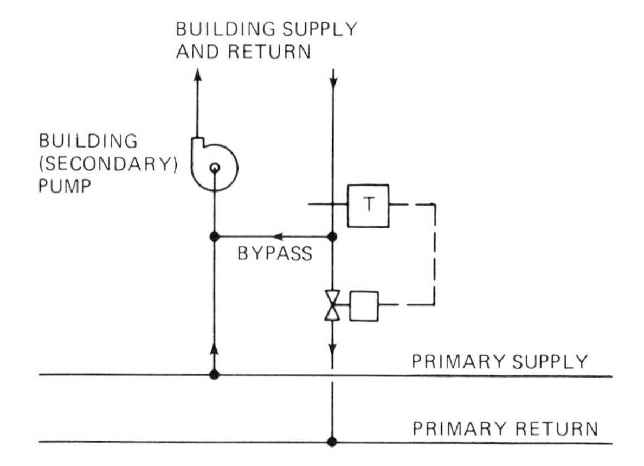

Figure 12-13 Building interface with secondary pump.

There are a number of secondary connections in common use. Not all of these meet the above criterion. One which does is shown in Figure 12–13. The thermostat modulates the control valve to maintain a constant return water temperature as the load varies. The control valve "sees" only the pressure differential between primary supply and return since the bypass line has no restrictions. The bypass line provides hydraulic isolation between primary and secondary systems.

With this arrangement the building water temperature will increase as building load decreases. Where a low dew point is required for humidity control this may not be satisfactory. The return water temperature can be reset as a function of building humidity to satisfy this requirement.

12.7 SUMMARY

Existing central plant systems use many arrangements and control methods. Not all of these are satisfactory in performance and most use excessive pumping and equipment energy. Obviously, there is ample opportunity for improvement in system design and control. This chapter is by no means a comprehensive study and the reader should refer to other more detailed references such as the ASHRAE HANDBOOK, SYSTEMS volume, and references 23 and 29 in the Bibliography.

13 Retrofit of Existing Control Systems

13.1 INTRODUCTION

The "energy crisis" has created an awareness of the inefficiency of many existing HVAC systems and their controls. The majority of the systems installed in the 50s, 60s and early 70s were concerned only with control for comfort, and minimization of energy consumption was seldom considered. Also, many existing systems do not perform in a satisfactory manner for either comfort or economy.

This chapter will consider some of the more common existing HVAC systems and show how they can be most easily retrofit to save energy while still providing adequate comfort control. These are simply suggestions, not guaranteed cures. Each individual situation requires careful study of the system, the building and the application. Then the basic principles of HVAC system and control design can be applied. Because most existing control systems are pneumatic, this discussion will use pneumatic devices for illustration. Pneumatic indicators are shown in the figures. These would be used in control panels. Positive positioners for valves and dampers are recommended, though not always shown.

13.2 ECONOMIC ANALYSIS

It will often be necessary to make an economic analysis of the cost-effectiveness of any HVAC System/Control retrofit. Methods of making these analyses are described in many references, including the ASHRAE HANDBOOK. There are a number of computer programs available from public and private sources. These programs do not always model the specific control strategies desired, so care must be taken in using them. "Bin method" analyses are adequate in many cases.

13.3 DISCRIMINATORS

Many of the diagrams which follow suggest the use of discriminators. When using discriminators keep in mind the problems which may be encountered as described in paragraph 7.3.1.

13.4 CONTROL MODES

Most existing systems will be using proportional controllers. It is almost always beneficial to replace them with proportional plus integral controllers. An improvement in both energy efficiency and control accuracy should result. See Chapter 1 for a discussion of control modes.

13.5 ECONOMY CYCLE CONTROLS

A surprising number of HVAC systems are operating with fixed outside air quantities. In some special application situations or because of geometry this may be necessary, but, in general, the addition of economy cycle controls will save energy. Remember that relief must be provided for outside air in excess of that needed for exhaust and pressurization.

Economy cycle with reset of the low-limit temperature will usually conserve the most energy over the season. Other factors may require modification or limitation of this approach. This scheme is discussed in Chapter 7 and included in the descriptions which follow.

13.6 SINGLE-ZONE SYSTEMS

A single-zone HVAC system with a traditional economy cycle (Figure 13–1) is typical of many existing systems. Energy consumption is low

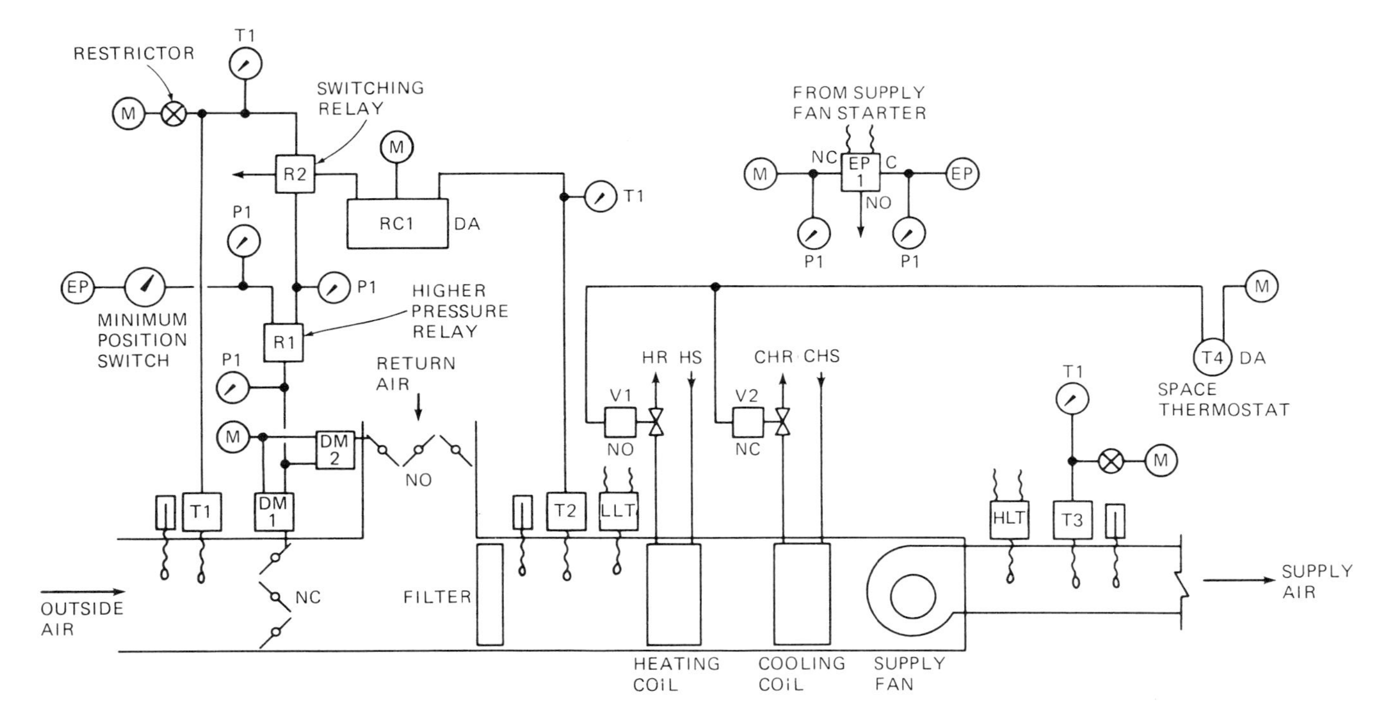

Figure 13-1 Single-zone HVAC system.

compared to multizone or reheat systems. Energy consumption can be decreased by any or all of several improvements shown in heavy lines in Figure 13–2. These include:

a. Sequencing of heating and cooling control valves to avoid overlap (simultaneous heating and cooling). Most, but not all, systems are designed in this way, accomplished by adjusting or replacing springs in the valves. The heating valve should be normally open (NO) with a 3–8 psi spring. The cooling valve should be normally closed (NC) with an 8–13 psi spring range. Then as the thermostat output goes through the range from full heating to full cooling (0–13 psi) the heating valve will gradually close and, in sequence, the cooling valve will gradually open. When the thermostat is "satisfied" it will have an output of 8 psi and both valves will be closed.

b. Replace the existing standard room thermostat with a dead-band thermostat. This will broaden the temperature range of the "satisfied" condition, allowing the space temperature to "float" over the dead-band range without the use of either heating or cooling energy.

c. Add reset from room temperature to the low-limit mixed-air controller. This will allow the mixed-air temperature to rise or fall with space heating or cooling demand, and, in particular, will minimize heating energy use.

d. Add supply fan variable air volume (VAV) control using inlet dampers as shown or motor speed control. Fan speed should be maximum at maximum cooling load, reducing to a minimum of about 40 to 50 percent at "thermostat satisfied" condition. The minimum should be retained for heating, and must be large enough to provide adequate heating.

13.7 REHEAT SYSTEMS

A typical reheat system is shown in Figure 13–3. Reset of supply temperature from outside air is sometimes, but not always, provided. Reheat systems are inherently wasteful of energy. However, much can be done to improve them, as shown in Figure 13–4.

a. Replace existing thermostats with dead-band thermostats. This will minimize reheat but will not, by itself, minimize cooling energy use.

b. Provide a discriminator relay (output equal to highest of several input pressures) to reset the supply air temperature to satisfy the zone with the greatest cooling demand. This will minimize cooling energy use. Obviously, there will still be reheat, with energy waste, in some zones.

c. Reset the low-limit mixed air temperature from the discriminator relay. This will minimize preheating and save some cooling as well, especially in mild weather (50°–70°F outside).

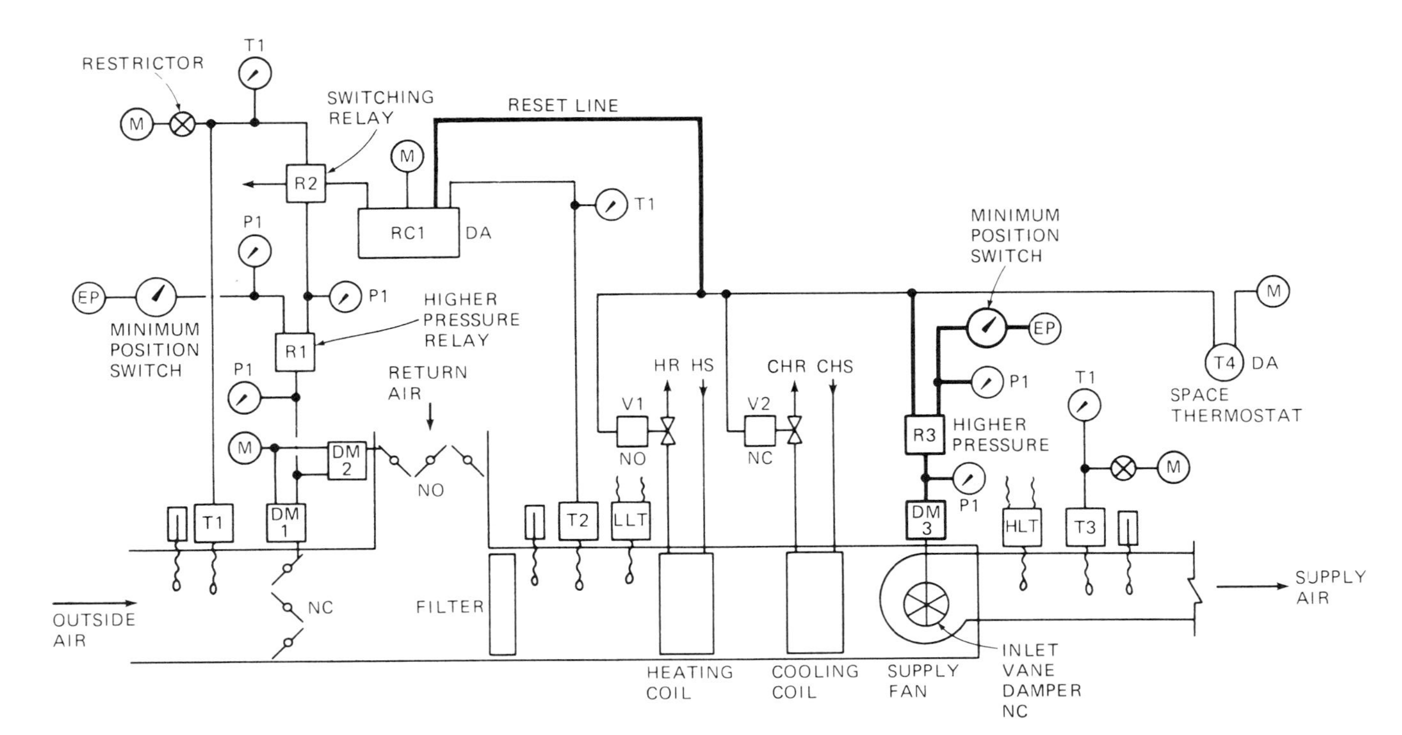

Figure 13-2 Single zone HVAC system; retrofit.

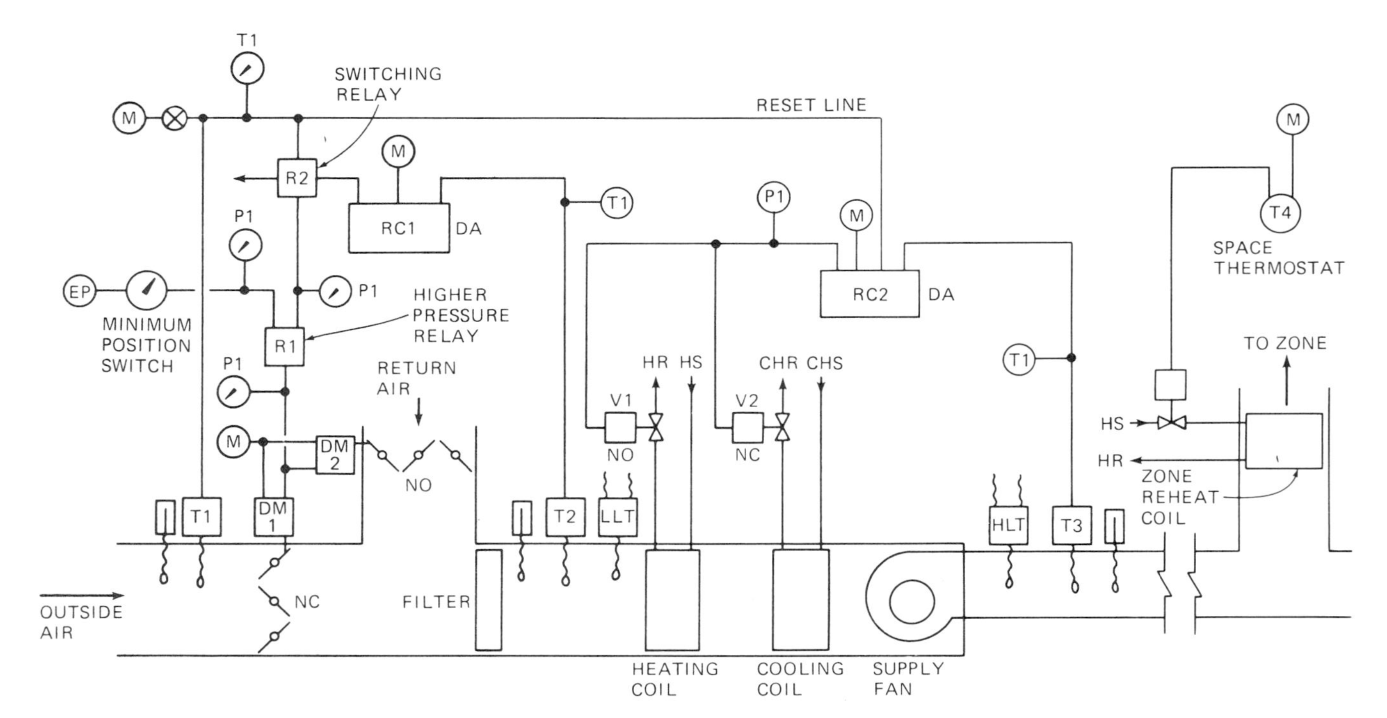

Figure 13-3 Zone reheat HVAC system.

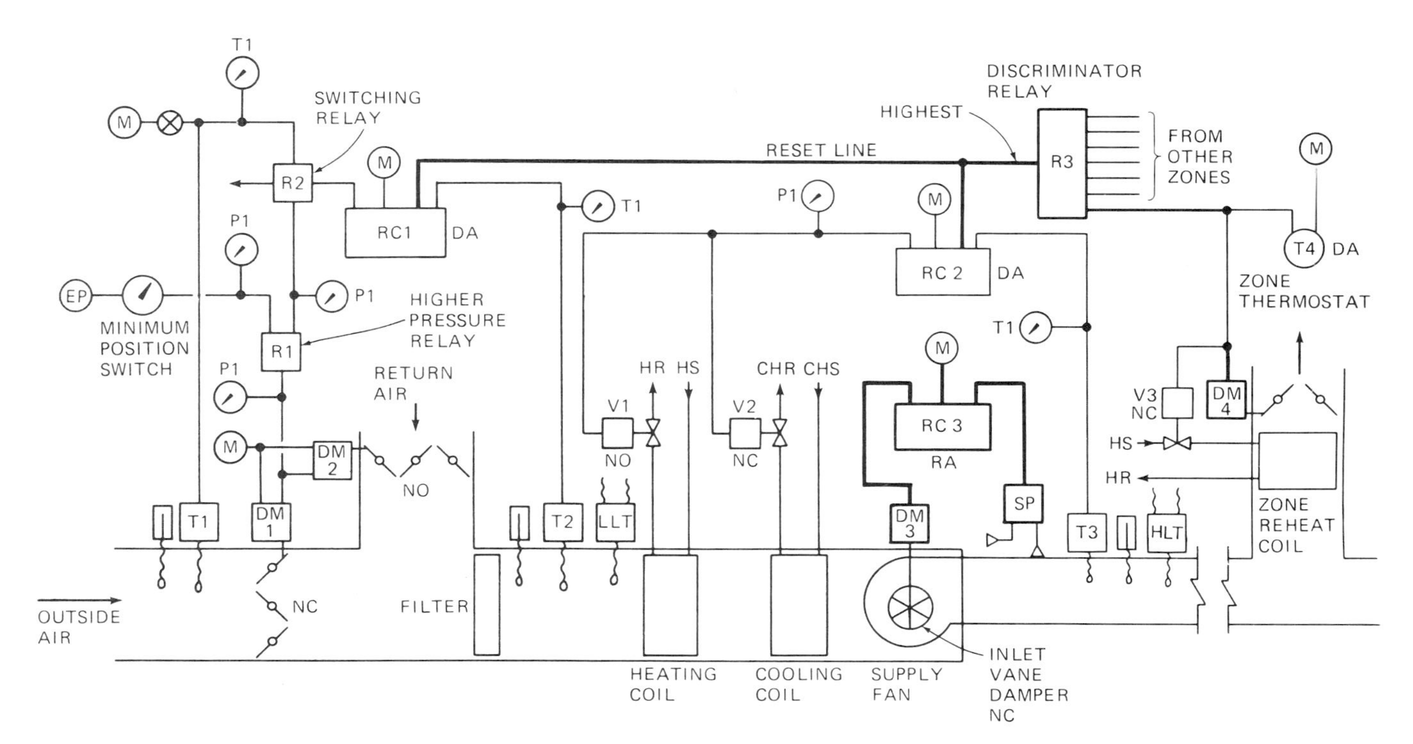

Figure 13-4 Zone reheat HVAC system; retrofit.

d. Remodel the system to become a VAV system. This will require adding variable volume boxes or dampers at all zones, with a duct static pressure control to control supply fan air volume (using damper or speed control). It will often require a larger supply fan and motor and higher fan speed to provide the extra pressure needed for the VAV boxes. There will still be a net saving in fan motor energy use over the conventional system but the cost-benefits of this revision require careful study.

13.8 MULTIZONE SYSTEMS

A typical multizone system with economy cycle is shown in Figure 13–5. Space temperature control is obtained by mixing hot and cold duct air through zone mixing dampers. Because one motor drives both dampers simultaneously (constant volume) the use of dead-band thermostats would tend to position the dampers in the 50-50 mixing position and would not conserve energy. The use of summer-winter thermostats is better here. Typically, thermostats are reset manually from heating to cooling as the seasons change.

These systems are effective and use minimum energy at or near design heating and cooling conditions. At these times the unneeded heating or cooling coil may be manually shut off and the bypassed mixed air provides the reheating or recooling needed. However, at least 50 percent of HVAC operating hours occur with mild outside air temperatures, 55°F to 70°F, when both heating and cooling energy may be required.

Under these conditions, and especially with the fixed low-limit mixed air and cold plenum control set points, multizone systems are notorious energy wasters. Some procedures are helpful.

a. Add disciminator relays for reset of hot and cold plenum set points (Figure 13–6). Notice that the lowest pressure output is used for hot plenum reset. This satisfies the zone with the greatest heating demand. The highest pressure output is used for cold plenum reset to satisfy the zone with greatest cooling demand. There will still be reheating and wasted energy but it will be limited to that required to meet zone control requirements.

b. Reset the low-limit mixed air controller as a function of cooling demand (Figure 13–6). This will minimize reheating requirements.

c. Remodel the system for VAV, as described below for dual-duct systems. This is not a simple task, and may not always be cost-effective.

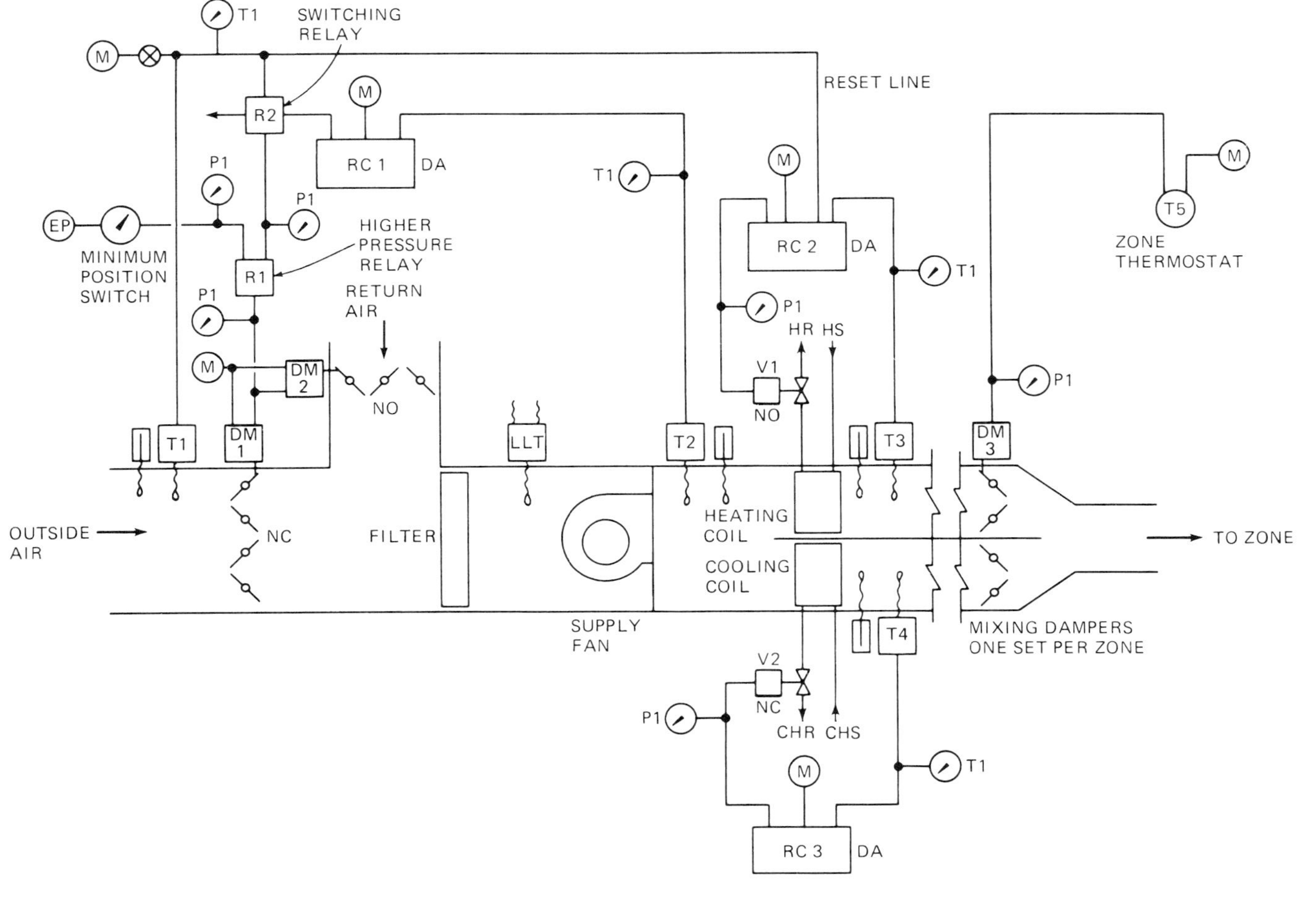

Figure 13-5 Multizone HVAC system.

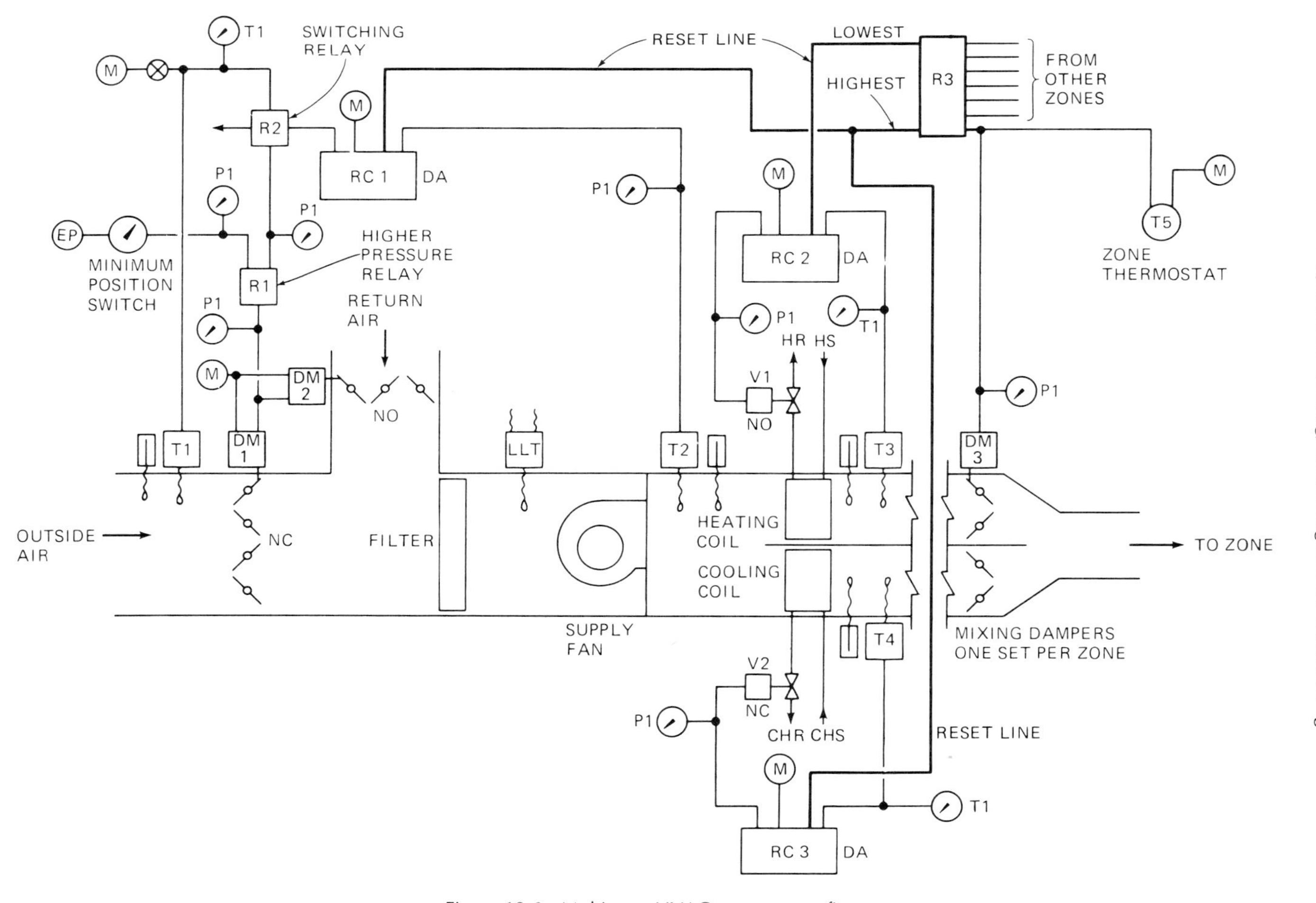

Figure 13-6 Multizone HVAC system; retrofit.

13.9 DUAL-DUCT SYSTEMS

Dual-duct systems are simply multizone systems with hot and cold plenums extended and mixing dampers located near the zone. They have the same problems and solutions as multizone systems:

a. When adding discriminator controls it is not usually practical to provide signals from all zones to the discriminator relay. A few typical zones may be selected.

b. Dual-duct systems may be converted to variable volume as shown in Figure 13–7. Most existing systems use single motor mixing boxes. It is necessary to provide separate motors for hot and cold dampers. Then the dampers are controlled in sequence as shown in Figure 13–8. At maximum cooling the cold damper is full open and the hot damper is closed. As cooling demand decreases the cold damper modulates toward the closed position while the hot damper remains closed (variable volume). At some minimum position of the cold damper—usually 25 to 30 percent open—the hot damper begins to open. As the heating demand increases, the cold damper continues to close and the hot damper modulates toward the open position. When the hot damper is 25 to 30 percent open, the cold damper is closed. There is some overlap and mixing of hot and cold air but it occurs at minimum air flow.

Supply fan volume controls must be provided (either dampers or motor speed controls) using the lower of the static pressures in hot and cold ducts.

Dead-band thermostats may be used.

c. Dual-duct systems may also be converted to "standard" VAV systems, by using the hot and cold ducts in parallel as though they were a common supply duct. The mixing boxes will be replaced with VAV boxes (with reheat coils if needed) and the hot plenum coil will be replaced with a cooling coil. A preheat coil may be required. This is a major remodeling project but is very effective where air supply might otherwise be inadequate.

The typical dual-duct system is designed with a high pressure drop and a change to VAV may produce dramatic savings in fan power requirements.

13.10 SYSTEMS WITH HUMIDITY CONTROL

Systems serving spaces requiring close control of relative humidity do not easily lend themselves to energy conservation. Close control of space temperature always goes along with humidity control. The simplest way to achieve humidity control is by controlling the dew point temperature at the HVAC unit and this, inevitably, requires reheat. Chemical dehumidi-

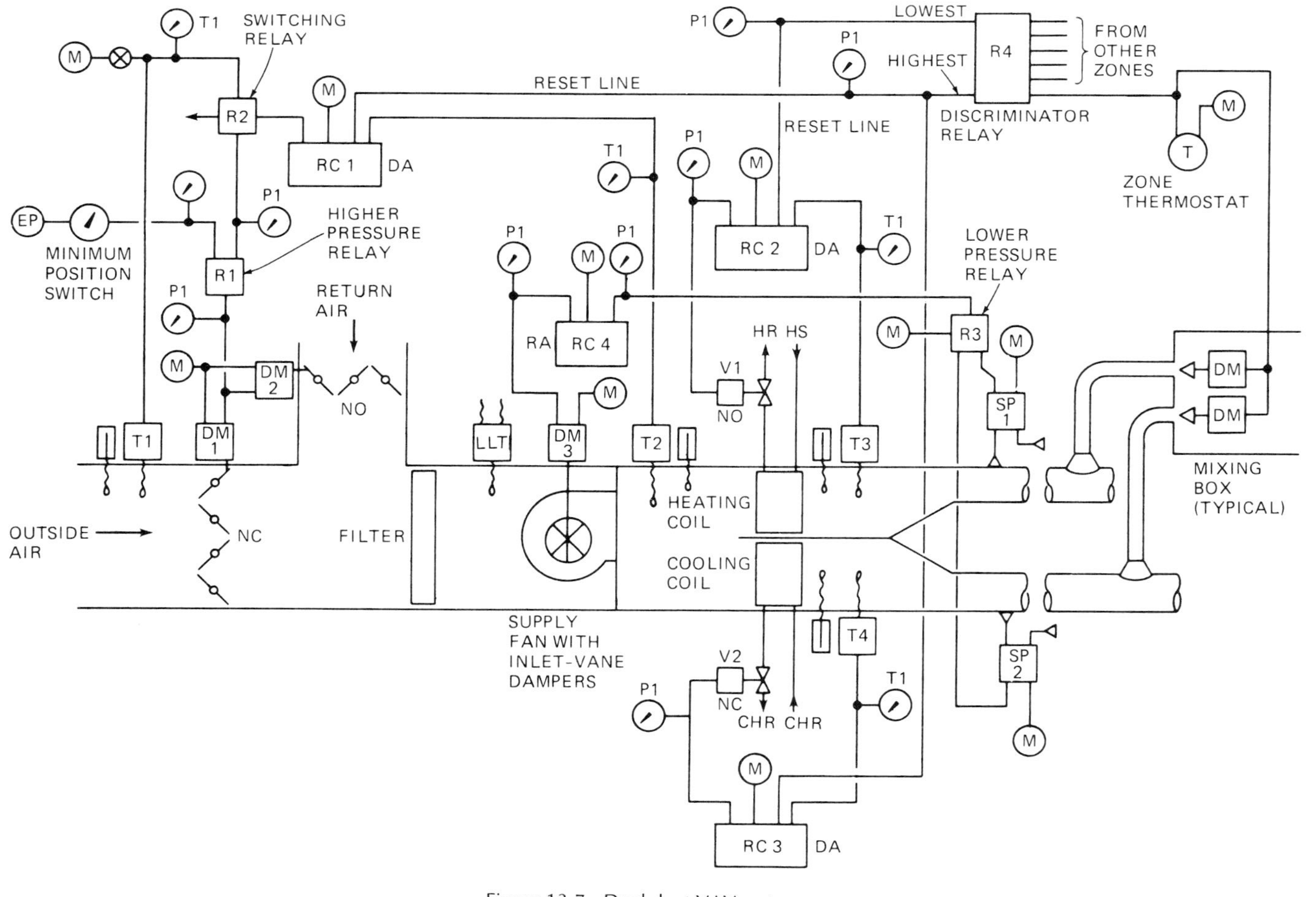

Figure 13-7 Dual-duct VAV system.

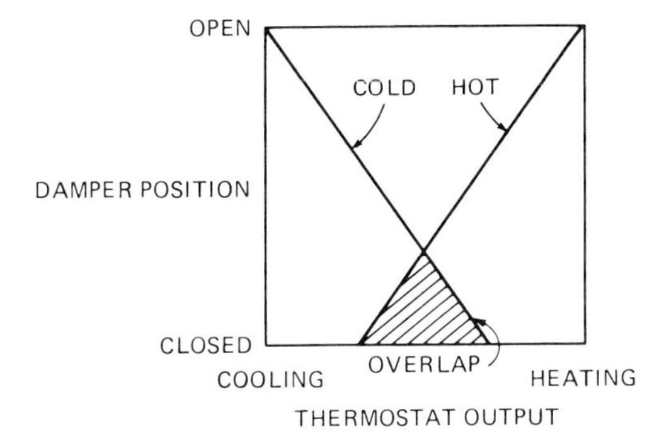

Figure 13-8 Damper relationships; dual-duct VAV.

fication requires heat energy for regeneration and additional cooling to remove the heat added to the air stream by the dehumidification process.

There is one method of saving energy: heat reclaim at the HVAC unit. This system has been described in detail in Bibliography reference 26 and is shown in Figure 13–9. The runaround system provides some or all of the required reheat by precooling the mixed air. This limits the outside air economy cycle somewhat, but, overall, results in considerable saving in energy use. (See paragraph 7.7.4.)

13.11 CONTROL VALVES AND PUMPING ARRANGEMENTS

Many chilled and hot water piping distribution systems are designed using 3-way control valves at the HVAC units. In a small system with only one chiller (or boiler) normally in use, this is satisfactory. When more than one chiller (or boiler) is needed, then pumping energy can be saved by using straight-through control valves and taking advantage of diversity as described in chapter 12 and Bibliography reference 29.

In the typical retrofit situation, the coil and valve are piped as shown in Figure 13–10(A) (cooling) or 13–10(B) (heating). The bypass valve shown will not always exist, and in some systems the bypass line may be extremely short. There are three ways of changing the system to straight-through control:

a. If the bypass valve exists, simply close it off.
b. Remove the bypass piping and plug the bypass port of the control valve.
c. Replace the three-way valve with a straight-through valve.

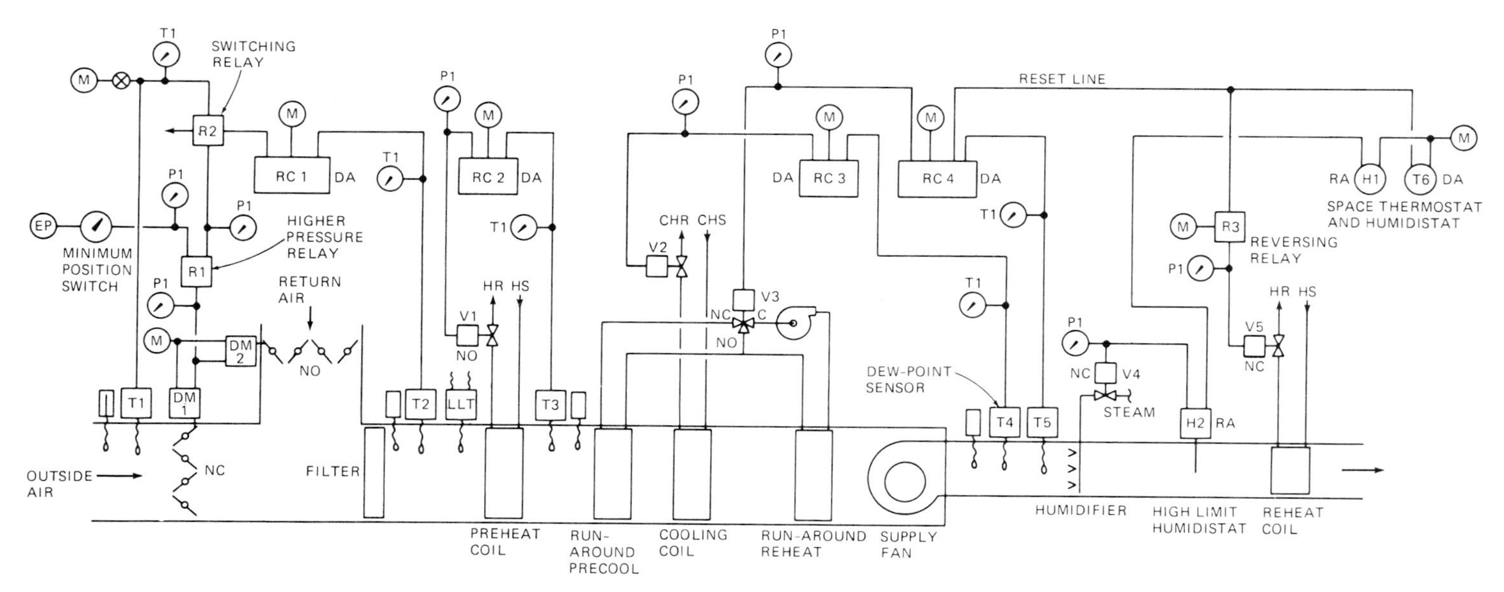

Figure 13-9 Humidity-control HVAC system with runaround for reheat.

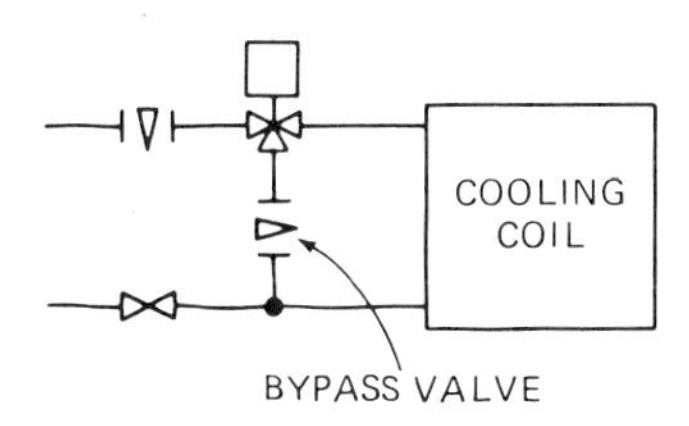

Figure 13-10(A) Three-way valve; cooling coil.

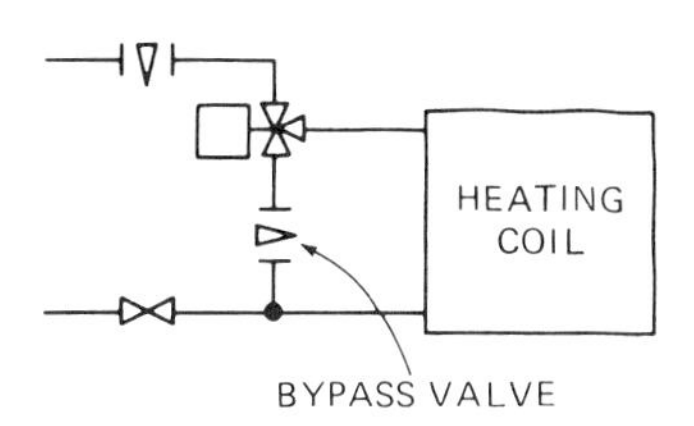

Figure 13-10(B) Three-way valve; heating coil.

In any case, the change from three-way to straight-through valves will create some system pressure problems unless a pressure controlled bypass is provided at the chillers or boilers and some provision is made for sequencing pumps with load. It is also useful to retain the three-way valve on the HVAC unit at the end of the piping system to provide a bypass to keep the line cold (or hot).

13.12 SUMMARY

As noted at the beginning of this chapter, these are suggestions for solving some of the more common retrofit problems. No system modification should be undertaken without carefully applying one's knowledge of HVAC and control fundamentals to determine the economic and technical feasibility of any proposed changes. In other words, use engineering principles rather than a "cookbook" how-to-do-it procedure.

13.13 CONCLUSION

We have discussed control systems for heating, ventilating and air conditioning systems, beginning with elements and ending with large supervisory systems. It is hoped that you have a better understanding of HVAC controls and, more importantly, control philosophy. Some of the ideas presented here may conflict with what you have learned previously. You may see what appear to be better ways of doing certain things. By all means, try something new if it looks possible. A great deal of the information presented here was learned "by mistake." Control design is interesting, challenging and fun. Enjoy it!

Control Bibliography

1. *Terminology for Automatic Control* (ASA C85,1–1963.)
2. D. M. Considine, et al.: *Process Instruments & Controls Handbook* (McGraw Hill Book Co., New York, 1957).
3. American Society of Heating Refrigeration and Air Conditioning Engineers: *HANDBOOK series* (four volumes: *FUNDAMENTALS, EQUIPMENT, SYSTEMS* and *APPLICATIONS*)
4. N. Peach: "Motor Control," *Power*, December 1962, pp. S–1–S–20.
5. E. J. Brown: "How to Select Multiple Leaf Dampers for Proper Air Flow Control," *Heating Piping and Air Conditioning*, April 1960, pp. 167–168.
6. W. M. Marcinkowski: "How to Select Controls for End Use Exchangers in HTW Systems," *Heating Piping and Air Conditioning*, March 1964, pp. 141–156.
7. "Understanding, Applying Solid State Devices," *Heating Piping and Air Conditioning*, November 1968, pp. 115–162.
8. J. E. Haines: *Automatic Control of Heating and Air Conditioning* (McGraw Hill Book Co., New York, 1961, 2nd ed.).
9. R. C. Mott: "Building Blocks for Fluidic Control," *Heating Piping and Air Conditioning*, June 1968, pp. 103–106.
10. R. E. Wagner: "Fluidics—A New Control Tool," *IEEE Spectrum*, November 1969, pp. 58–59.
11. P. M. Kintner: "Interfacing a Control Computer with Control Devices," *Control Engineering*, November 1969, pp. 97–101.
12. "Fluidic Output Interfaces," *Automation*, September 1969, pp. 54–58.
13. Honeywell, Inc.: *Basic Electricity*.

14. Johnson Service Co.: *Field Training Handbook*, Fundamentals of Pneumatic Control.
15. Ibid., *Fundamentals of Electronic Control Equipment*.
16. Ibid., *Engineering Report No. 685*, "Electric Heat and Its Control."
17. Barber Coleman Co.: *Control Application Data*, manual Nos. CA–5–1, CA–18, CA–19–1, CA–24, CA–27, CA–28.
18. Ibid., *Automatic Controls Bulletins*, Nos. EN–54–1 and EN–61–1.
19. Ibid, *Pneumatic Controls Practices and Typical Devices* (F–13955).
20. National Environmental Systems Contractors Assn.: *Automatic Controls for Heating and Air Conditioning Systems*.
21. R. W. Haines: "The Economy Cycle, How Economical?" *Heating Piping and Air Conditioning*, June 1969, pp. 86–88.
22. Roger W. Haines: "How to Pipe Recirculating Pumps for Preheat Coils," *Heating Piping and Air Conditioning*, July 1971, pp. 56–60.
23. Bahnfleth D. R.: "Pumping Interfaces Between the Chiller Plant Distribution System and the Buildings." *Purdue University Central Chilled Water Conference Proceedings*, 1976.
24. F. A. Govan and H. T. Kimball: "Combustion Controls for Industrial Boilers," *Heating Piping and Air Conditioning*, November 1981, pp. 66–75.
25. Trane Co.: *Fans in Air Conditioning*, 1970.
26. R. W. Haines: "Control for Low Humidity," *Heating Piping and Air Conditioning*, August 1980, pp. 88–89. (For Figure 6 see October issue, p. 70.)
27. C. J. Bell and L. R. Hester: "Electric Motors," *Heating Piping and Air Conditioning*, December 1981, pp. 51–56.
28. R. W. Haines: "Stratification." *Heating Piping and Air Conditioning*, November, 1980, pp. 70–71.
29. R. W. Haines: "Central Chilled Water Plant Control," *Heating Piping and Air Conditioning*, December 1981, pp. 68–69, and January 1981, pp. 135–136.
30. R. W. Haines: "Supply and Return Air Fan Control in a VAV System," *Heating Piping and Air Conditioning*, February 1981, pp. 75–76.
31. R. W. Haines:"Economy Cycle Control," *Heating Piping and Air Conditioning*, April 1981, pp. 111–113.
32. R. W. Haines: "Discriminator Controls," *Heating Piping and Air Conditioning*, May 1981, pp. 94–95.
33. R. W. Haines: "Air Flow Balance in a Laboratory," *Heating Piping and Air Conditioning*, November 1981, pp. 159–160.
34. "Design of Smoke Control Systems for Buildings," *National Bureau of Standards*, September 1983.
35. D. C. Hittle and D. L. Johnson: "Energy Efficiency Through Standard Air Conditioning Control Systems," *Heating Piping and Air Conditioning*, April 1986, pp. 79–94.

Abbreviations Used in This Book

A Amperes
AC Alternating current
Btu British thermal units
C Common
CB Circuit breaker
CHR Chilled water return
CHS Chilled water supply
CHW Chilled water
CWR Condensing water return
CWS Condensing water supply
D Derivative control mode
DA Direct-acting
DC Direct current
DB Dry bulb temperature
DP Dew point temperature
DPDT Double-pole, double-throw
DX Direct-expansion
EP Electric-pneumatic
ft Feet
h Enthalpy
HOA Hand-off-auto
hp Horsepower
HTWR High-temperature water return
HTWS High-temperature water supply
HVAC Heating, ventilating and air conditioning
HWR Hot water return
HWS Hot water supply
IC Instantaneous closing
in. Inches
IO Instantaneous opening
KW Kilowatt
MAX Maximum
MIN Minimum
min minutes
NC Normally closed
NO Normally open
OL Overload
P Proportional control mode
PB Pushbutton
PD Pressure drop
PE Pneumatic-electric
PI Proportional plus integral control mode
PRV Pressure-reducing valve
PSIA Pounds per square inch, absolute
PSIG Pounds per square inch, gage
RA Reverse-acting
RG Refrigerant gas (discharge)
RH Relative humidity
RL Refrigerant liquid
RS Refrigerant suction
SCFM Standard cubic feet of air per minute
SCIM Standard cubic inches of air per minute
sec Seconds
SP Static pressure
SPST Single-pole, single-throw
SS Start-stop
TC Timed closing
TO Timed opening
TR Tower return
TS Tower supply
V Volts
VAV Variable air volume
WB Wet bulb temperature
w Humidity ratio or specific humidity

Symbols Used in This Book

Capacitor

Centrifugal fan

Check valve

CB
Circuit breaker

Coil for solenoid valve

Contact, or point of force application

Contact, NC

Contact, NO

Contactor or controller

RC Receiver-controller

(M) Control air supply

(EP) Control air supply from EP relay.

(X) Restrictor in control air supply.

C NC NO Control valve, three-way

Control valve, two-way

DM Damper motor

Diaphragm

DP Differential pressure sensor or controller

EP Electric-pneumatic relay

Electromagnetic coil in starter or relay with identifier. All contacts actuated by this coil have the same identifier.

Fire safety switch

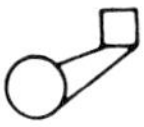

Float switch

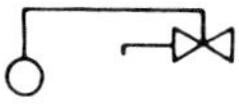

Float valve

Flow switch

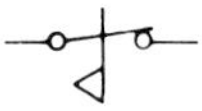

Flow switch, NC

Flow switch, NO

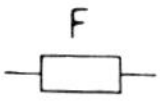

Fuse

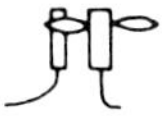

Gas pilot flame with thermocouple

Globe valve

Ground connection

Hand-off-auto (HOA) switch

Heater (heating element) or resistor

Humidistat, room

Humidistat, duct

Industrial-type recording controller

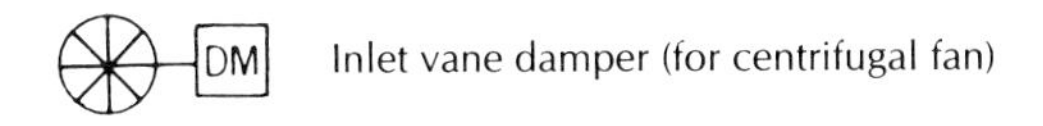

Inlet vane damper (for centrifugal fan)

Limit switch

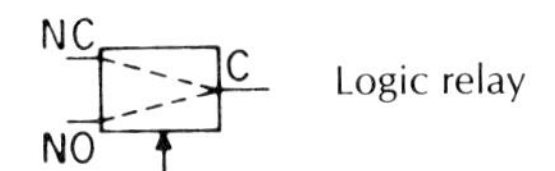

Logic relay

Manual switch

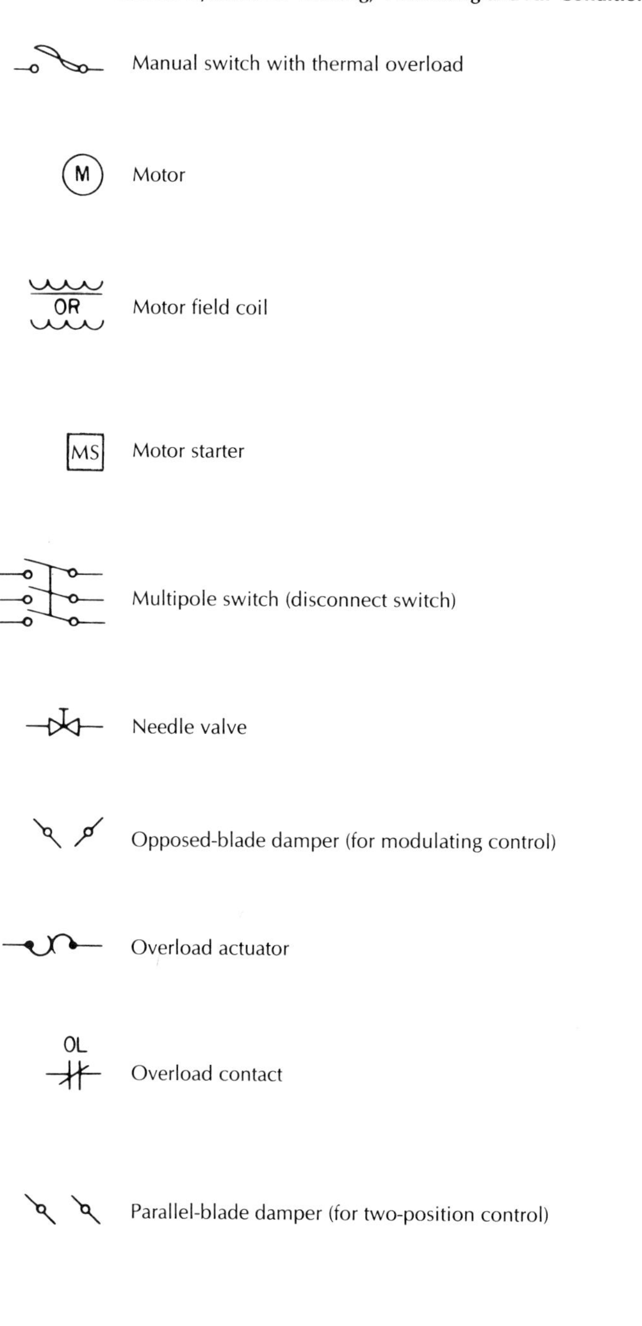
Manual switch with thermal overload
M
Motor
OR
Motor field coil
MS
Motor starter
Multipole switch (disconnect switch)
Needle valve
Opposed-blade damper (for modulating control)
Overload actuator
OL
Overload contact
Parallel-blade damper (for two-position control)
Pilot light, color indicated by initial

Plug valve

Point of solid contact, as to a device case or baseplate

Pressure gage

P1 Pressure indicator at control panel

T1 Temperature indicator at control panel

Pressure regulator (pressure-reducing valve)

P Pressure switch or sensor

Pressure switch, NC

Pressure switch, NO

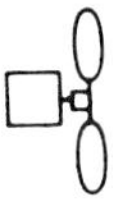

Propeller fan and motor

PC Proportioning controller, solid-state

Pump

Pushbutton, normally closed (PB, NC)

Pushbutton, normally open (PB, NO)

R Relay

OR Relay coil

Relay or starter contact, NC

Relay or starter contact, NO

Relief valve

Resistor

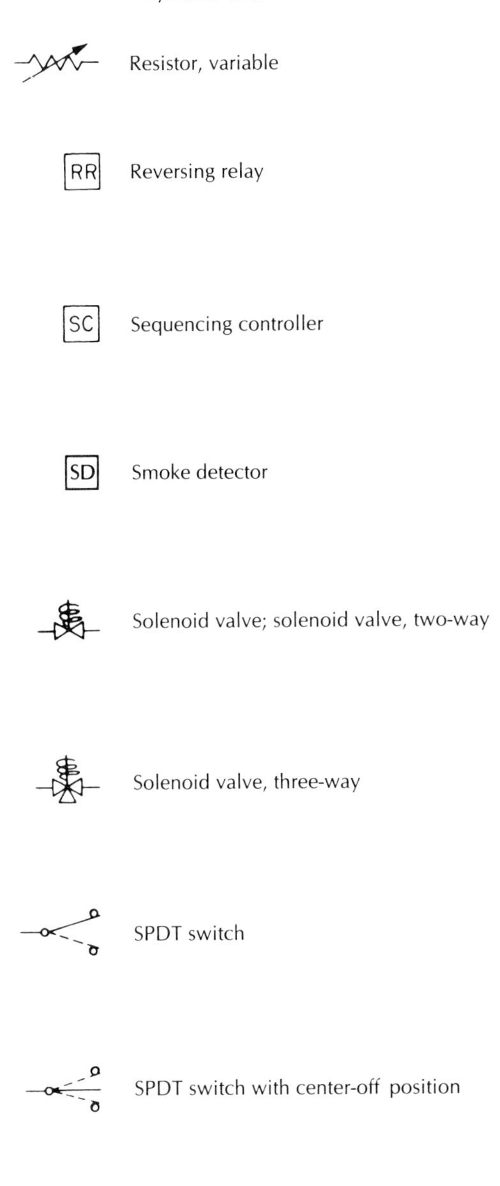
Resistor, variable
RR
Reversing relay
SC
Sequencing controller
SD
Smoke detector
Solenoid valve; solenoid valve, two-way
Solenoid valve, three-way
SPDT switch
SPDT switch with center-off position
Spray nozzle
Spring (where identified as such)

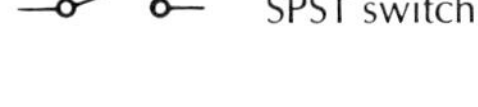

SPST switch

Static pressure controller

Steam trap

Thermal expansion valve, thermostatic expansion valve

Thermal switch, NC

Thermal switch, NO

Thermometer, remote-bulb or insertion type

Thermostat or temperature sensor, insertion type.

Thermostat or temperature sensor, remote-bulb, duct or pipe, or insertion type

Low temperature safety cutout

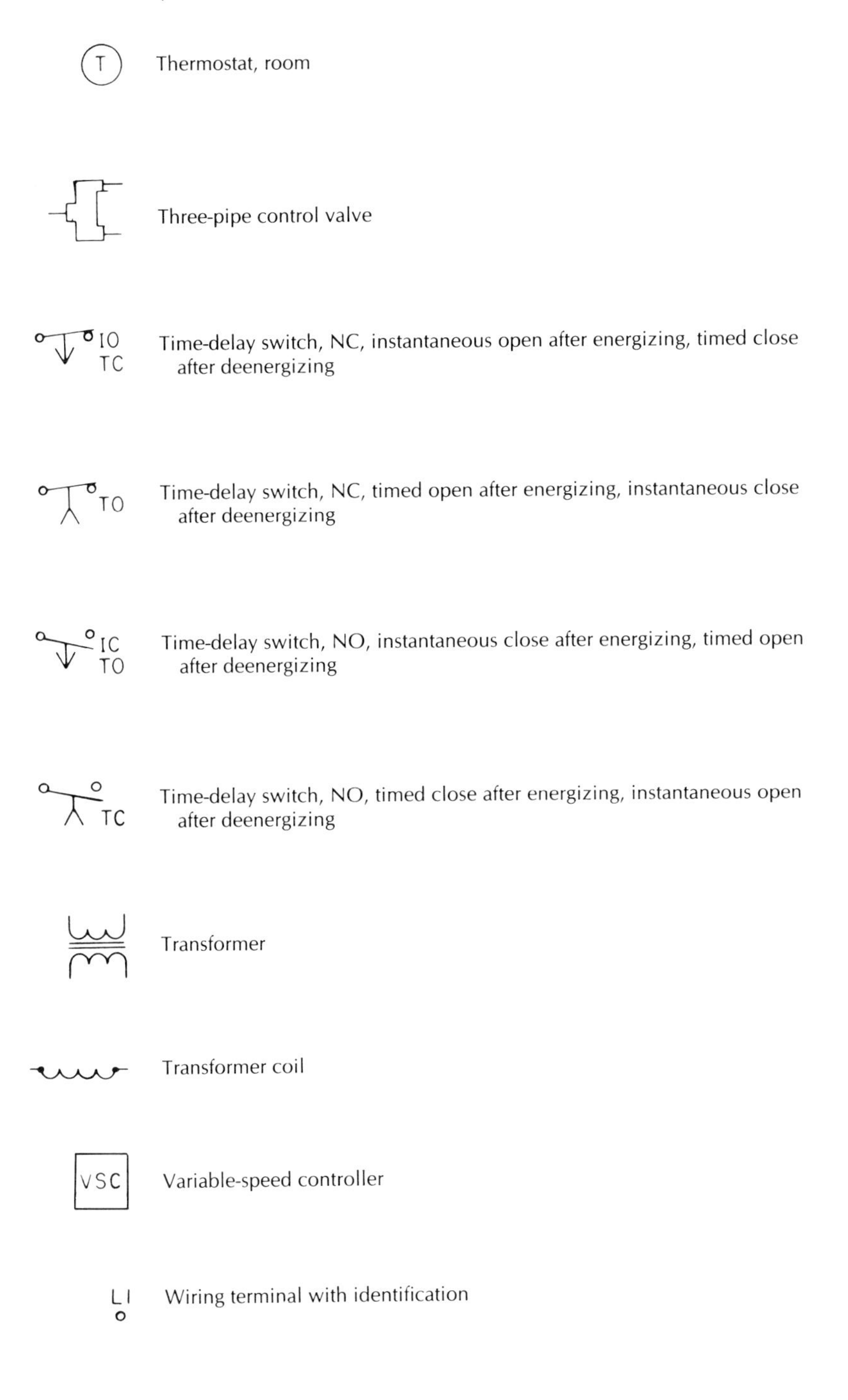

Thermostat, room

Three-pipe control valve

Time-delay switch, NC, instantaneous open after energizing, timed close after deenergizing

Time-delay switch, NC, timed open after energizing, instantaneous close after deenergizing

Time-delay switch, NO, instantaneous close after energizing, timed open after deenergizing

Time-delay switch, NO, timed close after energizing, instantaneous open after deenergizing

Transformer

Transformer coil

Variable-speed controller

Wiring terminal with identification

Index

Index